T0251287

Behavioral Biology:

Neuroendocrine Axis

Behavioral Biology:
Neuroendocrine Axis

edited by

Trevor Archer

Stefan Hansen

Gothenburg University

Psychology Press
Taylor & Francis Group

New York London

First Published by
Lawrence Erlbaum Associates, Inc., Publishers
365 Broadway
Hillsdale, New Jersey 07642

Transferred to Digital Printing 2009 by Psychology Press
270 Madison Ave, New York NY 10016
27 Church Road, Hove, East Sussex, BN3 2FA

Copyright © 1991 by Lawrence Erlbaum Associates, Inc.
    All rights reserved. No part of this book may be reproduced in
    any form, by photostat, microform, retrieval system, or any other
    means, without the prior written permission of the publisher.

Library of Congress Cataloging-in-Publication Data

Behavioral biology : neuroendocrine axis / [edited by] Trevor Archer, Stefan Hansen.
    p. cm.
    "Outcome of an international symposium . . . held in Gothenburg from 13-15 December, 1989, in
honor of Knut Larsson, Professor Emeritus"--Pref.
    Includes bibliographical references and index.
    ISBN 0-8058-0790-X
    1. Psychobiology--Congresses. 2. Neuroendocrinology--Congresses.   I. Archer, Trevor. II. Han-
sen, Stefan. III. Larsson, Knut.
QP360.B428   1991
152--dc20

                                                                    91-14480
                                                                    CIP

Publisher's Note
The publisher has gone to great lengths to ensure the quality of this reprint
but points out that some imperfections in the original may be apparent.

Contents

Preface

This book is the outcome of an international symposium in Biological Psychology, held in Gothenburg from 13-15 December, 1989, in honor of Knut Larsson, Professor Emeritus. The contents are, therefore, heavily flavored by the areas of research that he was involved in so successfully. We have attempted to present some important features of necessary paradigms for the analysis and study of experimental psychology within the biological perspective. It is with some surprise that one considers the divergent areas that Knut Larsson either instigated or became associated with: the neurobiology of sexual behavior and sexual differentiation, aspects of functional neuroanatomy, behavioral endocrinology, psychopharmacology, development and the ontogeny of behavior, clinical aspects of psychobiology, neurorethological determinants of critical behavior patterns, the biological dynamics of the essential parent-offspring relationships, and, not least, the generalizable aspect of comparative behavior analyses. Truly it was a wonderful thing that Knut Larsson started and is continuing - all over the world. But, the most wonderful thing was his substrate for these endeavors - *friendship.*

We are extremely grateful for very generous financial support and encouragement to the Council for Humanities and Social Sciences and the Medical Research Council in Sweden. Associate Professor Gösta Jonsson of Astra Research Center, Södertälje, gave both financial and scientific support, as did Dr. Thorsten Klint of Pharmacia Leo Therapeutics AB, Malmö. The dedicated efforts of Claery Persson, throughout, and Esther Ronn at the meeting and for the book are much appreciated. Lena Dahlgren, Claudia Fahlke, Elizabeth Wallin, Christina Harthon, Josefa Vega-Matuszczyk, and Lennart Svensson also provided excellent assistance.

Trevor Archer

List of Contributors

Sven Ahlenius, Department of Psychology, Astra Research Center, S-151 85 Södertälje, Sweden.

Mansour Al-Hazmi, King Abdul-Aziz University, Saudi Arabia.

Trevor Archer, Department of Psychology, University of Gothenburg, Box 14158, S-400 20 Gothenburg, Sweden.

Carlos Beyer, CIRA, Apdo Postal 62, Tlaxcala 90000, Tlaxcala, Mexico.

Paul Brain, School of Biological Sciences, University College of Swansea, SA2 8PP, United Kingdom.

Anders Broberg, Department of Psychology, University of Gothenburg, Box 14158, S-400 20 Gothenburg, Sweden.

Sven Carlsson, Department of Psychology, University of Gothenburg, Box 14158, S-400 20 Gothenburg, Sweden.

Annika Dahlström, Department of Histology, University of Gothenburg, Box 33031, S.400 33 Gothenburg, Sweden.

Ivan Divac, Institute of Neurophysiology, Panum Institute, Blegdamsvej 3, DK-2200, Copenhagen, Denmark.

Jörgen Engel, Department of Pharmacology, University of Gothenburg, Box 33031, S-400 33 Gothenburg, Sweden.

Barry Everitt, Department of Anatomy, University of Cambridge, Downing Street, Cambridge CB2 3DY, United Kingdom.

Elizabeth Factor, Institute of Animal Behavior, Rutgers, The State University of New Jersey, 101 Warren Street, Newark, NJ 07102, U.S.A.

Alonso Fernandez-Guasti, Seccion de Terapeutica Experimenatal Departamento de Farmacologia, Apdo Postal 22026, Mexico 14000 D.F., Mexico.

Gabriela González-Mariscal, CIRA, Apdo Postal 62, Tlaxcala 90000, Tlaxcala, Mexico.

Jean-Marie Guastavino, Ethology Universite Paris, Avenue J.B. Clement, F-93430, Villetaneuse, France.

Ernest Hård, Department of Psychology, University of Gothenburg, Box 14158, S-400 20 Gothenburg, Sweden.

Cheryl Harding, Department of Psychology, Hunter College, CUNY, 695 Park Avenue, New York, NY 10021, U.S.A.

Viveka Hillegaart, Astra Research Center, S-151 85 Södertälje, Sweden.

Mikael Heimann, Department of Psychology, University of Gothenburg, Box 14158, S-400 20 Gothenburg, Sweden.

C. Philip Hwang, Department of Psychology, University of Gothenburg, Box 14158, S-400 20 Gothenburg, Sweden.

Thorsten Klint, Pharmacia Leo Therapeutics AB, Box 839, S-201 80 Malmö, Sweden.

Barry R. Komisaruk, Institute of Animal Behavior, Rutgers, The State University of New Jersey, 101 Warren Street, Newark, NJ 07102, U.S.A.

Knut Larsson, Department of Psychology, University of Gothenburg, Box 14158, S-400 20 Gothenburg, Sweden.

Gerard Marek, Departments of Pharmacology and Physiological Sciences, University of Chicago, 947 East 58th Street, Chicago, IL 60637, U.S.A.

Anne Mayer, Institute of Animal Behavior, Rutgers, The State University of New Jersey, 101 Warren Street, Newark, NJ 07102, U.S.A.

Michael Leon, Department of Psychology, University of California, Irvine, CA 92717, U.S.A.

Bengt Meyerson, Department of Medical Pharmacology, Biomedicum, Box 594, S-75124, Uppsala, Sweden.

Paul Micevych, Departments of Anatomy and Cell Biology, UCLA School of Medicine, Los Angeles, CA 90024-1763, U.S.A.

Berend Olivier, CNS-Pharmacology, Duphar B.V., P.O. Box 900, 1380 DA Weesp, The Netherlands.

O. Picazo, Seccion de Terapeutica Experimenatal Departamento de Farmacologia, Apdo Postal 22026, Mexico 14000 D.F., Mexico.

Nanne van de Poll, Netherlands Institute for Brain Research, Meibergdreef 33, 1105 AZ Amsterdam ZO, The Netherlands.

B. Roldán-Roldán, Seccion de Terapeutica Experimenatal Departamento de Farmacologia, Apdo Postal 22026, Mexico 14000 D.F., Mexico.

Jay Rosenblatt, Institute of Animal Behavior, Rutgers, The State University of New Jersey, 101 Warren Street, Neward, NJ 07102, U.S.A.

A. Saldivar, Seccion de Terapeutica Experimenatal Departamento de Farmacologia, Apdo Postal 22026, Mexico 14000 D.F., Mexico.

Lewis Seiden, Departments of Pharmacology and Physiological Sciences, University of Chicago, 947 East 58th Street, Chicago, IL 60637, U.S.A.

Per Södersten, Psychiatric Clinic, Huddinge Hospital, S-14186 Huddinge, Sweden.

Frank Ödberg, State University of Ghent, Laboratory of Animal Husbandry, Heidestraat 19, B-9220 Merelbeke, Belgium.

1 Introduction: My Journey into Biological Psychology

Knut Larsson
University of Göteborg

I entered the University of Göteborg just after the end of World War II and decided to devote the first year to study philosophy with Gunnar Aspelin, a well-known professor of philosophy. I had become interested in the problem of mind and matter during my philosophy studies in high school and wanted to know what 'mind' really meant before I began to study zoology or psychology. I was quickly confronted with the concept of the gestalt psychologists of an isomorphy between the phenomenon perceived and its underlying physiological substrates. Köhler (1929), the leading gestalt psychology theorist, told me that experience and matter are only different sides of the same coin. Since experienced order is a representation of a corresponding physiological order, a true description of the experience would also give access to the physiological processes determining this order. I was intrigued by this idea of an isomorphy between matter and experience, although somewhat surprised that the laws of nature would be so easily accessible.

I soon discovered, however, that the thought of an isomorphy between mind and matter was not new but had occupied philosophers for centuries (Kaila, 1944). Further, the verbal description of the immediately-given experience turned out to be far from an easy task to accomplish. For instance, when, participating in the laboratory course of psychology, I was asked to describe what I perceived as immediately presented to me in the periphery of retina, I found this task quite difficult, because the phenomenon perceived altered with the language I was using. In fact, the language did not seem to accommodate the communication of such types of experiences.

After much reading, discussion, and thinking, I concluded that the concept of an isomorphy between behavior and its physiological substrate cannot be maintained. I began, vaguely, to grasp the basic truth that each method used to investigate nature must be applied according to its own rules, having its own limitations as well as advantages.

The problem of mind and matter did not stop troubling me and I continued my readings, now confronted with the American learning theorists, at that time, the leading theorists in psychology. I read Tolman, Hull, and Skinner, and other behaviorists. Tolman called his book "Purposive Behavior in Animals and Man". He described the learning of rats in the maze as a formation of cognitive maps. However, the problem was, as it appeared to me, not to find the purpose in the behavior of the rat but rather the mechanisms determining the behavior. In vain, I looked for such mechanisms in the works of Tolman. Hull's book, "Principles of Behavior", was no less promising than that of Tolman, and, in one respect, this approach was easier to understand. A physiological language was used which attracted me, but, again, I discovered a gap between the empirical data presented and the concepts used. On the other hand, there was Skinner, who had written a book entitled "The Behavior of Organisms". He, an ardent critic of the bulk of theorizing by his fellow behaviorists, could not easily be accused of making the same blunder as they did. He claimed that the task of behavioral research was to achieve control of the environment and having reached this goal as Skinner certainly had, by using bar-pressing behavior, the goal of the scientific analysis set by him was apparently achieved. But then, I wondered, why not perform the behavioral analysis directly on man? Obviously, rats and doves are not equal to man, and what allows one to generalize from animal to man?

Although the work of the American animal psychologists appealed to me in terms of the intention to understand the organization of behavior, I was deeply dissatisfied with their discussions and concepts. Either the author was a Tolman, a Hull, or a Skinner; nothing more was offered the reader, so it seemed, than descriptions of the physical world in terms of turns to left or right in correct successions performed by the rat. A task to master the physical environment, not exceeding the capacity of the animal, must, by definition, result in organized behavior according to my understanding. Yet, this conformity was of only secondary importance, our scientific object being to reveal the mechanisms behind the behavior. Only by knowing these mechanisms would we be able to generalize from animals to man. These mechanisms, I believed, must be of a biological nature. I was surprised and confused to see that the American behaviorists all presented their own closed universe of concepts and data, largely ignoring studies of animal behavior performed outside their own laboratories, and with complete disregard to the information available, even at that time, of morphology and physiology of the organism.

I decided to leave Copenhagen, where I had spent some years listening to the lectures of Edgar Rubin, a gestalt psychologist famous for his discovery of the figure-ground relationship, and went to Paris to Henri Pieron, the leader of biological psychology in Europe at that time. I spent a Spring semester at the Sorbonne, a tumultuous experience for me, not only because I was exposed for the first time to a biological-oriented teaching of psychology of high standard but also because in Paris, after the war, I saw misery which was beyond all my imagination. I met a Russian physiologist, Popov, a former student of Pavlov. He had a small laboratory of his own in the basement of the Sorbonne where he made electrophysiological recordings of conditioned reflexes. I also became acquainted with Russian work on conditioning of autonomic reflexes by reading a number of translations into French and English. Bykov and his associates conditioned various reflexes in the urinary and intestinal functions. (see Bykov 1953). The Russian research fascinated me, but the unorthodox methods used in analyzing the data and the political jargon in which the results were wrapped made me uncertain about this work. Contributing to my confusion was the lack of analysis of either side on the performances of others. The exception was a book - "Conditioned Reflexes and Neuron Organization" - written by a Polish investigator, Konorski (1948), in which he tried to integrate Pavlovian concepts with those of Sherrington. It was a delight for me to read his book; I almost learned it by heart.

I returned to Copenhagen and spent another year there occupied by attempts to grasp and integrate the different views of behavior, guided and encouraged by my philosophy teacher, Gunnar Aspelin in Lund, to whom I returned now and then and to whom I wrote many letters. After four years of study, I finished my Master's thesis in philosophy, calling it "The Problem of Generalization in the Study of Behavior of Animals and Man". I concluded that nobody, so far, had approached these problems adequately. I was deeply discouraged by what I had learned, and, even worse, I did not know how to find a way out of this confusion.

Among European psychologists, only the gestalt psychologists had been interested in experimental work on animals and most of them had fled Europe long before the war. In Europe, at that time, in the beginning of the fifties, the behavior of animals was studied mainly by the ethologists, Lorentz and Tinbergen, in particular. I knew of this work but read it with disdain, burdened as it was by the heavily criticized concepts of instinct and inborn releasing mechanisms. I was, at this stage of my development, not prepared to understand the significance of this kind of work.

It was then I remembered that in Norway there had been a remarkable man, Schelderup-Ebbe, who in the early thirties had made a study, often cited, of the social life of hens in a hen yard. I heard, even then in 1952, that a small animal colony was maintained by a psychologist, Wulff Rasmussen. I, myself, had never had a rat in my hands. I decided to go to Oslo. Rasmussen studied rat sexual behavior in a Warden obstruction box, a method which, of course, I dismissed as an expression for American learning theory. I was, instead, assigned a task of my own, namely, to study the estrous cycle in the mouse. These studies gave me the key to my paradise. Each night I went to the small colony room for mice and tested their estrous. cycle.

I was fascinated by what I saw; I felt I was glancing right into Nature itself. Behavioral changes were unfolding before my eyes: estrus followed by anestrus, and this by estrus. The importance of gonadal hormones for the behavior under study was wonderfully demonstrated to me, a complete novice in the study of biology of reproduction. The laboratory situation gave me an example of the importance of environmental factors. When I entered the laboratory before 10 o'clock in the evening in the Spring, when sunset was late, none of the animals was in estrus. One hour later, after sunset, the rats showed full estrus. Thus, the neuroendocrine events determining the behavior were, themselves, adapted to the physical environment.

At last, I felt I had found a solution to my problem of mind and matter. The mouse estrous cycle belonged to the animal's repertoire of inborn behaviors, a part of the morphological and physiological characteristics of the animal itself, depending on the hormones acting upon the brain and environmental stimuli as well. The estrous cycle was just a window for me, letting me observe a behavior that seemed to be a direct expression of the forces governing the organism. For the first time, I had a tool with which I could continue my search for relationship between behavior and physiology.

Naturally, not all the scientific problems were solved for me. Names like Calvin Stone or Frank Beach were still unknown to me. Reproductive biology was an area of research entirely new to me and nobody could guide me into this literature. The estrous cycle fascinated me, indeed, but here, so I believed, no problems remained and I had to go elsewhere. In this situation, I had the good fortune to meet a French neurologist and physiologist, André Soulairac. He studied masculine sexual behavior in rats and had performed exciting studies on effects of hormones and drugs in sexual behavior. He reported that distinct behavioral traits were related to specific physiological events, i.e., he seemed to have answers to exactly those kinds of questions which I had, in vain, been looking for when reading Hull and Tolman. Soulairac invited me to his laboratory in Paris. I spent the Spring of 1953 in Paris, less confused than I had been four years earlier, learning some elementary skills which I needed.

I then returned to Sweden, obtained permission from the head of the Department of Psychology, John Elmgren, to use part of the cellar in the Psychology Department as a colony room and laboratory, bought some rats from a private rat dealer in the outskirts of Copenhagen, where rats were cheaper than in Stockholm - in Gothenburg such animals were unavailable - and then began to study sexual behavior in rats. My intention was to study the effects of cortical lesions upon masculine sexual behavior, but the months went by and I was confronted with new methodological problems which I felt must be solved before going any further. I studied the influence of environmental factors, like the light/dark periodicity, and the importance of the testing situation, including social grouping. I studied the importance of sexual experience and the interaction of sexual behavior with conditioned responses. I investigated how often I could test each individual rat so that I would not exhaust them sexually. Carefully, I observed the unfolding sexual responses, trying to uncover stimulatory and inhibitory factors determining the appearance of the pattern. Since I had no money to buy new rats, they aged during the course of the experimentation. Thus, a chapter was added, describing age changes in the behavior. After two years, my funds were depleted. I closed the lab, wrote down the results, and sent them to the printer. I entitled the book "Conditioning and Sexual Behavior", and soon after I became a Doctor of Philosophy.

It may be proper, at this stage, to give a picture of the research going on in my field in the latter half of the fifties. In the U.S., there were two dominating figures, Frank Beach and William Young. In Europe, nobody, other than Soulairac, worked in this field, as far as I knew. Beach and his students had studied most aspects of reproductive behavior in male and female rodents since the end of the thirties. He had written a timely book on hormones and behavior, and every other month another important contribution left his lab. Young was recognized as the leading expert on the ovarian regulation of the estrous cycle in rodents, problems with which he had been working since the early thirties. Now, in the fifties, he, together with his brilliant students, Gerall, Goy, Phoenix, Valenstein, Riss and others, was deeply engaged in studying the hormonal regulation of sexual behavior, preparing for a major contribution on the hormonal regulation of sexual differentiation. Beach had been working at the Museum of Natural History in New York during the forties, and there a laboratory had been established for the study of reproductive behavior which included, among others, T.C. Schneirla, Lester Aronson, Dan Lehrman, Jay Rosenblatt, and Barry Komisaruk. The study of species-specific behavior analysis as an approach to the study of physiological substrates of behavior was the focus of interest. Schneirla and Lehrman were the central figures in these discussions around inborn and experiential factors determining behavior. In contrast to Lorentz and Tinbergen, they maintained that the manifested behavior was neither an expression for inborn, nor for acquired factors, but rather for an interaction between these both classes of factors, the relative importance of which only being accessible by way of experimental analysis. The studies of the reproductive behavior of the ring dove performed by Dan Lehrman (1955) were viewed as a model for a study of the interaction of environmental and hormonal factors in the behavior.

Having completed my doctoral thesis in 1956, I had to make new decisions about my future. My experimental work was meant to be a contribution to the study of conditioning. As a phenomenon of conditioning, I considered, for instance, the delay in the reversal of the circadian cyclicity occurring after a reversal of the light/dark stimulation. I had studied the interaction between a conditioned operant response and sexual behavior and made some observations of what the ethologist would call displacement effects. In vain, I had tried to condition the enforced interval effect, i.e., the reduction of the number of intromissions to ejaculation produced by experimental prolongation of the inter-intromission intervals in the masculine sexual behavior pattern. I was curious about the relationship between sexual behavior and pain, having noticed that sexual behavior might increase the pain threshold. On the other hand, the neuroendocrine regulation of masculine sexual behavior itself called for a study, as did feminine sexual behavior, apart from its tool to study conditioning. In the journals, studies of the physiological mechanisms of sexual behavior began to appear. Paul MacLean (1959) reported localization of genital functions using electrical stimulation by implanted electrodes, James Olds (1958) reported effects of testosterone on brain self stimulation behavior, and Harris (1955) studies of the brain regulation of the pituitary gland. Sawyer, Everett, and their associates combined morphological, endocrine, neurophysiological, and morphological techniques in studying the mechanisms controlling ovulation in rats and rabbits (Sawyer, 1959).

As I hesitated between these many different ways to continue my research, I received an invitation from Frank Beach to be with him for the year of 1959 at Berkeley. Ron Rabedeau and Dick Whalen were graduate students in Beach's laboratory and later we were joined by Thomas McGill. I traveled extensively during this year, visiting several laboratories in the United States engaged in physiological behavior research. With initial help from Karl Pribram and Gilbert French, I performed a study of effects of neocortical lesions on the sexual behavior. I returned to Sweden where, in the meantime, I had received new laboratory locations in the Anatomy Department. My hesitation was gone and my intention now was to study the neural regulation of the masculine sexual behavior, using morphological and pharmacological methods.

I was now also ready to take on students. My first student was Ernest Hård. A former school teacher, he was interested in behavior ontogenesis: he observed how the rat mothered her pups, licked them, lactated them, defended them, and warmed them. He chose one of the components: licking. Licking, like mating, was a species-typical behavior, depending upon the bodily needs of liquid and food. It was also a behavior highly sensitive to taste stimuli and greatly dependent upon experience. Like mating, it was a behavior which one should be able to record and quantify, at least in principle. Ernest soon got into difficulties when trying to record suckling in newborn pups, and instead began to study suckling in the adult mother. A method was developed for the microanalysis of the licking behavior. A very important branch of research, concerned with behavior ontogenesis and taste preference behavior, began to grow. In addition to these studies, he and I worked together on the ontogenesis of various behaviors, a subject almost entirely undeveloped at that time. One of our ideas was to establish a time table for the behavioral development which then could be used for studies of teratogenic influences on development.

My second student was Sven Carlsson, a former medical student. He was interested in the problems of arousal and activity levels, problems in fashion at that time, after the discovery of the role of the reticular system in the brain stem. Sven developed a method to study arousal by measuring habituation of the startle response. Unfortunately, he was more attracted to human beings than rats and left us to make a pioneering contribution to problems of psychosomatics.

A third person soon appeared in the laboratory, Lennart Heimer, a medical student and a devoted teacher in neuroanatomy. We immediately became close friends. We designed a stereotaxic instrument of our own, developed a technique for localizing the electrode with X-ray measurements, and Lennart began to work on his own variant of silver impregnation, later called the Fink-Heimer method. Together we explored the brain of the rat for its role in regulating sexual behavior. Our collaboration continued for almost ten years and ended when he left for the United States of America.

During these years, in the early sixties, I met Arvid Carlsson and Nils-Åke Hillarp, and their students: Kjell Fuxe, Annika Dahlström, Nils Erik Anden, Thomas Hökfelt, Urban Ungerstedt, Lars Olsson, Georg Thieme, and others. All of them were engaged in working out the new fluorescence techniques for tracing monoaminergic pathways in the brain. At this time, I also met Lewis Seiden, who started a behavioral laboratory in the Pharmacology Department. In the Department of Physiology, a psychobiology laboratory was started by Ulf Norrsell, and we immediately became good friends. Heimer and I had discovered that lesions in the medial mesencephalon resulted in drastic changes in the coital pattern of the rat. The rats ejaculated after a few intromissions and after ejaculation they did not wait for a couple of minutes, as they usually did, but almost immediately remounted the female. It seemed that an inhibitory mechanism had been destroyed by the lesion. We scrutinized the literature then available but were unable to find any reasonable answer to why the behavior had become distorted in this way. Perhaps these newly discovered monoamine pathways, including cellbodies in the brainstem and terminals in the forebrain, might give an explanation for these remarkable behavioral effects. I quickly joined efforts with this enthusiastic group of people, and the friendship then established has lasted over the years.

I entered the field of biology of reproduction as a philosopher with a minimal insight into this field. Naturally, the treatment of the problems, experimentally and conceptually, were amateurish in many respects. It may be of interest, from both a historical and a factual point of view, to look upon the problems, as they then were treated, in the light of present day knowledge.

My ultimate goal was to disclose the physiological substrates of the masculine sexual behavior. To that purpose a detailed description of the behavior was needed. The sexual behavior of the rat consists in a series of intromissions culminating in ejaculation which, in turn, is followed by a period of refractiveness to sexual stimulation, the postejaculatory refractory period. The male may then resume copulating, achieving another ejaculation, etc. The waves of sexual activity and the periods of quiescence between them follow a strict temporal pattern. The latency of the first ejaculation is longer than the immediately following ones, while the postejaculatory intervals are prolonged by each successive ejaculation with extreme regularity (Larsson, 1956). These alterations in the coital pattern call for neural mechanisms operating with great precision. How much do we know, at present, of these mechanisms? One may assume that the repeated intromissions produced a state of rising excitation resulting in ejaculation. We tried to determine the duration of these excitatory effects and found that they lasted for a few hours only (Larsson, 1960). Besides the excitatory effect of an intromission, we postulated an inhibitory effect. By prolonging the intromission intervals beyond their normal length we prevented the series of copulations to be negatively affected by this inhibitory effect, as evidenced by a reduction in the number of intromissions preceding ejaculation (Carlsson & Larsson, 1962). The duration of this postulated inhibitory effect is only a fraction of a minute. The inhibitory effect of repeated ejaculations as reflected in the postejaculatory refractory periods seems to have a very long duration of days and even weeks. These different durations of effects induced by the intromissions and ejaculations support the pharmacological findings suggesting different physiological mechanisms involved in controlling these behavioral components. We know that electrical stimulation in the medial preoptic area may reduce the number of intromissions to ejaculation, and shorten the postejaculatory refractory periods (Van Dis & Larsson, 1970). Brain lesions in the medial mesencephalon may shorten the postejaculatory intervals, as was indicated above. Generally, it is as if normal inhibitory regulatory mechanisms can be set out of play by lesions or brain stimulation. However, at present, the nature of these processes and the neural circuits involved are unknown.

When Heimer and I started our lesion studies, there was much uncertainty of the neural circuits involved in the sexual behavior. Some lesion work had been performed and it was generally believed that the hypothalamus and the limbic brain had an important function in mediating the behavior, although empirical evidence largely failed. Heinrich Kluver had reported dramatic effects of lesions in the limbic brain causing an abnormal increase in sexual behavior, and a possible inhibitory function of the brain in the sexual behavior was well recognized. Further indications of a role of the limbic brain for male sexual behavior was given by MacLean's finding of penile erection in his studies of electrical stimulation of the brain in squirrel monkeys (MacLean, Robinson, & Ploog,1959). When Heimer and I found that lesions in the medial preoptic-anterior hypothalamic area abolished the sexual behavior, we did not further question this observation (Heimer & Larsson, 1966/67). The sexual behavior and also sexual desire were eliminated, or so we believed, even though many of our males continued to follow the females, sniffing them in the back. Repeating these studies many years later, Stefan Hansen made the same observation and, in addition, found that the lesioned rats displayed excessive grooming and scratching, and, if given an opportunity to drink, also excessive drinking (Hansen, & Drake af Hagelsrum, 1984). This displacement activity suggested to them that the animals may have been sexually motivated but were unable to link this motivation with sexual motor behavior. Treatment with lisuride restored normal copulatory behavior, giving further support to the idea that the preoptic lesions did more than eliminate sexual behavior. As we have seen, Everitt and his associates (see chapter in this volume) have reported evidence supporting this hypothesis in a most elegant manner.

In male rats, I found marked changes in the sexual behavior accompanying increasing age. Around puberty, the male needs an exceedingly high number of intromissions to achieve ejaculation, and in old age, the ejaculation latency greatly increases. These variations do not seem to be directly dependent on variations in testosterone levels. My observations were later confirmed in other laboratories, but, somewhat surprisingly, the study of aging in sexual activity is still a neglected area of research.

There is one aspect of this research which I feel may be of particular potential interest, namely, premature aging. Working with the staggerer mouse, Guastavino observed that males, after having initiated copulatory activity at approximately normal age, permanently ceased copulating after only a week (Larsson, Guastavino, & Ly, 1986). Similarly, females began to cycle and sire young at around 3 months of age. They continued to cycle for an additional 3 or 4 months and then became permanently unfertile. Thus, in both males and females, sexual activity may initially be fairly normal, but abnormal features appear after a while. It may be that diseases and abnormalities are expressed by premature aging, and to discover such behavior variations, long term observations of the behavior are needed. An interesting series of studies has been performed by Joseph Knoll (1990), showing that the male rat's life expectancy and sexual appetite can, surprisingly, be enhanced by daily doses of the monoamine oxidase-b inhibitor, L-Deprenyl.

I believed that the sexual behavior analysis, like the Pavlovian analysis of the conditioned salivary reflex, would give us access also to cognitive factors determining the behavior. Although this has been more of an aspiration than an achievement, it may be proper, finally, to discuss some of our observations of the role of cognitive factors in the sexual behavior.

One aspect of the sexual activity is its circadian variation. This variation is adapted to environmental cues by the cycle of light and darkness. Since a reversal of the environmental light/dark cycle results in a reversal of the circadian sexual rhythmicity, but this reversal is not completed before several weeks, I suggested that a conditioned dynamic stereotype had been established. This interpretation was suggested to me from reading Russian physiological literature which began to be available in translation to West European languages during the fifties. The importance of sexual experience for the rise of sexual activity at puberty was another issue which we studied in various ways without, however, obtaining a satisfactory answer. It appears that intact rats readily respond to sexual stimulation with a minimum of learning. It is possible, however, that a role of learning can be more easily uncovered in individuals which are deprived of the many compensatory mechanisms characteristic to the normal animal. This was demonstrated by a series of experiments in which sexually inexperienced male rats were deprived of the sense of smell (Larsson, 1975). When anosmia was induced before the animals had acquired sexual experience, they never copulated. When, however, the same operation was performed in experienced rats, it had only a minor effect on their sexual activity. We suggested, on the bases of these observations, that intact animals may learn to respond sexually to non-odoferous stimuli, but the sense of smell is essential for such learning to take place.

A series of articles by Barry Everitt and his associates has interested me much (Everitt, Cador, & Robbins, 1989; Everitt, Fray, Kostarczyk, Taylor, & Stacey,1987, Everitt, & Stacey, 1987). They have indicated that there is a striking contrast between the interest the psychologists have given to cognitive factors in regulating food and liquid intake and sexual behavior, although in both cases we are dealing with basic bodily needs. Because of my interest in the research on conditioned reflex activity by Russian investigators after the war, as a contrast to the American way of studying learning, I studied the interaction between a Skinnerian pedal-pulling response for food reward and sexual reflexes. Some of the observations may be worthwhile to cite because they need to be reexplored in a more proper conceptual context. I found, for instance, that ejaculation prevented the elicitation of the operant response for only fractions of a minute, intromission having no inhibitory effect as all. Secondly, I observed that the operant and the sexual responses had a mutually activating influence upon one another. Sometimes, a strange "switching" occurred, and the two mutually excluding responses alternated with extreme rapidity. Observations of this kind led me later to observe for effects upon sexual behavior, hoping to use sexual activity as an analgesia. I was never able to integrate these observations in a conceptual schedule and altogether abandoned this approach, although here clearly was a path worth traveling, as has been demonstrated so well by Barry Komisaruk.

The studies of Komisaruk pertain, at present, largely to monoamine involvement. This neuropharmacological approach has even been applied by us in the analysis of male rat sexual behavior. It is well established that the ejaculation latency can be shortened to a few seconds by treating rats with either the serotonin agonist, 8-OH-DPAT, the dopamine agonist, quinperole, or lisuride, a compound that is assumed to exert an agonistic influence upon both serotonin and dopamine receptors (Ahlenius & Larsson, 1987). On the bases of these observations, we suggested that the brain normally exerts an inhibitory influence upon the ejaculatory reflex and that both 5-HT and DA may be involved as neurotransmitters in these mechanisms. This inhibition can be antagonized by using compounds interfering with the monoaminergic neurotransmission. Unfortunately, the mechanisms of action of these substances are still unclear. For instance, 8-OH-DPAT, although primarily a 5-HT agonist, may serve as a serotonin antagonist, as well. It is possible that 8-OH-DPAT, like lisuride and quinperole, acts as 5-HT and/or DA antagonists, thereby facilitating the occurrence of ejaculation. DA, presumably, is important also in mechanisms stimulating sexual behavior, since central DA activates sexual behavior according to much evidence. Interestingly, treatment with 8-OH-DPAT or any of the other substances that facilitate the occurrence of ejaculation does not influence the length of the postejaculatory intervals, suggesting that different neural mechanisms control these two aspects of the sexual behavior pattern. The postejaculatory intervals may be under the influence of GABAergic neural mechanisms, since local infusion of the GABA antagonist bicuculline in the medial preoptic area will shorten the intervals to one or a few minutes (Fernandez-Guasti, Larsson, & Beyer, 1986).

I began this chapter relating my struggle to find a way through the confusion of concepts of cognitive psychology. Now, I find myself in need of some of these concepts. I realize that we have underestimated the capacity of an animal to store and use its own experience. I am not intimidated by the concept of sexual motivation. Sexual orientation and sexual preference cannot be equalized with copulation but calls for an analysis of its own, and, naturally, the brain, with all its potentials, is used in searching for sexual reward as for food and liquid. Masculine or feminine sexual behavior are only concepts representing phenomena observed under experimental conditions, abstractions of real events relating to social interaction between males and females. These events may with equal right be called responses of aggression, or expressions for pain and reward.

REFERENCES

Ahlenius, S., & Larsson; K. (1987). Evidence for a unique pharmacological profile of 8-OH-DPAT by evaluation of its effects on male rat sexual behavior. In C.T. Dourish, S. Ahlenius, and P. Hutson (Eds.), *Pharmacology of Central 5-HT Receptors* (pp. 185-198). Chichester: Ellis Horwood.

Bolles, R.C. (1985). Short-term memory and attention. In L-G- Nilsson and T. Archer (Eds.), *Perspectives on Learning and Memory* (pp. 137-146). Hillsdale, New Jersey: Lawrence Erlbaum Assoc., Inc., Publ.

Bykov, K.M. (1953). *Grosshirnrinde und Innere Organe.* Berlin: VEB Verlag Volk und Gesundheit.

Carlsson, S.G., & Larsson, K. (1962). Intromission frequency and intromission duration in the male rat mating behavior. *Scand. J. Psychol., 3,* 189-191.

Dahlgren, I.L., Eriksson, C.J.P., Gustafsson, B., Harthon, C., Hård, E., & Larsson, K. (1989). Effects of chronic and acute ethanol treatment during prenatal and early postnatal ages on testosterone levels and sexual behaviors in rats. *Pharmacol. Biochem. Behav., 33,* 867-873.

Dallo, J., Lekka N., & Knoll, J. (1985). Age dependent decrease of copulatory activity and its correction by (-) Deprenyl in male rats. *Pharmacol. Soc. Budapest., 3,* 35-38.

Everitt, B.J., Cador, M., & Robbins, T.W. (1989). Interactions between the amygdala and ventral striatum in stimulus-reward associations: studies using a second-order schedule of sexual reinforcement. *Neuroscience 30,* 63-75.

Everitt, B.J., Fray, P.J., Kostarczyk, E., Taylor, S., & Stacey, P. (1987). Studies of instrumental behavior with sexual reinforcement in male rats (Rattus norvegicus): I. Control by brief visual stimuli paired with a receptive female. *J. Comp. Psychol., 101,* 395-406.

Everitt, B.J., & Stacey, P. (1987). Studies of instrumental behavior with sexual reinforcement in male rats (Rattus norvegicus): Effects of preoptic area lesions, castration and testosterone. *J. Comp. Psychol., l0l,* 407-419.

Fernandez-Guasti, A., Larsson, K., & Beyer, C. (1986). GABAergic control of masculine sexual behavior. *Pharmacol. Biochem. Behav., 24,* 1065-1070.

Hansen, S., Köhler, C., Goldstein, M., & Steinbusch, H.W.M. (1982). Effects of ibotenic acid-induced neuronal degeneration in the medial preoptic area and lateral hypothalamic area on sexual behavior in the male rat. *Brain Res., 239,* 213-232.

Hansen, S., & Drake af Hagelsrum, L.J.K. (1984). Emergence of displacement activities in the male rat following the thwarting of sexual behavior. *Behav. Neurosci., 98,* 868-883.

Heimer, L., & Larsson, K. Impairment of mating behavior in male rats following lesions in the preoptic-anterior hypothalamic continuum. *Brain Res., 3,* 248-263.

Hull, C.L. (1943). *Principles of Behavior.* New York: Appleton-Century.

Kaila, E. (1944). Logik und Psychophysik. *Theoria, 10,* 91-119.

Köhler, W. (1929). *Gestalt Psychology.* New York: Liveright Publ. Corp.

Knoll, J. (1990). Effects of L-Deprynyl on age and sexual behavior in male rats. *Polish J. Pharmacol.,* in press.

Konorski, J. (1948). *Conditioned Reflexes and Neuron Organization.* Cambridge University Press.

Larsson, K. (1956). *Conditioning and Sexual Behavior in the Male Albino Rat.* Acta Psychologica Gothoburgensia I, J. Elmgren (Ed.). Stockholm: Almqvist & Wiksell.

Larsson, K. (1960). Excitatory effects of intromission in mating behavior of the male rat. *Behaviour, 16,* 66-73.

Larsson, K. (1975). Sexual impairment of inexperienced male rats following pre- and postpuberal olfactory bulbectomy. *Physiol. Behav., 14,* 195-199.

Larsson, K., Guastavino, J-M., & Ly, K.A. (1986). Altered reproductive behavior of the neurological mutant mouse staggerer. In J. Medioni and G. Vaysse (Eds.), *Genetic Approaches to Behavior* (pp. 69-79). Toulouse: Privat, I.E.C.

Lehrman, D.S. (1953). A critique of Konrad Lorentz' theory of instinctive behavior. *Quart. Rev. Biol., 28,* 337-363.

Lehrman, D.S. (1955). The physiological basis of parental feeding behavior in the ring dove (Streptopelia risoria). *Behaviour, 7,* 16-286.

MacLean, P.D., Robinson, B.W., & Ploog, D.W. (1959). Experiments on localization of genital function in the brain. *Trans. Amer. Neurol. Assoc.,* pp. 105-109.

Olds, J. (1958). Effects of hunger and male sex hormone on self-stimulation of the brain. *J. Comp. Physiol. Psychol., 51,* 320-324.

Sawyer, C.H.H. (1959). Effects of brain lesions on estrous behavior and reflexogeneous ovulation in the rabbit. *J. Exp. Zool., 142,* 227-245.

Skinner, B.F. (1938). *The Behavior of Organisms.* New York: Appleton-Century.

Soulairac, A. (1952). Etude experimentale du comportement sexual male. Independence relative des divers elements moteurs chez le rat male normal. *Ann. d'Endocrinologie, 13,* 775-780.

Tinbergen, N. (1951). *The Study of Instinct.* Oxford Univ. Press.

Tolman, E.C. (1932). *Purposive Behavior in Animals and Man.* New York: Century.

Van Dis, H., & Larsson, K. (1970). Seminal discharge following intracranial electrical stimulation. *Brain Res., 23,* 381-386.

I Developmental Psychobiology

Jean-Marie Guastavino
Universite Paris Nord
Villetaneuse
France

Ethogenesis through normal and mutant animals

This section deals with the ontogenesis and ethogenesis of the individual and is assembled into four chapters. Jay Rosenblatt (Chapter 2) demonstrates the importance of hormonal versus non-hormonal factors influencing maternal behavior in female rats. He indicates that maternal behavior, as measured by aggression directed towards intruders, is dependent upon the hormones estradiol and progesterone, until just after parturition. After 1 or 2 days, maternal aggression is no longer dependent upon these hormones but is instead supported by sensory stimulation from the pups. On the other hand, Michael Leon (Chapter 3) discusses the importance of early learning in rat pups and the correlation with both physiological (relative uptake of 2-DG: 14C labeled 2-deoxyglucose) and morphological parameters (number of neurons in the neuropil). Early learning results in changes in both these parameters. An increase in cellular activity takes place, as revealed by elevated levels of 2-DG uptake. This elevation is associated with an increase in the number of neurons in the neuropil composing the glomeruli in the olfactory bulb. Cheryl Harding (Chapter 5) reports on studies of the role of hormone-sensitive neural mechanisms controlling male social behavior in birds. Both androgenic and estrogenic metabolites are indispensable for normal courtship and aggressive behaviors. The hormonal control of singing behavior varies according to the social context in which the song is taking place. Finally, Ernest Hård (Chapter 4) presents data revealing mechanisms underlying ontogenic development of vocalization in rat pups. He demonstrates the involvement of serotonergic, dopaminergic, noradrenergic, and GABAergic systems in the ontogenesis of ultrasound communication with the mother. These chapters all testify to the importance of environmental and internal factors for the ontogenesis. Species-typical behaviors like parental and mating behavior are required for the survival of the individual and of the species. These behaviors are supported by neuroendocrine circuits, which, like all morphological and physiological features of the organism, have their roots in the genes.

Attempts to reveal the genetic basis of behavior and its physiological substrates have largely been directed towards a global analysis of the role of different gene pools. This is well demonstrated in domestic animal breeding, as this probably has been practiced since thousands of years, but remains as true in present day genetic analysis. The approach taken here is to compare groups having identical gene pools with groups of a different gene pool. However, by this methodology we can only relate differences in behavior to differences in gene pools. While entirely legitimate, this tactic will reveal nothing of the role of individual genes. Further, since the link between differences in behavior and differences in neural structures remains mostly unknown, the physiological mechanisms accounting for the behavioral differences may remain entirely obscure. In order to understand the role of single genes, methods must be used where populations are compared which differ exclusively with respect to one single gene.

The advantage of studying the mutant animal is that it allows us to perform such studies. After the mutant has been identified by its phenotypical abnormality and its genetic basis assured, one may begin to ask for physiological mechanisms accounting for these abnormalities (Sidman, 1968). The neurological mutant gives us an example by Nature of a unique relationship between a single gene, a peculiarity in the nervous system, and the behavior. This provides an opportunity to study neural functions in a way different from that usually practiced in experimental neurology, e.g., by injuring the brain. In the mutant, the lesion is there from the very beginning of the development, progressing with the growth of the animal, precisely localized, without accompanying bleeding, and with no variability. Even the very progression of the lesion is advantageous. Knowing the normal course of development, and systematically changing the environmental conditions for growth, one may evaluate the role of environmental stimulation in the development (Guastavino & Larsson, 1985). This, in turn, will provide a basis for an understanding of the interaction between genetical and environmental factors in determining the behavioral development. It may be argued that the mutant does not allow a choice between different targets. On the other hand, it should be observed that a choice can be made among many different mutants. The rapidly increasing list of neurological mutants now includes about 200 species (Sidman, Green, & Apple, 1965). I will illustrate the use of mutants in this research by demonstrating the existence of alterations in the food intake and in the temporal pattern of the feeding activity of the *staggerer*. Already at the very beginning of our research on the mutant, we observed that these mice were engaged in feeding activity much more frequently than normal mice. Indeed, under standard rearing conditions, undernourishment seemed to be one of the causes of early death in young *staggerer* mice (Guastavino, 1978). We have shown that young *staggerers* continued suckling their mothers up to the age of 60 days, versus 18-20 days for normal pups (Guastavino, 1978). However, in spite of such prolonged care by the mother and access to food ad libitum, the body weight of the mutant pups remained about only half of that of the normal sibs, indicating a possible metabolic abnormality induced by the *staggerer* gene (Guastavino, 1983, Guastavino & Larsson, 1986,). In fact, this observation of frequent feeding activity helped us to detect young mutants in a litter, as they stayed either in the bottom of the nest or close to the containers in which mashed food was delivered (Guastavino, 1976). However, we had not carried out any quantitative study of that behavior in adult *staggerers* until now.

In the present study, we investigated: 1) the existence in *staggerer* mice of the circadian pattern of feeding activity, according to the temperature in which the subjects were reared (28° C and 22° C), as compared to controls; 2) the quanitity of food intake for both *staggerers* and controls, according to the temperature in which the subjects were reared; 3) the existence of a physiological correlate: the activity of the brown adipose tissue of the *staggerer*.

Twenty grams of hydrated diet was delivered in a small plastic container. Every four hours, day and night, the plastic container was removed and weighed in order to determine the quantity of food eaten. About one week before the experiment began, the animals were placed in the experimental conditions, i.e., isolated in an individual cage, subjected to their group temperature (28° C or 22° C), and fed ad libitum with the hydrated diet. However, the food was delivered only twice a day: 8 a.m. and 8 p.m.

The *staggerer* mutation affects the feeding pattern of the *staggerer* mouse. This mutant is observed to eat the same quantities of food when subjected to different temperatures, while the normal mouse adapts its food intake to the ambient temperature. One likely explanation is that the mutant eats as much as possible at the temperature of 28° C and cannot increase its food intake at a lower temperature. Moreover, the day-night rhythmicity observed in the eating behavior of the normal mouse is absent: when subjected to the 22° C conditions, the mutant eats as much during the light phase as during the dark one. In contrast, the normal mouse eats much more during the dark phase, as laboratory rodents do. However, mutant rhythmicity is slightly apparent at the 28° C conditions.

Subjected to temperatures of either 28° C or 22° C, the staggerer is observed to eat more than the control (in g per g of body weight). The ratio is about 1 to 1.5. This hyper-food intake is in discrepancy with the low body weight of the *staggerer,* which represents a drastic deficit, weighing about half of the control. The normal mouse adapts its food intake to the ambient temperature. When the temperature decreases from 28° C to 22° C, the food intake of the mouse increases, whereas the staggerer eats the same amount of food regardless of thermal conditions. Similarly, the obese Zucker rat, with a comparable hyperphagia, does not increase its food intake when subjected to a low temperature (Bertin, De Marco, & Razanamaniraka, 1983; Bertin, De Marco, & Portet, 1984; Portet & De Marco, 1976).

It could be argued that some internal factors are involved. The energetic metabolism has been studied in both normal and mutant animals. Investigations concerning brown adipose tissue (BAT) revealed that BAT activity is stimulated in these animals. In mutant mice born and reared at 28° C, we observed a very high cytochrome oxydase activity (about twice as much as in the normal mice). This activity of the last enzyme of the respiratory chain in the inner membrane of the mitochondria reflects the intensity of the respiratory phenomenon. It shows that even when mutants are maintained close to thermal neutrality, their BAT is stimulated. When mice are reared at a temperature of 22° C, the resting metabolism of the staggerer is raised by 70% and the activity of the cytochrome oxydase by a ratio of 1.5. In both cases, the increased activity of BAT leads to an enhanced energy expenditure which prevents normal growth of the mice by a higher proportion of acoloric consumption.

These observations pose the problem of the relationship between the damaged brain structure on one hand and the altered physiology and behavior on the other. Rhythmicity and food intake are presumably not related to the cerebellum, nor is the physiology of the brown fat tissue. This rises the basic problem whether or not the effects of the mutation surpass the frame of the cerebellum or if this nervous structure is responsible for more functions than those usually known.

REFERENCES

Bertin, R., De Marco, F., & Portet, R. (1984). Effet de l'adaptation au froid sur le comportement alimentaire du rat Zucker obèse génétique. *J. Physiol. Paris, 79,* 361-364.

Bertin, R., De Marco, F., & Razanamaniraka, G. (1983). Effects of cold acclimation on feeding pattern and energetic metabolism of the genetically obese Zucker rats. *Comp. Biochim. Physiol., 74A,* 855-860.

Guastavino, J-M. (1978). Sur le developpement comportemental de la souris atteinte par la mutation staggerer. *C.R. Acad. Sc. Paris, 286,* 137-139.

Guastavino, J-M. (1983). Constraint of the mother with pups restores some aspects of the maternal behavior of mutant staggerer mice. *Physiol. Behav., 30,* 771-774.

Guastavino, J-M., & Larsson, K. (1985). Restorative plasticity in the behavior of the mutant mouse staggerer. In B.E. Will, P. Schmitt, and J.C. Dalrymple (Eds.), *Brain Plasticity Learning and Memory* (pp. 33-40). New York: Plenum Press.

Guastavino, J-M., & Larsson, K. (1986). Genetically determined impairment and environmental recovery in the maternal behavior of the staggerer mutant mouse. *Behav. Genet., 16,* 507.

Portet, R., & De Marco, F. (1976). Comportement alimentaire du rat adapté au froid. *J. Physiol. Paris, 72,* 115A.

Sidman, R.L. (1968). Development of interneural connections in brains of mutant mice. In F.D. Carlsson (Ed.), *Physiological and Biochemical Aspects of Nervous Integration* (pp. 163-193).

Sidman, R.L., Green, M.C., & Appel, S.H. (1965). Catalog of the neurological mutants of the mouse. Harvard University Press.

Sidman, R.L., Lane, P.W., & Dickie, M.M. (1962). Staggerer, a new mutation in the mouse affecting cerebellum. *Science, 137,* 610-612.

2 Aggressive Aspects of Maternal Behavior in the Rat

Anne D. Mayer, Elizabeth M. Factor, and Jay S. Rosenblatt
Institute of Animal Behavior
Rutgers, The State University of New Jersey
Newark, New Jersey

Until recently, the aggressive aspects of maternal behavior in the rat were neglected in our studies of the hormonal basis of this behavior pattern. They are an important feature of maternal behavior serving to protect the young from predator and conspecific attack. While aggression is an integral part of the pattern of maternal behavior, it differs from other components in important ways. Nursing, retrieving, and anogenital licking are directed at the young and provide directly for their care: aggression is directed at conspecific intruders and predators and only indirectly is concerned with their care. In this sense it is like nestbuilding which also is not directed at the young. We shall use the term maternal behavior or maternal responsiveness to refer to the former behavior patterns and reserve the term aggression for the offensive and defensive behavior patterns exhibited by the female in response to an intruder.

During the past five or six years, we have studied patterns of aggression displayed by females during late pregnancy and lactation (Mayer, Carter, Jorge, Mota, Tannu, & Rosenblatt, 1987). We had several questions in mind in initiating these studies: (1) What is the relationship between maternal behavior and aggression during the maternal behavior cycle? When does aggression arise during the cycle and how long is it maintained before it declines? How does this compare with the performance of maternal behavior? (2) Are the hormonal determinants, sensory stimuli, and brain regions mediating aggression the same as those which mediate maternal behavior?

REVIEW OF RESEARCH FINDINGS

Course of aggression and its relation to maternal behavior

Our initial studies were conducted on nulliparous females between 60 and 120 days of age that were mated and then gave birth 21.5 or 22.5 days later. They were given 10.minute tests for aggression with a strange adult male (Mayer, Reisbick, Siegel, & Rosenblatt, 1987) during which the females exhibited both offensive and defensive behavior patterns. These are shown in Table 1. Since then we have frequently used ovariectomized females as intruders; they elicit as much aggression as males.

TABLE 2.1

Behavioral Items Recorded in Aggression Tests	
Sniff Male:	Sniffs, nose close to the male without open mouth contact.
Sniff-Follow:	Maintains nose proximity to the male while following him at least three paw steps.
Rear-Sniff:	Sniffs male, forefeet raised.
Paws-On:	Rears and puts forepaws on male.
Fur-Nibble:	Sows mouth movements with male hair in mouth; occurs most frequently around the head.
Rear-Threat:	Faces male, forefeet elevated, fur sleeked, snout extended, moving only the snout to follow the male.
Attack:	Lunges suddenly at the male with mouth open; usually followed by a scuffle and/or male freezing. The typical male response suggests biting has occurred, but frequently it cannot be seen because of the rapidity of the action.
Pin-Male:	In contact with or hovering over male lying on his back (or side-back).
Rear-Defense:	Forefeet raised, fur ruffled; may squeak, usually looking away from male. Foreleg movements (boxing) sometimes seen when male also is upright.
Squeak:	High pitched, brief vocalization.
On-Back:	Female lies on back or side-back.
Pinned-By-Male:	On back or side, with male in contact or hovering over.
Kick:	Kicks in direction of male with a hind limb.

Our first study traced the rise and decline of aggression during gestation and lactation. Females were tested at intervals from gestation day 1 through lactation day 24. Aggression had a prepartum onset: offensive *attack* and *rear threat* began to increase between day 16 and 19 of gestation and continued to increase reaching peak frequencies on lactation day 1 (Fig. 1). These behavior patterns were maintained at a high level until lactation day 14 after which they abruptly declined. Defensive behavior patterns *on back* and *rear defense* were exhibited at low frequencies throughout gestation, then declined when offensive aggression increased; later, during lactation, *on back* increased in frequency again.

Our results during lactation were similar to those of Flannelly and Flannelly (1987), and Flannelly, Flannelly, and Lore (1986), but differed somewhat from those of Erskine, Barfield, and Goldman (1978) who found the peak frequency of aggression on lactation day 9. Different strains, housing and testing conditions may account for these differences.

The appearance and decline in aggression during lactation closely parallel the appearance and decline in pup nursing, retrieving, and anogenital licking (Fleming & Rosenblatt, 1974; Reisbick, Rosenblatt, & Mayer, 1975; Rosenblatt & Lehrman, 1963).

Late pregnant females exhibited aggression without of course first exhibiting maternal behavior. Maternal responsiveness also increases during the last week of gestation; from gestation day 14 onwards, pregnant females require fewer days of pup exposure before commencing maternal care than do nonpregnant females (Rosenblatt & Siegel, 1975). Females exposed to pups for the first time within a few hrs of delivery frequently retrieve, gather and lick them almost immediately (Mayer & Rosenblatt, 1984; Slotnick, Carpenter, & Fusco, 1973). Thus during pregnancy also, the rise in aggressiveness parallels a rise in maternal responsiveness.

To distinguish aggression displayed during pregnancy, when females are not exhibiting maternal care, from aggression shown after parturition when maternal behavior has been initiated, we have called the former *home cage aggression (HCA)* and reserved the term *maternal aggression (MA)* for the situation in which females are caring for young. The significance of this distinction is that HCA may be stimulated directly by hormones alone, while MA may require some combination of hormones and pup stimulation during the performance of maternal behavior.

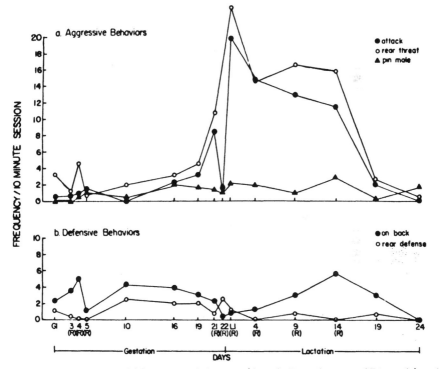

FIG. 2.1. Frequencies of (a) agonistic behaviors (Attack, Rear.threat, and Pin-male) and (b) defensive behaviors (On-back, Rear-defense) shown by females toward intruders as a function of day of Gestation or Lactation. Tests were conducted by introducing an unfamiliar male into the female's home cage and were 10 min in duration. Each point represents the mean score of 6-10 females. Repeated tests are identified by (R); other points are independent groups. By Mann-Whitney U-tests (independent groups), Attacks were less frequent on G10 than on G16, and L19 (P < .02); Attacks were more frequent on G16 and G19 than on L24 (P < .05); Attacks were more frequent on L19 than on L24 (P < .05). See text for analysis of repeated test data. (Reprinted with permission from Mayer, Reisbick, Siegel, and Rosenblatt, 1987.)

Hormonal factors influencing the onset of aggression

1. Pregnancy termination

The parallel between the rise and decline of aggression and maternal responsiveness to young suggested that hormones which stimulate maternal behavior may also stimulate aggression. Terminating pregnancies on day 16 by hysterectomizing (H) females or by hysterectomizing and ovariectomizing them and giving them estradiol benzoate (HO+EB), has proven effective in stimulating maternal behavior (Rosenblatt, Siegel, & Mayer, 1979). These procedures were used to elicit HCA and MA in further work. Table 2 shows the groups that were used and the results of tests of aggression prior to exposure to pups (HCA). The treatment we have found to be most effective in stimulating maternal behavior, HO+EB (20 μg/kg), was also highly effective in stimulating HCA. Pregnancy-termination without ovariectomy, and HO+EB treatment of nonpregnant females (given a higher dose of EB, 100 mg/kg) also raised levels of attack above the negligible levels of sham-treated nonpregnant females, but were much less effective than pregnancy-termination followed by HO+EB.

TABLE 2.2

Agonistic Behavior During Intruder Test 1 (Prior to Exposure to Pups)			
		Females attacking one or more times	
Treatment Group	N	N	%
Pregnant and pregnancy terminated			
Repeated test controls	13	5	
Sham-HO-Oil	15	5	37.8a
Sham-H	9	4	
HO-EB (20 μg/kg)	13	12	92.3b
HO-Oil	13	1	7.7c
H	13	6	46.2
Nonpregnant			
Repeated test controls	9	1	8.3
Sham-HO-Oil	15	1	
HO-EB (100 μg/kg)	17	5	29.4

a Greater than nonpregnant controls, $P < 0.02$
b Greater than Preg and Preg-terminated controls, $P < 0.001$; Preg HO-Oil, $P < 0.001$, Preg H, $P < 0.02$.
c Lesser than Preg and Preg-terminated controls, $P < 0.05$; Preg H, $P < 0.05$

This was the first direct evidence in the female rat that high levels of aggression can be stimulated by ovarian hormones, independent of pup stimulation.

An additional effect of pup stimulation, however, became evident when the females were exposed continuously to foster pups, and retested for aggression after they had initiated maternal behavior. After initiating maternal care, the pregnant group and all pregnancy-terminated groups showed increases in MA over HCA levels. Nonpregnant females, on the other hand, did not increase in aggressiveness while displaying maternal care (Table 3).

TABLE 2.3

Agonistic Responses to Unfamiliar Intruders after Initiation of Maternal Behavior

		Females attacking one or more times	
Treatment Group	N	N	%
Pregnant and pregnancy terminated			
Combined controls	20	18	90.0[a]
HO-EB (29 μg/kg)	13	13	100.0[b]
HO-Oil	13	11	84.6
H	13	12	92.3
Nonpregnant			
Combined controls	8	2	25.0
HO-EB (100 μg/kg)	14	6	42.8

Note: Data are based on females attacking at least once during either or both intruder tests 1 or 2.

a   Greater than nonpregnant controls, $P < 0.001$.
b   Greater than nonpregnant HO-EB, $P < 0.01$.

The results of this study revealed additional similarities between aggression and maternal behavior. We have proposed that the onset of maternal behavior is based upon hormonal stimulation, but pup stimulation is required to maintain this behavior after parturition (Rosenblatt et al, 1979). Moreover, there is an interval immediately following parturition when hormonal stimulation and pup stimulation combine to increase maternal responsiveness. Following this, pup stimulation alone can maintain maternal behavior. With respect to maternal aggression, stimulation from pups during the performance of maternal behavior perhaps also provides an additional impetus, augmenting the effects of hormones during a transition period postpartum.

This hypothesis is supported by the finding of Erskine, Barfield, and Goldman (1980b): maternal aggression can be stimulated nonhormonally in nonpregnant females just as they are able to be stimulated by pups to initiate maternal behavior (Rosenblatt, 1967). A lengthy period of continuous pup exposure was necessary, though once the females exhibited maternal aggression it was as intense as that of lactating females during the second week. It is unlikely that the hormonal factors responsible for aggressiveness during late pregnancy and the peripartum period continue to act throughout lactation, since the females' hormonal profiles are very much altered.

A second finding by these investigators is that hypophysectomy, which greatly reduces ovarian hormones as well as removing pituitary hormones, does not affect the maintenance of postpartum aggression (Erskine, Barfield, & Goldman, 1980a).

2. Estrogen and progesterone treatment

Prolonged treatment of ovariectomized females for 16 days with estradiol (E) and progesterone (P) followed by EB also stimulates short-latency maternal behavior (Giordano, 1987). In our next experiment we tested females for MA following this treatment after they had initiated maternal behavior (Mayer, Ahdieh, & Rosenblatt, in press). In Table 4, we have combined all of the groups that received the full treatment, those that did not receive the final EB injection, and those that received no hormone at all. Not all of the females of each group initiated maternal behavior during 4 days of exposure to pups; this gave us the opportunity to compare maternal aggression in maternal and nonmaternal females of each group.

TABLE 2.4

Number of Maternal and Nonmaternal Females Showing Aggression (More than 1 Attack, or 1 Attack Eliciting a Freezing Response) / Number per Treatment Group			
	Controls	E/P/Oil	E/P/EB
Maternal	1/3	9/11	9/9
Nonmaternal	0/7	2/5	5/7 [a]
All Females	1/10 (10%)	11/16 (69%) [b]	14/16 (88%) [c]

a Different from nonmaternal controls, $P < 0.03$, Fisher Exact Probability Test.
b Different from controls, $X^2 = 8.5$, $P < 0.01$.
c Different from controls, $X^2 = 15.1$, $P < 0.001$.

In the absence of hormone treatment, few of the females initiated maternal behavior within the four days of pup exposure provided by the experiment, and only one displayed aggression toward an intruder. Following E+P treatment, 69% to 88% were aggressive, with maternal females showing aggression more frequently than nonmaternal females (90% vs 58%).

In studies that are still in progress using a similar protocol, we are treating nonpregnant, ovariectomized females for 16 days with combinations of E, P, and EB and testing them for HCA *before* they are exposed to pups and also for MA *after* they have initiated maternal behavior (Mayer, Monroy, Siegel,& Rosenblatt, 1989; see Table 5). Statistical analyses of aggression scores, which are based on latencies to attack, number of attacks, and the presence or absence of intruder freezing, indicated that E+P+EB, E+P, and E+E+EB all significantly raised HCA. However, only females receiving E+P+EB showed elevated levels of MA, which was tested six days after the last hormone injection. At this time, 100% of females that had received E+P+EB were showing maternal care and 100% attacked intruders. To examine the importance of pup stimulation to this result, we tested additional females that received E+P+EB but were not exposed to pups; on the 6th day after EB injection, only 33% displayed any aggression, and their mean aggression score was 1.2 as compared to a mean of 9.7 for those females that had initiated maternal care.

TABLE 2.5

Home Cage and Maternal Aggression after 16-Day Treatment with Estradiol, Progesterone, and Terminal Injection of Estradiol Benzoate, Treatment with Vehicles Only, or Treatment with Fewer Components		
Treatment	HCA (%)	MA (%)
E+B+EB	100	100
E+P+oil	62	50
E+oil+EB	78	25
E+oil+oil	37	50
Blank+P+EB	25	33
Blank+P+oil	0	0
Blank+oil+EB		
Blank+oil+oil	10	14

Note: Females received intruder tests (10 min) for HCA two days
following termination of the 16-day treatment. Exposure to pups
commenced 15-min later and was continuous for 4 days. Intruder tests
for MA were conducted on the final day of pup exposure for (only) females that
had initiated maternal care.

These studies specify that the hormonal basis for aggression in maternal females is the action of the ovarian hormones estradiol and progesterone. The doses of estradiol and progesterone we have used had previously been shown to produce circulating levels similar to those occurring naturally during pregnancy. Contact with pups is *not* necessary for the appearance of HCA after hormone treatment, nor is it necessary in late pregnancy. However, in the absence of pup stimulation aggression wanes within a few days after either the 16-day treatment, or after parturition (unpublished data). The combination of prior hormone exposure and pup stimulation received during the performance of maternal behavior results in high levels of maternal aggression in both nonpregnant, hormone-treated females and postpartum females.

3. Role of pituitary hormones in aggression

A recent series of studies by Bridges and his associates (Bridges, Loundes, DiBase, & Tate-Ostroff, 1985) have implicated prolactin in the initiation of maternal behavior in the rat. For example, they have shown that hypophysectomy reduces maternal responsiveness in nonpregnant females produced by an 11-day treatment with progesterone and estradiol, and that the effectiveness of the progesterone-estradiol treatment can be restored by injecting bovine prolactin. More recently, they have shown that growth hormone also can increase maternal responsiveness (Bridges & Millard, 1988).

We have examined the role of the pituitary gland in the initiation of aggression in estrogen- and progesterone-treated females, using our paradigm for stimulating maternal behavior and aggression described above. At the start of the 16-day hormone treatment, the females were hypophysectomized (Hypx) or Sham-Hypx. They were tested for aggression both before and after four days of continuous exposure to pups, during which the majority initiated maternal behavior.

The groups studied are shown in Table 6. With regard to maternal responsiveness, the results replicated those of Bridges et al. (1985); whereas E+P+EB treatment stimulated short-latency maternal behavior in Sham-Hypx females, Hypx females that had received hormones did not initiate maternal behavior more rapidly than nonhormonally treated animals. Hypophysectomy, however, did not prevent the stimulation of either HCA or MA by hormone treatment. After receiving E+P+EB, 75% of Hypx females showed HCA, and 86% showed MA as compared to 100% of Sham-Hypx (both HCA and MA). No females that had not received hormones were aggressive either before or after initiating maternal care. Mean latencies to attack were significantly longer among Hypx females, and they attacked less frequently, deficiencies which we attribute to their loss of weight following Hypx and to the general physical debilitation which this operation produces. Hypx females lost between 15% and 22% of their body weight during the three weeks that followed the surgery regardless of whether they received ovarian hormones or not.

We conclude from these results that pituitary hormones are not directly involved in the onset of HCA or MA. Erskine et al. (1980a) have shown that the pituitary also is not necessary for the *maintenance* of maternal aggression and this is in agreement with findings on maternal behavior (Obias, 1957). Erskine et al. (1980) showed that females that were Hypx on lactation day 5 were not significantly less aggressive than Sham-Hypx females on day 9. Thus there appears to be a significant difference in the extent to which pituitary hormones influence aggression, and other components of maternal behavior during pregnancy and early lactation.

Sensory factors in maternal aggression

In the mouse, postpartum maternal aggression does not appear to be stimulated by the hormones of pregnancy and parturition, which earlier are responsible for the appearance of aggression during gestation (Svare & Mann, 1983). It is now established that suckling stimulation during the first 48 hr postpartum is required to initiate this behavior after which it can be maintained without such stimulation (Svare & Gandelman, 1976), although suckling continues to modulate the intensity of the female's aggressive behavior (Garland & Svare, 1988).

Gandelman and Simon (1980) have extended this idea to postpartum maternal aggression in the rat. They showed that aggression was absent if pups were removed at parturition and replaced 8 days later, whereas it was present if the pups were replaced after 4 days, although neither the longer nor the shorter interval without pups prevented the females from rapidly initiating maternal behavior. The loss of maternal aggression proved to be correlated with regression in nipple size, which diminished between 4 and 8 days after removal of the young. Additionally, they treated (ovariectomized) nonpregnant females with estradiol and progesterone to produce nipple growth and exposed them to pups, eliciting maternal care. After some days of exhibiting maternal behavior, the majority of the hormonally-treated females were aggressive toward intruders, whereas hormonally-treated females that had not been exposed to pups remained nonaggressive. The authors concluded that in the rat, as in the mouse, estradiol and progesterone facilitates maternal aggression indirectly through stimulating nipple development; maternal aggression that is mediated by suckling rather than by the action of these hormones on the central nervous system.

TABLE 2.6

Home Cage and Maternal Aggression in HYPX and SH-HYPX Females Treated with Estradiol-Progesterone for 16 Days and Given One Injection of Estradiol Benzoate

Group	N	HCA[1]	LAT	FREQ	MA[3]	LAT	FREQ
HYPX/E/P/EP	8	75%	261 ± 91^{2}	2.5 ± 0.82	86%	197 ± 83^{2}	3.6 ± 1.2
SH-HYPX/E/P/EB	8	100%	129 ± 43	5.7 ± 1.1	100%	67 ± 24	7.9 ± 0.4^{3}
HYPX/BLANK/OIL	8	0		0	0		
SH-HYPO/BLANK/OIL/OIL		8				0	

[1] Day 18 test before exposure to pups
[2] HYPX and SH-HYPX non significant differences (P < .10)
[3] After first 4+ score indicating initiation of maternal behavior
[4] SH-HYPX significantly greater than HYPX (P < 0.5)

TABLE 2.7

Maternal Behavior of Thelectomized vs Sham-thelectomized Primiparous Females: Latencies to Commence Retrieving and Time Required to Gather 10 Pups at Tests 30 min following Parturition and on Days 1-5 Postpartum (Repeated Tests of the Same Female)

| | Thelectomized (N = 6) | | Sham-thelectomized (N = 8) | |
| | Latency (sec) | Duration (sec) | Latency (sec) | Duration (sec) |
Day	Mean (SD)	Mean(SD)	Mean (SD)	Mean (SD)
0	121.7 (114.3)	239.3 (185.3)	150.8 (156.7)	107.6 (68.9)
1	22.5* (14.9)	214.0 (250.5)	7.5 (4.9)	163.6 (199.1)
2	18.7 (13.9)	159.5 (168.6)	15.5 (10.5)	81.4 (84.1)
3	38.7 (20.3)	128.5 (103.7)	19.4 (19.3)	113.7 (177.3)
4	18.0 (10.0)	196.8 (191.3)	20.6 (17.3)	102.1 (95.5)
5	21.7 (17.5)	212.3 (202.2)	63.6 (93.6)	236.6 (178.5)

Note: All females commenced retrieving during tests. Females failing to retrieve and gather all 10 pups within 10 min were assigned completion times of 601 sec. On Days 0,4, and 5, Sham-thelectomized N = 7. Data were subjected to square root transformation prior to statistical analysis.

*Different from Sham-thelectomized, t(12) = 2.26, p < 0.05.

We have attempted to test this hypothesis by using thelectomized females (i.e., females whose nipples have been removed surgically; (Mayer & Rosenblatt, 1987). Following the surgery, females were mated and tested for HCA on gestation days 18 and 21, and for MA on lactation days 3, 5, and 12 (some groups received repeated tests, while others were tested one time only). Females without nipples were not less likely to attack than sham.thelectomized females (Table 7). In a second experiment, 16-day pregnant females were hysterectomized-ovariectomized and then exposed to pups until they had shown maternal behavior for 1-2 days when they were tested for MA. All females attacked at least once, and again thelectomized and sham-thelectomized females were equally aggressive.

These results do not support the hypothesis that nipple stimulation is crucial for the initiation of maternal aggression in the rat. However, they do not rule out a role for ventral stimulation which is not eliminated by thelectomy, nor do they address the question of whether nipple stimulation normally plays a role in maintaining MA.

Although the importance of tactile stimulation by pups is unclear, it is evident that pup stimuli of some sort are necessary to maintain postpartum aggression since it declines rapidly if the pups are removed (Erskine et al, 1978). Indeed, Ferreira and Hansen (1986) found that maternal aggression declined significantly after as little as 4 hours of separation from the pups. They then found that suspending the pups in mesh plastic bags within the females' cages was sufficient to prevent the decline in attacks on intruders, indicating that direct contact between female and pups, ventral or otherwise, is not essential at least for the short-term maintenance of MA. The authors reported that only olfactory cues were readily available when pups were in the mesh bags; in other studies, they found that olfactory deafferentation achieved by surgically removing the olfactory epithelium, and lesions of either the mediodorsal thalamic nucleus or insular prefrontal cortex, brain areas which are part of the central olfactory system, significantly diminished the aggressiveness of females tested 6-7 days postpartum (Ferreira, Dahlöf, & Hansen, 1987). These findings suggest that olfactory stimuli are important, perhaps of primary importance, in the maintenance of MA during lactation.

We have studied the roles of olfaction in maternal aggression using intranasal injections of a zinc sulfate solution, which cause necrosis of the olfactory epithelium and therefore (a reversible) olfactory deficit (Mayer, Catania, & Rosenblatt, unpublished data). Lactating females were tested with intruders on either lactation day 4 or day 13, one day following intranasal treatment. In agreement with Ferreira and Hansen (1986), aggression was significantly reduced by olfactory loss at the later point in lactation. On day 4, however, anosmic females were as aggressive as sham-treated females. These findings suggest that the importance of olfactory stimuli develops gradually postpartum, perhaps as the hormonal stimulation of aggression wanes.

As yet, we do not know whether pup odors, the odors of the unfamiliar intruders, or both (perhaps via some comparative mechanism) are important in the mediation of MA.

In the mouse, it appears that the vomeronasal system may be more active in maintaining maternal aggression than the primary olfactory system. Surgical removal of the vomeronasal organs significantly reduces aggression in lactating mice, particularly if they have had little aggressive experience (Bean & Wysocki, 1989).

Recently, Stern and Kolunie (1989) have shown that acute perioral anesthesia of the mystacial pads markedly disrupts attacks on intruders by lactating females, suggesting that somatosensory stimulation in the region of the mouth and snout also plays a role in this behavior. Interestingly, perioral somatosensory cues also appear to be necessary to elicit normal nursing, retrieving, and licking of pups.

The peripeduncular nucleus of the midbrain and maternal aggression

Hansen et al. (1986) reported that bilateral electrolytic lesions in the lateral midbrain, localized to the region of the of the peripeduncular nucleus (PPN), disrupted lactational performance and also abolished postpartum aggression. They attributed both effects to the interruption of ascending somatosensory pathways which subserve oxytocin (and prolactin) release, and interpreted them as supporting the hypothesis that rats are similar to mice in depending upon suckling stimulation to initiate and modulate maternal aggression. Since they had earlier found that axon-sparingacid lesions of the PPN interrupted lactation (Hansen & Köhler, 1984), they proposed that cell bodies within this small site constitute part of that ascending pathway.

At the IAB, Elizabeth Factor has repeated this study to investigate 1) whether the above results can be replicated in our strain of rat, 2) whether lesioned females show deficits in maternal behavior when more refined observational methods are employed than those used by Hansen and Ferreira, 3) whether axon-sparing lesions of the PPN result in the same deficits of MA as electrolytic lesions, and 4) whether lesions of the PPN produced before parturition, or after parturition but earlier than day 8, have the same effects as the day 8 lesions studied by Hansen and Ferreira. This work is still in progress but has come far enough to report some interesting findings.

First, when females were treated on lactation day 8, lesions which damaged the PPN, whether produced by radiofrequency current or by axon-sparing microinjection of N-methyl aspartic acid (NMA), significantly reduced aggression toward intruders and interfered with lactational performance as measured by the daily weight changes of fostered litters. These findings replicate those of Hansen et al. (1986). However, whereas Hansen and Ferreira had reported that the maternal care-giving behavior of lesioned rats was not altered, detailed observations revealed that females with PPN lesions showed deficits in retrieving their pups from runway tunnels or from scattered points within their cages, although they demonstrated oral-motor competence by carrying candy which approximated the size and weight of a pup. On the other hand, nestbuilding and nursing were not different between sham-lesioned and lesioned groups. The significance of the retrieving deficit is not yet clear, but it does suggest that neural sites mediating pup care and MA are not completely independent.

The fourth question has been addressed by lesioning the PPN either prepartum or at one one of two points between parturition and day 8. Curiously, PPN lesions (radiofrequency) made on gestation day 18 has no effect on pregnancy-induced aggression (HCA) nor on any aspect of postpartum maternal behavior, maternal aggression or lactation throughout the first two postpartum weeks. PPN lesions made on the second or fifth postpartum day caused slight deficits in litter weight gain, but did not disrupt retrieving or other maternal behavior or reduce maternal aggression. These results suggest that the PPN gains functional significance during the postpartum period, and that its role may be reduced or eliminated if injury occur prior to about the eighth postpartum day.

CONCLUSIONS

With respect to the questions asked at the start of this research, the following answers can now be given: 1) The aggressive aspects of maternal behavior parallel the other components of maternal behavior during the maternal behavior cycle. They arise during gestation (HCA) and are maintained and enhanced during lactation (MA). Aggression declines at around the same time as retrieving, nestbuilding, and anogenital licking; nursing may continue somewhat longer. 2) Aggression and maternal behavior are stimulated by the same combination of estradiol and progesterone, although maternal behavior also appears to require pituitary hormones, especially prolactin, and aggression does not. With regard to both maternal responsiveness and aggression, it is likely that estradiol and progesterone act chronically to prepare the female to respond to the terminal rise in estrogen which precedes parturition.

Neither the prepartum rise in maternal responsiveness nor the home cage aggression displayed during late pregnancy require pup stimulation, although the former cannot be measured without exposing the females to test pups. During the postpartum period, however, both maternal behavior and maternal aggression require contact with pups to be maintained. There is a long-term component, which is dependent upon pup stimulation, underlying both maternal behavior and maternal aggression: 7-10 days without pups starting at birth results in the loss of maternal responsiveness (Orpen & Fleming, 1987) and also maternal aggression (Gandelman et al. 1980). Maternal aggression has a short-term component as well: 4-24 hrs without pups leads to a temporary loss of aggression which is rapidly restored when pups are reintroduced (Ferreira et al. 1986).

It seems likely, therefore, that aggressive components of maternal behavior conform to the larger maternal pattern with respect to the hormonal onset and nonhormonal maintenance of the behavior, with a transition period around parturition joining the two.

ACKNOWLEDGEMENTS

The research in this article was supported by NSF Grant BNS 8808662 to JSR. EMF was partially supported by a grant from Sigma Xi. We wish to acknowledge the contribution of several undergraduate students who were supported by a Minority Biomedical Research Grant (JSR). Our appreciation to Winona Cunningham for secretarial assistance and Cindy Banas for graphic art. Publication number 504 of the Institute of Animal Behavior.

REFERENCES

Bean, N.J., & Wysocki, V.J. (1989). Vomeronasal organ removal and female mouse aggression: the role of experience. *Physiol. Behav., 45,* 875-882.

Bridges, R.S., Loundes, D.D., DiBase, R., & Tate-Ostroff, B.A. (1985). Prolactin and pituitary involvement in maternal behavior in the rat. In R.M. MacLeod, M.O. Thorner, & U. Scapagnini (Eds.), *Prolactin: Basic and Clinical Correlates, Fidi Research Series, Vol. I* (pp. 591-599). Livinia: Padua.

Bridges, R.S., & Millard, W.J. (1988). Growth hormone is secreted by ectopic pituitary grafts and stimulates maternal behavior in rats. *Horm. Behav., 22,* 194-206.

Erskine, M.S., Barfield, R.J., & Goldman, B.D. (1978). Intraspecific fighting during late pregnancy and lactation in rats and effects of litter removal. *Behav. Biol., 23,* 206-218.

Erskine, M.S., Barfield, R.J., & Goldman, B.D. (1980a). Postpartum aggression in rats. I. Effects of hypophysectomy. *J. Comp. Physiol. Psychol., 94,* 484-494

Erskine, M.S., Barfield, R.J., & Goldman, B.D. (1980b). Postpartum aggression in rats: II. Dependence on maternal sensitivity to young and effects of experience with pregnancy and parturition. *J. Comp. Physiol. Psychol., 94,* 495-505.

Ferriera, A., Dahlöf, L-G., & Hansen, S. (1987). Olfactory mechanisms in the control of maternal aggression, appetite, and fearfulness: Effects of lesions to olfactory receptors, mediodorsal thalamic nucleus and insular prefrontal cortex. *Behav. Neurosci., 101,* 709-717.

Ferreira, A., & Hansen, S. (1986). Sensory control of maternal aggression in Rattus norvegicus. *J. Comp. Physiol. Psychol., 100,* 173-177.

Flannelly, K.J., & Flannelly, L. (1987). Time course of postpartum aggression in rats (Rattus norvegicus). *J. Comp. Psychol., 1,* 101-103.

Flannelly, K.J., Flannelly, L., & Lore, R.K. (1986). Postpartum aggression against intruding male conspecifics by Sprague-Dawley rats. *Behav. Proc., 13,* 279-286.

Fleming, A.S., & Rosenblatt, J.S. (1974). Maternal behavior in the virgin and lactating rat. *J. Comp. Physiol. Psychol., 86,* 957-972.

Gandelman, R., & Simon, N.G. (1980). Postpartum fighting in the rat: nipple development and the presence of young. *Behav. Neural Biol., 28,* 350-360.

Garland, M., & Svare, B. (1988). Suckling stimulation modulates the maintenance of postpartum aggression in mice. *Physiol. Behav., 44,* 301-305.

Giordano, A.L. (1987). *Relationship between nuclear estrogen receptor binding in the preoptic area and hypothalamus and the onset of maternal behavior in female rats.* Ph.D. Thesis, Rutgers University.

Hansen, S., & Ferreira, A. (1986). Food intake, aggression, and fear behavior in the mother rat: control by neural systems concerned with milk ejection and maternal behavior. *Behav. Neurosci., 100,* 64-70.

Hansen, S., & Köhler, C. (1984). The importance of the peripeduncular nucleus in the neuroendocrine control of sexual behavior and milk ejection in the rat. *Neuroendocrinology, 39,* 563-572.

Mayer, A.D., Ahdieh, H.B., & Rosenblatt, J.S. (In press). Effects of prolonged estrogen-progesterone treatment and hypophysectomy on the stimulation of short-latency maternal behavior and aggression in female rats. *Horm. Behav.*

Mayer, A.D., Carter, L., Jorge, W.A., Mota, M.J., Tannu, S., & Rosenblatt, J.S. (1987). Mammary stimulation and maternal aggression in rodents: thelectomy fails to reduce pre- or postpartum aggression in rats. *Horm. Behav., 21,* 501-510.

Mayer, A.D., Monroy, M.A., Siegel, H.I., & Rosenblatt, J.S. (1989). *Stimulation of home-cage and maternal aggression in nonpregnant female rats by prolonged treatment with estrogen.* Presented at the 21st meeting of the Conference on Reproductive Behavior, Saratoga, Springs, New York, Abst. (p. 60).

Mayer, A.D., Reisbick, S., Siegel, H.I., & Rosenblatt, J.S. (1987). Maternal aggression in rats: changes over pregnancy and lactation in a Sprague-Dawley strain. *Aggr. Behav., 13,* 29-43.

Mayer, A.D., & Rosenblatt, J.S. (1984). Prepartum changes in maternal aggressiveness and nest defense in Rattus norvegicus. *J. Comp. Psychol., 98,* 177-188.

Mayer, A.D., & Rosenblatt, J.S. (1987). Hormonal factors influence the onset of maternal aggression in laboratory rats. *Horm. Behav., 21,* 253-267.

Obias, M.D. (1957). Maternal behavior of hypophysectomized gravid albino rats and development and performance of their progeny. *J. Comp. Physiol. Psychol., 50,* 120-124.

Orpen, B.G., & Fleming, A.S. (1987). Experience with pups sustains maternal responding in postpartum rats. *Physiol. Behav., 40,* 47-54.

Reisbick, S., Rosenblatt, J.S., & Mayer, A.D. (1975). Decline of maternal behavior in the virgin and lactating rat. *J. Comp. Physiol. Psychol., 89,* 722-732.

Rosenblatt, J.S. (1967). Nonhormonal basis of maternal behavior in the rat. *Science, 156,* 1512-1514.

Rosenblatt, J.S., & Lehrman, D.S. (1963). Maternal behavior of the laboratory rat. In H.L. Rheingold (Ed.), *Maternal Behavior in Mammals* (pp. 8-57). New York: Wiley.

Rosenblatt, J.S., & Siegel, H.I. (1975). Hysterectomy-induced maternal behavior during pregnancy in the rat. *J. Comp. Physiol. Psychol., 89,* 685-700.

Rosenblatt, J.S., Siegel, H.I., & Mayer, A.D. (1979). Progress in the study of maternal behavior in the rat: hormonal, nonhormonal, sensory, and developmental aspects. In J.S. Rosenblatt, R.A. Hinde, C.G. Beer, and M.-C. Busnel (Eds.), *Advances in the Study of Behavior, Vol. 10* (pp. 225-311). New York: Academic Press.

Slotnick, B.M., Carpenter, M.L., & Fusco, R. (1973). Initiation of maternal behavior in pregnant nulliparous rats. *Horm. Behav., 4,* 53-59.

Stern, J.M., & Kolunie, J.M. (1987). *Perioral stimulation from pups: role in Norway rat maternal retrieval, licking, crouching, and aggression.* Conference on Reproductive Behavior, Tlaxcala, Mexico, June.

Svare, B., & Gandelman, R. (1976). Postpartum aggression in mice. *Horm. Behav., 7,* 407-416.

Svare, B., & Mann, M.A. (1983). Hormonal influences on maternal aggression. In B. Svare (ed.), *Hormones and Aggressive Behavior* (pp. 91-104). New York: Plenum Press.

3 Learning Early in Life: A Neurological Analysis

Michael Leon
University of California
Irvine, California

INTRODUCTION

In order to survive, altricial mammals must seek out and maintain contact with their mother. In the species that have been studied, olfactory cues appear to be used by young mammals to identify their mother. Mothers typically produce an individualized odor and the young must learn to approach it postnatally (Leon, 1983). In the following review, I will discuss the role of such early experiences in altering the olfactory preference behavior of human and rat neonates. I will also discuss the neural changes evoked by early learning in developing rats.

Early olfactory learning in human infants

In the days after birth, human infants come to be attracted to the odor of their mother's breast and will prefer it to that of a strange mother (Macfarlane, 1975; Russell, 1976; Schaal, 1988). Infants will also develop a preference for a nonmaternal odor (Balogh & Porter, 1986). We have recently shown that this kind of preference may be due to an associative process (Sullivan, Taborsky-Barbar, Mendoza, Itino, Leon, Cotman, Payne, & Lott, in press). Specifically, newborn infants were exposed to a citrus odor that was paired with tactile stimulation designed to mimic maternal contact. On the second day of life, the infants were tested for a preference to the odor that could be demonstrated by either a head turn or a conditioned increase in body movement.

Infants that were concurrently exposed to both citrus odor and the tactile stimulation, subsequently, had both an increase in head turns and an increase in body movement in response to that odor. The infants exposed sequentially to the odor and tactile stimulation (and therefore unable to associate the stimuli) had no change in response to the odor. Similarly, infants exposed to the odor alone or to tactile stimulation alone on the first day of life, increased neither body movement nor head turns toward the odor when reexposed to the odor on the second day of life. The conditioned olfactory responses were then shown to be independent of arousal state and were specific to the conditioned odors. These data indicate that the development of this early olfactory preference may be due to an associative process and not to either a simple familiarity or an imprinting process.

Early olfactory learning in rats

Young rats also come to approach the odor of their mother (Leon, 1983). Since this odor can vary with the diet of the mother (Leon, 1983), pups must acquire a preference for the maternal odor postnatally. In addition to maternal odors, young rats can also acquire a preference for nonmaternal odors, particularly if that odor is paired with tactile stimulation of the kind that mimics maternal contact (Coopersmith & Leon, 1986; Sullivan & Leon, 1986). The development of such a preferential response appears to be due to an associative process, rather than to simple exposure to the odor (Sullivan, Wilson, & Leon, 1989b). Only those pups experiencing concurrent odor and tactile stimulation develop a preference for the odor; odor exposure alone, or sequential exposure to tactile stimulation and odor does not induce an olfactory preference.

Neural correlates of early olfactory learning

To examine the neural changes accompanying early olfactory learning, we determined the relative uptake of ^{14}C-labeled 2 deoxyglucoses (2-DG) in the olfactory bulb in response to the conditioned odor. This technique allows one to survey differential cellular activity in the brain by differential uptake of this radioactively labeled glucose analogue. Since the 2-DG is incompletely metabolized, it accumulates in brain cells. The assumptions here are that brain activity depends on glucose utilization and that 2-DG uptake reflects glucose metabolism. In the olfactory bulb, there are different focal patterns of 2-DG uptake in the glomerular layer for different odors (Sharp, Kauer, & Shepherd, 1975). The individual glomeruli are relatively large balls of neuropil in which the olfactory receptor neurons make synaptic contact with the second-order neurons in the olfactory system. Restricted regions of the mammalian brain are therefore revealed by this technique for quantitative, systematic investigation.

When postnatal day 19 pups that have been injected with [14C]2-DG are exposed to the maternal odor with which they had been raised, they show an enhanced focal uptake of the glucose analogue compared to the glomerular uptake of pups raised with mothers whose dominant maternal odor had been suppressed (Sullivan, Wilson, Wong, Correa & Leon, in press). Rat pups therefore normally develop an enhanced neural response to their mother's odor.

Pups will also develop an enhanced glomerular response to a nonmaternal odor when it has been experienced while their mother's dominant odor has been suppressed (Sullivan et al., submitted). When peppermint odor was introduced to the pups while their mothers were caring for them from postnatal day (PND) 1-18, the pups developed both a behavioral preference and an enhanced focal glomerular uptake of 2-DG in response to that odor.

Rat pups will even develop a behavioral preference and an enhanced neural response to nonmaternal odors that have been paired with tactile stimulation designed to mimic maternal contact. Coopersmith and Leon (1984) trained rat pups in this manner for 10 min/day on PND 1-18. On PND 19, the pups were injected with 2-DG and then exposed to the training odor. Those rats that received the olfactory stimulation that had been paired with tactile stimulation developed a preference for the odor and had an enhanced focal 2-DG uptake in the glomerular layer of the olfactory bulb in response to the trained odor (Coopersmith & Leon, 1984). Those pups that received odor alone, tactile stimulation alone, or sequential presentation of odor and tactile stimulation, developed neither the behavioral preference nor the enhanced focal 2-DG uptake in response to the trained odor (Sullivan & Leon, 1986; Sullivan, Wilson & Leon, 1989b). These data indicate that the neurobehavioral response to the trained odor is due to an associative process, rather than to an imprinting or familiarization process. Indeed, the successful training procedure constitutes a classical conditioning paradigm.

This neurobehavioral response can be acquired during the first postnatal week, but not subsequently (Woo & Leon, 1987). Training during PND 1-4 is also insufficient to produce a change in brain or behavior on PND 19. At the same time, a single brief training episode on PND 6 produces a behavioral preference and an enhanced 2-DG uptake on PND 7 (Sullivan & Leon, 1987). We do not know whether the effects of a single training trial persist past PND 7.

Although the neurobehavioral consequences of early olfactory preference conditioning develop during a sensitive period early in life, these changes are not transient. Rats trained on PND 1-18 and tested on PND 90 retain their enhanced 2-DG uptake in the olfactory bulb glomerular layer (Coopersmith & Leon, 1986). Adult rats also show increased behavioral responsiveness to odors conditioned early in life (Fillion & Blass, 1986).

The neurobehavioral response appears to be limited to a specific developmental period, but it is not limited to a single nonmaternal odor. Training with different odors produces specific behavioral preferences and enhanced neural responses in different areas of the bulb (Coopersmith, Henderson, & Leon, 1986). The glomerular 2-DG response is also specific to the conditioned odor; training with one odor does not enhance the glomerular responsiveness to another odor (Coopersmith, Henderson, & Leon, 1986). These data speak to the specificity of the neural consequences of early olfactory conditioning.

Differential respiration

One possible mechanism underlying the enhanced neural response is a behavioral modification in response to the conditioned odor. That is, pups that have developed a preference for an odor may sample it more than controls by increasing their respiration rate in its presence. This behavior pattern would increase olfactory stimulation at the level of the olfactory receptor neuron, thereby increasing the glomerular 2-DG response. However, when we monitored pup respiration during 2-DG testing, we found no difference in respiration rate on PND 19 (Coopersmith & Leon, 1984; Coopersmith & Leon, 1986; Coopersmith, Henderson, & Leon, 1986; Sullivan & Leon, 1986; Woo & Leon, 1987; Sullivan, Wilson, & Leon, 1988a). The exception to this finding is the PND 7 pups, conditioned on PND 6, who increased their respiration in response to the trained odor (Sullivan & Leon, 1987). This finding raises the interesting possibility that different mechanisms underlie similar functional changes in the bulb at different stages of development.

An additional, critical experiment was performed to evaluate the role of respiration in these pups during the PND 19 2-DG test. Anesthetized pups were tracheotomized and artificially respired during the 2-DG test. While all pups had an identical number of respirations imposed on them, only the conditioned pups had an enhanced 2-DG uptake in response to the trained odor (Sullivan, Wilson, Kim & Leon, 1988). These data make it very unlikely that differences in respiration produce the enhanced 2-DG response in pups trained on PND 1-18.

Anatomical changes

We then considered the possibility that the increase in 2-DG uptake in conditioned pups could be a result of an increase in the number of neurons associated with the enhanced 2-DG foci. To address this possibility, we counted and measured the size of the glomerular-layer neurons associated with the 2-DG foci of conditioned and control pups (Woo & Leon, submitted). While there was no difference in the soma size of the neurons, there was a 19% increase in the number of neurons associated with the 2-DG foci. The increase in

31

glomerular-layer neurons is largely due to an increased number of periglomerular neurons in the 2-DG foci. There were no differences between groups outside the 2-DG foci. This increase could be due to differential survival of these neurons, changes in migration patterns of neurons during development, or even the production of additional neurons destined for those glomerular regions. We are currently evaluating these possibilities.

An increase in the neuropil composing the glomeruli associated with the enhanced 2-DG foci accompanies the increase in cell number in conditioned pups. The olfactory bulbs of conditioned and control pups were tested for 2-DG uptake in response to the conditioned odor and then alternate sections were reacted for succinic dehydrogenase to reveal the glomerular neuropil (Woo, Coopersmith, & Leon, 1987). The stained sections were then aligned with the 2-DG autoradiographs to determine the location of the 2-DG foci with respect to the glomerular regions. Conditioned pups have a glomerular region 26% wider than controls in the focal 2-DG regions and the groups do not differ in the nonfocal glomerular regions.

Part of the reason for the increase in the width of the glomerular layer is that there is a 20% increase in the size of individual glomeruli in the 2-DG foci of conditioned pups. In addition, there is also an increase in the area occupied by cell bodies in the glomerular layer of conditioned pups that reflects the increase in cell number.

While it is not clear in which cells the 2-DG uptake increases, the increase in neuropil and cell bodies in the 2-DG foci could account for the enhanced uptake observed as a consequence of learning. It is also possible that increased glial uptake could contribute to the enhanced uptake. Another possibility is that there are long-lasting changes in the pathways for glucose metabolism that underlie the enhanced 2-DG uptake. Finally, it is possible that there is an increased firing of the external tufted cells in response to the conditioned odor. Indeed, there is some indirect evidence supporting this last possibility. There is an increase in glycogen phosphorylase activity in a restricted region, deep to the conditioned 2-DG foci that increases in response to the conditioned odor (Coopersmith & Leon, 1987). This region coincides with the projection of axon collaterals in a subclass of tufted cells (Orona, Rainer, & Scott, 1984).

Changes in olfactory bulb output signal

Given the increase in periglomerular neurons associated with the conditioned 2-DG foci and the inhibitory role of such neurons at the glomerular level (Shepherd, 1972), it seemed possible that there would be a suppression of the output neurons of the bulb that are associated with these glomeruli. Increased firing of the tufted cell axon collaterals might also suppress mitral cell responses. We therefore recorded from the output neuron population (the mitral and tufted cells) associated with the 2-DG foci of conditioned and control pups. Indeed, there is a striking change in the neurophysiological response pattern to the conditioned odor (Wilson, Sullivan, & Leon, 1987; Wilson & Leon, 1988a). The cells in the regions of the enhanced 2-DG uptake in the bulbs of conditioned pups have significantly more inhibitory responses and significantly fewer excitatory responses to the conditioned odor than control group responses. This pattern of response is specific to both the trained odor and the output neurons associated with the region of the enhanced 2-DG foci. There are no differences in the number of responses or in the magnitude of either the excitatory or inhibitory responses to the trained odor. There are also no respiratory differences between trained and control pups during neurophysiological recordings. Sequential presentation of odor and stroking during the training sessions did not affect the subsequent neurophysiological output signal of the bulb (Sullivan, Wilson, & Leon, 1989b). The olfactory bulb therefore sends what may be a unique signal to the olfactory projection regions to indicate the presence of a learned odor.

Pharmacological mechanisms

The type of early learning that I have described depends on concurrent exposure to both the odor and the reinforcing tactile stimulation. To understand the cellular bases of the neural changes associated with this kind of learning, we thought it more profitable to start our study of the cellular aspects of early learning with a consideration of tactile stimulation.

The effects of the tactile stimulation that we studied appear to be modulated by the noradrenergic system. The specific kind of tactile stimulation that is reinforcing to pups activates the noradrenergic neurons in the locus coeruleus, a response that is limited to early life (Nakamura, Kimura, & Sakaguci, 1987). Locus coeruleus neurons are also electrotonically coupled early in life (Christie, Williams, & North, 1987), perhaps amplifying the effect of such stimulation.

The neurons in the locus coeruleus are noradrenergic and send a massive projection to the olfactory bulb (Shipley, Halloran, & De la Torre, 1985). This projection is present (McLean & Shipley, 1987) and functional (Wilson & Leon, 1988b) in the first week of life. It therefore seemed possible that if the stroking stimulation was facilitating the development of the neurobehavioral response pattern by the release of norepinephrine, it should be possible to block those changes by blocking noradrenergic receptors during training (Sullivan, Wilson, & Leon, 1989a).

We first blocked beta noradrenergic receptors with the drug propranolol during the PND 1-18 conditioning sessions. PND 19 testing revealed a suppression of the behavioral preference, the enhanced 2-DG response, and the altered neurophysiological response of the bulb output neurons (Sullivan et al., 1989a). We then substituted a beta noradrenergic agonist for the tactile stimulation, pairing it with olfactory stimulation on PND 1-18. This substitution proved successful, as the pups developed the behavioral preference, the enhanced 2-DG response, and the suppressed mitral/tufted cell response to the odor. Moreover, the dose/response curve for the agonist described an inverse "U" shape that is characteristic of the facilitatory effects of norepinephrine on memory in adult rats (McGaugh, 1983).

While some or all of these noradrenergic effects could be due to direct neural responses, it is also possible that noradrenergic modulation of early learning is due to indirect effects. One candidate for an indirect action involves the very high levels of glycogen in the olfactory bulb (Hussong, Coopersmith, & Leon, 1989). This is accompanied by high activity of glycogen phosphorylase, the rate-limiting enzyme involved in the mobilization of glycogen to glucose (Coopersmith & Leon, 1987). Noradrenaline can mobilize glucose from olfactory bulb glycogen (Hussong et al., 1989) and it can also increase circulating glucose from peripheral stores of glycogen in young rats (Coopersmith & Leon, unpublished observations). Many of the facilitatory effects of epinephrine on memory formation can be mimicked by glucose (Gold, 1986; Messier & White, 1984) and it is possible that some of the effects of noradrenaline on early learning is mediated by its effects on either central or peripheral stores of glucose.

The excitatory amino acids are also involved in early olfactory learning. The putative neurotransmitters glutamide and N-aspartylglutamate are found in the mitral and tufted cell populations in the rat olfactory bulb (Anderson, Henderson, Cangro, Namboodiri, Neale, & Cotman, 1987; Saito, Kumoi, & Tanaka, 1986) and glutamate receptors of the N-methyl-D-aspartate (NMDA) type are found in the olfactory bulb and its projection sites (Monaghan & Cotman, 1985). We found that blocking these receptors with APV during the daily conditioning sessions also blocks both the behavioral preference and the enhanced 2-DG uptake in the olfactory bulb glomerular layer in response to the preferred odor (Lincoln, Coopersmith, Harris, Cotman, & Leon, 1988).

CONCLUSIONS

The analysis of learning in the developing brain has allowed us to reveal dramatic localized changes that have been difficult to find in similar investigations of adults. The unique plasticity of the young brain may amplify critical experiences into major changes in local neuronal circuitry and such changes may be large and permanent enough to be evident to the investigator. The relatively simple nature of the learning, combined with the relatively simple nature of the neural network underlying the altered neural response make this model an exciting one for the investigation of the neural basis of learning and memory.

ACKNOWLEDGEMENTS

The research presented here has been supported in part by a grant from NICHD (HD24236) and a Research Scientist Development Award from NIMH (MH00371).

REFERENCES

Anderson, K.J., Henderson, S.R., Cangro, C.B. Namboordiri, M.A., Neale, J.H., & Cotman, C.W. (1987). Localization of N-aspartylglutamate-like immunoreactivity in selected areas of the rat brain. *Neurosci. Lett., 72,* 14-20.

Balogh, R.D., & Porter, R.H. (1986). Olfactory preference resulting from mere exposure in human neonates. *Infant Behav. Dev., 9,* 395-401.

Christie, M.J., Williams, J.T., & North, R.A. (1987). Synchronous activity in locus coeruleus neurons from neonatal rats. *Soc. Neurosci. Abst., 13,* 538.

Coopersmith, R., Henderson, S.R., & Leon, M. (1986). Odor specificity of the enhanced neural response following early odor experience in rats. *Dev. Brain Res., 27,* 191-197.

Coopersmith, R., & Leon, M. (1984). Enhanced neural response following postnatal olfactory experience in Norway rats. *Science, 225,* 849-851.

Coopersmith, R., & Leon, M. (1986). Enhanced neural response by adult rats to odors experienced early in life. *Brain Res., 371,* 400.403.

Coopersmith, R., & Leon, M. (1987). Glycogen phosphorylase activity in the olfactory bulb of the young rat. *J. Comp. Neurol., 261,* 148-154.

Fillion, T.J., & Blass, E.M. (1986). Infantile experience with suckling odors determines sexual behavior in male rats. *Science, 231,* 729-731.

Gold, P. (1986). Glucose modulation of memory storage processing. *Behav. Neural Biol., 45,* 342-349.

Hussong, M., Coopersmith, R., & Leon, M. (1989). Alpha-1 adrenergic mediation of noradrenaline-stimulated glycogenolysis in the rat olfactory bulb. *Soc. Neurosci. Abst., 15,* 1315.

Leon, M. (1983). Chemical communication in mother-young interactions. In J. Vandenbergh (Ed.), *Pheromones and Mammalian Communication* (pp. 39-77). New York: Academic Press.

Lincoln, J., Coopersmith, R., Harris, E.W., Cotman, C.W., & Leon, M. (1988). NMDA receptor activation is required for the neural basis of early learning. *Dev. Brain Res., 39,* 309-312.

Macfarlane, A.J. (1975). Olfaction in the development of social preferences in the human neonate. *Ciba Found. Symp., 33,* 103-117.

McGaugh, J.L. (1983). Hormonal influence on memory. *Annual Rev. Psychol., 34,* 297-323.

McLean, J.H., & Shipley, M.T. (1987). Postnatal development of noradrenergic afferents to the olfactory bulb. *Assoc. Chemoreception Sci. Abst., 58,* 510-524.

Messier, C., & White, N. (1984). Contingent and non-contingent action of sucrose and saccharin reinforcers: Effects of taste preference and memory. *Physiol. Behav., 32,* 195-203.

Monaghan, D.T., & Cotman, C.W. (1985). Distribution of NMDA-sensitive L-^{3}H glutamate binding sites in the rat brain as determined by quantitative autoradiography. *J. Neurophysiol., 5,* 2902-2919.

Nakamura, S., Kimura, F., & Sakaguchi, T. (1987). Postnatal development of electrical activity in locus coeruleus. *J. Neurophysiol., 58,* 510-524.

Orona, E., Rainer, E.C., & Scott, J.W. (1984). Dendritic and axonal organization of mitral and tufted cells in the rat olfactory bulb. *J. Comp. Neurol., 226,* 346-356.

Russell, M.J. (1976). Human olfactory communication. *Nature, 260,* 520-522.

Saito, N., Kumoi, K., & Tanaka, C. (1986). Aspartate-like immunoreactivity in mitral cells of the rat olfactory bulb. *Neurosci. Lett., 65,* 89-93.

Schaal, B. (1988). Olfaction in infants and children: developmental and functional perspectives. *Chem. Senses, 13,* 145-190.

Sharp, F., Kauer, J., & Shepherd, G.M. (1975). Local sites of activity related glucose metabolism in rat olfactory bulb during olfactory stimulation. *Brain Res., 98,* 596-600.

Shepherd, G. (1972). Synaptic organization of the mammalian olfactory bulb. *Physiol. Rev., 52,* 864-916.

Shipley, M.T., Halloran, F.J., & De la Torre, J. (1985). Surprisingly rich projection from locus coeruleus to the olfactory bulb in the rat. *Brain Res., 329,* 292-299.

Sullivan, R.M., & Leon, M. (1986). Early olfactory learning induces an enhanced olfactory bulb response in rats. *Dev. Brain Res., 27,* 278-282.

Sullivan, R.M., & Leon. M. (1987). One-trial olfactory learning enhances olfactory bulb responses to an appetitive conditioned odor in 7-day-old rats. *Dev. Brain Res., 35,* 307-311.

Sullivan, R.M., Taborsky-Barbar, S., Mendoza, R., Itino, A., Leon, M., Cotman, C., Payne, T., & Lott, I. Olfactory classical conditioning in neonates. *Pediatrics,* in press.

Sullivan, R.M., Wilson, D.A., Kim, M.H., & Leon, M. (1988). Behavioral and neural correlates of postnatal olfactory conditioning: Effect of respiration on conditioned neural responses. *Physiol. Behav., 44,* 85-90.

Sullivan, R.M., Wilson, D.A., & Leon, M. (1989a). Norepinephrine and learning-induced plasticity in infant rat olfactory system. *J. Neurosci., 9,* 3998-4006.

Sullivan, R.M., Wilson, D.A., & Leon, M. (1989b). Associative processes in early olfactory preference formation: neural and behavioral consequences. *Psychobiology, 17,* 29-33.

Sullivan, R.M., Wilson, D.A., Wong, R., Correa, A., & Leon, M. Modified behavioral and olfactory responses to maternal odors in preweanling rats. *Brain Res.,* in press.

Wilson, D.A., & Leon, M. (1988a). Spatial patterns of olfactory bulb single-unit responses to learned olfactory cues. *J. Neurophysiol., 59,* 1770-1782.

Wilson, D.A., & Leon, M. (1988b). Noradrenergic modulation of olfactory bulb excitability in the postnatal rat. *Dev. Brain Res., 42,* 69-75.

Wilson, D.A., Sullivan, R.M., & Leon, M. (1987). Single unit analysis of postnatal learning: modified olfactory bulb output response patterns to learned attractive odors. *J. Neurosci., 7,* 3154-3162.

Woo, C.C., & Leon, M. (1987). Sensitive period for neural and behavioral response development to learned odors. *Dev. Brain Res., 36,* 309-313.

Woo, C.C., & Leon, M. Early learning increases a neuronal population in the olfactory bulb. Submitted.

Woo, C.C., Coopersmith, R., & Leon, M. (1987). Localized changes in olfactory bulb morphology associated with early olfactory learning. *J. Comp. Neurol., 263,* 113-125.

4 Ontogeny of Ultrasonic Vocalization in the Rat: Influence of Neurochemical Transmission Systems

Ernest Hård and Jörgen Engel
University of Göteborg

In many animal species , including man, separation of offspring from their mother provokes an array of reactions in the child, collectively designated separation distress anxiety (Hofer & Shair, 1987; Levine, 1986; Shirley & Poyntz, 1941). In the rat, removal of the pup from its nest, mother, and littermates, elicits crying in the ultrasonic frequency range , 30 - 70 kHz (Noirot, 1966). Ultrasonic vocalization is strongly controlled by cold stimuli and is also easily elicited by novel smells (Allin & Banks, 1971; Conely & Bell, 1978; Lyons & Banks, 1982; Okon, 1970, 1971). The function of ultrasonic crying is to call the attention of the mother, induce her to search for and retrieve the pup to the nest (Allin & Banks, 1972; Jans & Leon, 1983; Noirot, 1972). The amount of crying shows a characteristic developmental pattern with small amounts of vocalizations during the first days of life, increasing to peak level at about 9-11 days of age (Hård, Engel, & Musi, 1982; Noirot, 1968), thereafter gradually disappearing at about 16-18 days of age (see Figure 1 for a schematic overview of the developmental trend). At that age, several other developmental landmarks are reached: functional thermoregulation (Adolph, 1957), eye-lid opening, and air-righting (mastering of turning from a supine position during free fall in the air) (Hård & Larsson, 1975).

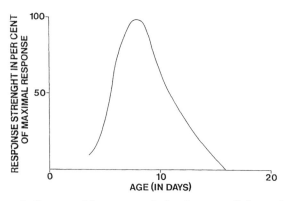

FIG. 4.1. Schematic diagram of the ontogenetic development of ultrasonic vocalization.

The eliciting conditions, the kind of behavior displayed, and the terminating events suggest the inclusion of ultrasonic crying in the category of fear-related behaviors, and more specifically, separation distress/anxiety reactions. It is well known that behaviors of this category are attenuated by various anxiolytic drugs such as benzodiazepines and barbiturates. The usefulness of rat pup crying for studies on neurochemical mechanisms in fear/anxiety reactions should, therefore, be enhanced if anxiolytics exerted similar attenuation of ultrasonic crying. As a first step in the series of studies reviewed here, we therefore investigated the effects of a benzodiazepine diazepam on distress crying. Before reporting the results of this study and the following ones, we will first give a brief account of the experimental protocol used in our studies (for a more detailed description, see Hård &Engel, 1988).

GENERAL PROCEDURE

Prior to treatment with the drug, the whole litter was removed from the mother and placed in a plastic cage immersed in a water bath, 37° C, where it was left undisturbed for at least 15 min. After treatment with drugs, the pups were returned to the rest of the litter, staying in the warm waiting cage. At the time of testing the rat was removed from the waiting cage and placed in a circular jar (11 cm diameter).

The floor of the circular jar consisted of a aluminium plate into which four photodetectors were placed. The photodetectors were connected to an electronic counter, cumulating the number of passages above the detectors during the testing period, giving a *measure of locomotor activity.*

On the inner side of the jar, the wall was covered by a sheet of aluminium, with its lower line separated from the bottom plate. The bottom plate and the wall sheet were connected to the input circuit of an amplifier. Each time the pup rose on its hindlegs and made contact with the sheet the electrical circuit was closed. The cumulative duration of all such contacts during the testing period gave a *measure of rearing behavior.*

The vocalizations of the pup were picked up by a microphone for the detection of ultrasonic callings up to 100 kHz. The microphone signals were fed through a pre-amplifier connected with an electronic counter estimating the *cumulative duration of all callings emitted during the test period* (5 min).

Within each litter, animals of different treatment groups (A-D) were tested in ABCDDCBA sequence with different permutations of order for various litters. After the test, each pup was placed alone for 10 min in a separate cage, placed in a 37° C water bath, before return to the litter. If not otherwise stated, the pups were tested at 10-11 days of age.

Influence of anxiolytics (pharmacological validation)

As mentioned above, the series of studies reported here began with the investigation of the effects of conventional anxiolytics on distress crying. For this purpose we studied the effect of a benzodiazepine, diazepam, on this behavior (Engel & Hård, 1987). The effects of various doses of diazepam on ultrasonic vocalization and motor activity are depicted in Figure 2. The results show a dose dependent decrease of the amount of ultra sonic callings. The attenuation of crying was not accompanied by a reduction of motor activity, indicating that the effect on crying was not due to a general sedation. The result therefore fulfills one requirement in the process of pharmacological validation.

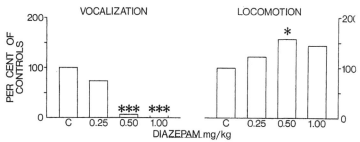

FIG. 4.2. Effects of various doses of diazepam on ultrasonic vocalization and locomotion in rat pups, 10 days of age.
Statistically significant differences between controls and various experimental groups are indicated as follows:
* P<0.05; *** P<0.001.

It is well established that the anxiolytic action of benzodiazepines (BZD) is mediated by interference with specific binding sites on the GABA-BZD-chloride-ionophore-receptor-complex (Hommer, Skolnick, & Paul, 1987). To test the hypothesis that the action of diazepam on ultrasonic crying is mediated by this mechanism we tested if the specific benzodiazepine receptor antagonist benserazide (Ro 15-1788) (Hunkeler, Mohler, Pieri, Polc, Bonetti, et al, 1981) could reverse the effect of diazepam (Engel & Hård, 1987). The results presented in Figure 3 show that the attenuating effect of diazepam is completely antagonized by benserazide.

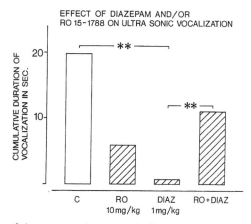

FIG. 4.3. Effects of diazepam or benserazide (Ro 15-1788) alone, or a combination of diazepam with benserazide on ultrasonic vocalization in rat pups 10 days of age. Statistical differences between various experimental groups are indicated as follows: ** P<0.01.

Conclusion

The results of these experiments, aimed at a pharmacological validation, lend support to the contention that ultrasonic crying in rat pups may be assigned to a category of fear/anxiety behaviors sensitive to benzodiazepines. Similar conclusions have been reached by other laboratories (Gardner, 1985a; Insel Hill, & Mayor, 1986; Sales, Cagiano, De Salvia, Colonna, Racagni, & Cuomo, 1986).

Influence of the GABA system

It is well established that the BZD sites are functionally related to the GABA neurotransmitter system (Hommer et al., 1987). An involvement of GABA in distress crying was tested by using drugs affecting the activity of the GABA-system. To accomplish a decrease of GABA-ergic influence we used the drugs bicuculline and picrotoxin, both diminishing GABA receptor activation, although in different ways. The former is a direct GABA receptor antagonist, and the latter is a blocker of the chloride channel (Lloyd & Morselli, 1987). In order to induce an enhanced GABA receptor activation, we treated the pups with the GABA transaminase inhibitor sodium valproate (ibid.). As shown in Figure 4, both bicuculline and picrotoxin dose-dependently increased the amount of ultrasonic callings, reaching very high levels at the highest doses. The range of doses used were clearly below those necessary to induce convulsions. In contrast to these results, sodium valproate decreased the amount of crying in a dose-dependent fashion (Figure 5).

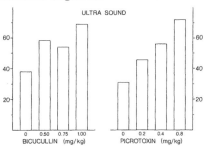

FIG. 4.4. Effects of varied doses of bicuculline and picrotoxin on ultrasonic locomotion in rats 10 days of age.

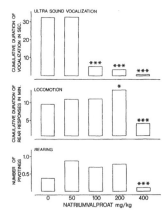

FIG. 4.5. Effects of various doses of sodium valproate on ultrasonic vocalization in rats 10 days of age.

Conclusion

The results support the hypothesis that the GABA-system is involved in distress crying in rat pups. Since decrease of GABA-ergic activity by picrotoxin and bicuculline increases crying, and an increase of GABA-ergic activity by sodium valproate decreases crying, the results suggest that the GABA-system exerts a tonic, inhibitory influence in the regulation of ultrasonic vocalization.

Influence of monoamine systems

Both human and animal studies on neurochemical mechanisms in fear/anxiety reactions and disorders imply influence of monoamine systems in these states. Specifically, the noradrenaline system (NA), but also the serotonin system (5-HT), have been accredited important roles in generalized fear/anxiety and panic attack disorders (Broekkamp & Jenck, 1989; Gardner, 1985b; Lader, 1974; Redwood, 1977; Soubrié, 1986)). Therefore, we studied the consequences of manipulating the NA- and 5-HT-systems for ultrasonic crying.

As tools for investigating the effects of dysfunction of these two systems on crying, we used neurotoxins producing chemical lesions specific for each of the neurotransmitter systems. Lesioning of the NA-system was accomplished by the neurotoxin 6-hydroxydopamine (6-OHDA), whereas 5,7-dihydroxytryptamine (5,7-DHT) was used to lesion the 5-HT system (Breese, Vogel, & Mueller, 1978; Smith, Cooper, & Breese, 1973). Both toxins were administered i.c.v. to neonatal rats on the day after birth and produced, at the dosages used (100 mg 6-OHDA or 25 mg 5,7-DHT), selective and persistent lesions of the NA and 5-HT systems, respectively (for details, see Hård, Ahlenius, & Engel, 1983). As shown in Figure 6 , animals treated with 6-OHDA showed a selective, 60 % decrease of whole brain levels of NA. Treatment with 5,7-DHT selectively decreased the 5-HT level by 50% accompanied by a small (10-15 %) increase of the level of dopamine (DA).

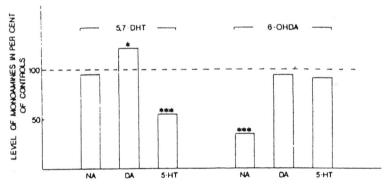

FIG. 4.6. Whole brain levels of monoamines as per cent of controls in 35-day-old rats neonatally (Day 1) treated with intracisternal injections of 5,7-DHT or 6-OHDA.
Statistical analyses performed by t-test (two tailed):
* <0.05; *** P<0.001.

The effects of the neurotoxin treatments on the ontogenetic development of ultrasonic crying are shown in Figure 7. The control animals displayed the normal development of the amount of distress crying with a conspicuous peak occurring on Days 8 - 12. The 5,7-DHT treated pups, on the other hand, did not display any peak, they were instead silent at all ages. By contrast, the pups treated with 6-OHDA did not show any consistent deviation from the control pups. Further experiments proved that the silence exhibited by the 5,7-DHT treated pups was not restricted to cold stimuli only. Thus, compared to normal pups, they were silent even when placed on a hot plate, 55° C, or when lifted by the tail or by the scruff of the neck. The locomotor activity of the 5,7-DHT treated pups did not differ from control pups. In contrast, 6-OHDA treated pups displayed an increase of locomotor activity from Day 8 onwards compared to controls.

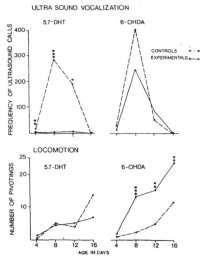

FIG. 4.7. Effects of neonatal i.c.v. treatment with 5,7-DHT (25 mg/kg) and 6-OHDA (100 mg/kg) on ultra sound vocalization and locomotor activity at various ages. The animals were tested at an ambient temperature of 2° C.
Statistically significant differences between controls and experimentals at indicated ages:
*** P<0.001; ** P<0.01; * P<0.05.

Since treatment with 5,7-DHT selectively reduced the level of 5-HT, these results suggest that the developing 5-HT system is involved in the ontogeny of ultrasonic calling. The experiment did not give any indications for a role of NA in this behavior.

Further support for the serotonin hypothesis was furnished by results from experiments performed by using subchronic treatment with para-chloro-phenylalanine (PCPA), which preferentially inhibits the synthesis of 5-HT by blocking the first step in the synthesis of 5-HT: the hydroxylation of the amino acid tryptophan to 5-hydroxytryptophan (5-HTP). This treatment will therefore result in a temporary depletion of 5-HT. Similar to the effects of 5,7-DHT treatment, PCPA decreased the amount of distress crying, as shown in Figure 8. Restoration of the 5-HT levels by administration of the precursor 5-HTP, thus bypassing the PCPA inhibition, dose-dependently antagonized the effect of PCPA (Hård et al., 1982).

The results of this experiment thus support the contention that ultrasonic vocalization is under control of the serotonergic system. This contention was further tested by using the drug 8-hydroxy-2-(di-n-propylamino)tetralin (8-OH-DPAT; see Hjort, Carlsson, Lindberg, Sanchez, Wikström, Arvidsson, Hacksell, & Nilsson, 1982). This drug presumably stimulates presynaptic serotonin receptors of the 5-HT$_{1A}$ type, thereby inhibiting the activity of the serotonin system (Hjort, Magnusson, & Carlsson, 1986). The results , shown in Figure 9, indicate a strong decrease of ultrasonic crying even at the lowest dose used, 7.5 mg/kg. The decrease of crying was not accompanied by any changes in motor activity, nor could it be attributed to changes in body temperature induced by 8-OH-DPAT (Hård & Engel 1988).

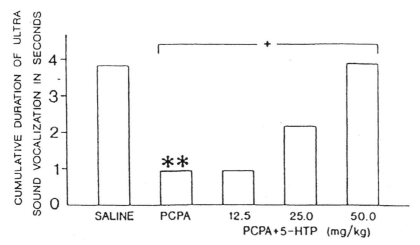

FIG. 4.8. Effects on ultra sound vocalization of treatment with PCPA (100 mg/kg/day during 3 days) alone and in combination with various doses of 5-HTP injected 45 min before test. The test was performed 24 hrs after the last PCPA-injection. 1 hr 15 min before test the animals were tested with the peripheral decarboxylase inhibitor, benserazide HC1 (25 mg/kg). At the test the animals were 9 days of age. All the tests were performed at an ambient temperature of 2° C.

Statistically significant trend over indicated range of doses: + P<0.05.

Statistically significant differences between animals treated with saline only and those treated with PCPA: ** P<0.01.

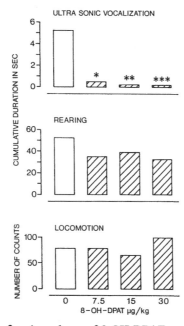

FIG. 4.9. The effects of various doses of 8-OH-DPAT on ultrasonic vocalization and motor activity in 10-day-old rats.

*, **, and *** denote statistically significant differences from the controls at the 0.05, 0.01, and 0.001 probability levels.

Conclusion

The results of the experiments reported in this section are all consistent with the hypothesis that the serotonin system is crucial for the manifestation of ultrasonic crying in rat pups. This contention is also supported by results from studies by Gardner (1985b), and Mos and Olivier (1989).

Influence of the catecholamine systems: Dopamine

As mentioned in the introduction, many studies implicate the participation of catecholamine systems in fear and anxiety. However, in the experiment with the catecholaminergic neurotoxin 6-OHDA, *vide supra*, we observed no significant effect on distress crying. In that experiment, however, the dose of 6-OHDA used was deliberately chosen to affect the NA system only, leaving the DA system intact. For a broader approach to the issue of catecholaminergic involvement in ultrasonic vocalization we performed an experiment by using the drug a-methyl-tyrosine (a-MT) (Hård et al., 1982). This drug inhibits the synthesis of CA by blocking the first step in the synthesis of CA: the hydroxylation of the amino acid tyrosine to DOPA, the immediate precursor of DA. a-MT was administered to rat pups of various ages to assess its effects on the ontogenetic development of distress callings. As shown in Figure 10, a-MT caused an increase of distress crying at all ages with the largest increase occurring on Day 12. The results thus suggested that that the CA's exerted an inhibitory influence on ultrasonic vocalization. Further support for this suggestion was afforded by the findings that replenishment of the CA levels by administration of L-DOPA, thus bypassing the a-MT inhibition, counteracted the effect of a-MT in a dose dependent fashion.

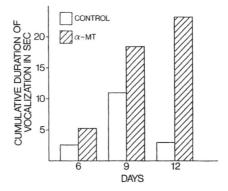

FIG. 4.10. The effects of a-methyl-tyrosine (250 mg/kg) on ultrasonic vocalization in rat pups at various ages.

The results thus support the hypothesis that one of the CA's exert an inhibitory function in the production of ultrasonic callings. Since both NA and DA levels are decreased by a-MT and increased by L-DOPA, the results are not conclusive with regard to the individual roles of the CA's. Since, however, selective depletion of NA by 6-OHDA was without effect or slightly decreased the amount of crying, DA seemed to be the most probable candidate responsible for the inhibitory influence on this behavior. This hypothesis was tested with the selective DA-receptor agonist apomorphine (Hård et al., 1982). As shown in Figure 11, apomorphine dose-dependently decreased the amount of crying.

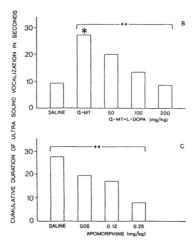

FIG. 4.11. Effects on ultra sound vocalization of the following treatments. Upper figure: a-methyl-tyrosine (250 mg/kg: 2 hrs before test) alone and in combination with various doses of L-DOPA injected 45 min before test. 1 hr 15 min before test the animals were treated with the peripheral decarboxylase inhibitor, benserazide HCl (25 mg/kg). Lower figure: Various doses of apomorphine injected 10 min before test. All the tests were performed at an ambient temperature of 2° C. The animals were 12 days of age.
Statistically significant trend over indicated range of doses: ++ P<0.01. Statistically significant differences between animals treated with saline only and those treated with a-MT: * P<0.05.

Conclusion

The results reported in this section suggest that the dopamine system exerts an inhibitory influence on ultrasonic crying in rat pups.

Influence of catecholamines: Noradrenaline

As reported above, neonatal treatment with the catecholaminergic neurotoxin 6-OHDA exerted no significant effect on the developmental trend of ultrasonic vocalization. The dose of 6-OHDA used in that study reduced brain levels of NA by about 60%, thus sparing an appreciable amount of functional NA neurons. Since the activity of the remaining NA neurons may maintain normal ultrasonic vocalization, the results of that study can not conclusively exclude a role for NA in this behavior. A renewed approach to this issue was therefore undertaken, this time with a-adrenoceptor active drugs (Hård, Engel, & Lindh, 1988).

Several reports indicate that clonidine might ameliorate feeling of anxiety during abstinence from morphine- or alcohol-abuse and also in anxiety disorders (Gold, Redmond, & Kleber, 1978; Hoehn-Saric, Merchant, Keyser, & Smith, 1981; Liebowitz, Fyer, McGarath, & Klein, 1981; Wålinder, Balldin, Bokström, Karlsson, & Lundström, 1981). The anxiolytic effects of clonidine are attributed to its agonistic action on presynaptic a_2-receptors, whose activation inhibits the presumed overactivity of NA neurons during anxiety (Aghajanian & Van der Maelen, 1982; Svensson, Bunney, & Aghajanian, 1975). Against this background of findings we found clonidine suitable as a probe for further studies on the possible involvement of NA in distress crying.

The results of the first study, using various doses of clonidine, were surprising. Instead of an expected decrease of crying, we found a dose-dependent increase, very intense at the highest dose, as depicted in Figure 12. The effects of clonidine were observed not only during cold stimulation from the testing cage, but also in pups held by their tails in head-down hanging position. This effect could not be attributed to the fall of body temperature normally induced by clonidine.

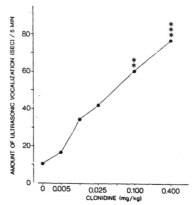

FIG. 4.12. The effects of various doses of clonidine on the amount of ultrasonic vocalization emitted by rat pups, 12 days of age, exposed to an ambient temperature of 4° C for 5 min.
*** P<0.001; ** P<0.01.

An ontogenetic study, performed by testing different animals at various ages, indicated that clonidine uniformly stimulated crying at all ages from Day 4 up to and including Day 20, an age at which distress crying normally has subsided (Figure 13).

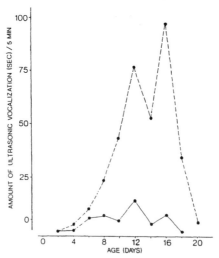

FIG. 4.13. The effects of 0.400 mg/kg clonidine (• - - - - - •) compared to vehicle (• _____ •) on the amount of ultrasonic vocalization emitted by rat pups of different ages from day 2 up to day 20.

From Day 21, marking the beginning of the weaning period, clonidine could no longer induce sustained crying. Instead, clonidine from that age induced the sedative effect well known from studies on adult animals (Nomura & Segawa, 1979). Since the sedation, caused by clonidine, is thought to be due to its presynaptic action, this result suggested that the presynaptic a_2-receptor might be functionally immature before the weaning period. (Nomura, Hoki, & Segawa, 1980). Consequently, the stimulatory action of clonidine on distress crying before the weaning period might be mediated by postsynaptic a-adrenoceptors.

To test this hypothesis we administered clonidine to pups whose presynaptic NA neurons were pharmacologically deactivated. This was accomplished by pretreatment with the monoamine depleting agent, reserpine, together with the inhibitor of CA synthesis, a-MT. This treatment almost completely abolished distress crying (Figure 14). Treatment with clonidine restored the vocalization in a-MT-treated pups to the same amount observed in intact pups treated with clonidine alone. The results thus lend additional support to the hypothesis that the stimulation of distress crying by clonidine is mediated by postsynaptic a-adrenoceptors. Furthermore, these data suggest that noradrenaline exerts an excitatory action on distress crying in developing rats.

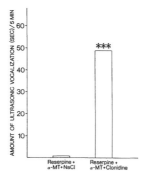

FIG. 4.14. The effect of clonidine on the amount of ultrasonic vocalization emitted by rat pups, 12 days of age, pretreated with reserpine (5 mg/kg) s.c. 6 hrs prior to test, and with a-methyl-tyrosine (200 mg/kg) s.c. 2 hrs before test. 20 min prior to test, one half of the animals were s.c. treated with clonidine (0.100 mg/kg) and the other half with 0.9% saline. During the 5 min test the pups were exposed to an ambient temperature of 4° C.
*** P<0.001.

In additional studies, we tried to establish if the stimulatory effect of clonidine on crying was mediated by postsynaptic a_1- or a_2-adrenoceptors. This was performed by trying to block the effects of clonidine by antagonists selective for each type of a-receptor: prazosin for the a_1-receptor (Menkes, Baraban, & Aghajanian, 1981), idazoxan for the a_2-receptor (Doxey, Roach, & Smith, 1983). Results of a first experiment, performed on pups 12 days of age, indicated that the effect of clonidine, a presumed a_2-agonist, as expected, was blocked by the a_2-antagonist idazoxan, but, unexpectedly, also by the a_1-antagonist prazosin (Figure 15).

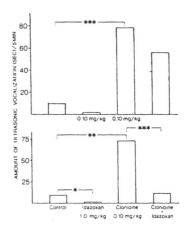

FIG. 4.15. The effects of the following treatments on the amount of ultrasonic vocalization emitted by rats pups, 12 days of age, exposed to an ambient temperature of 4° C. Upper figure: The effects of prazosin (0.5 mg/kg) or clonidine (0.1 mg/mg/kg) alone, or in combination. Lower figure: The effects of idazoxan (0.1 or 0.5 mg/kg) or clonidine (0.1 mg/mg/kg) alone, or in combination.
*** P<0.001; ** P<0.01; * P<0.05.

Thus, clonidine seemed to exert a less selective receptor activation than in other experimental situations. This suggested the hypothesis that the a-adrenoceptor population is not fully differentiated into functionally separate a_1- a_2-receptor populations at 12 days after birth. This hypothesis was tested by repeating the same experimental protocol on pups 18 days of age. At that age, only idazoxan blocked the action of clonidine, whereas prazosin did not (Figure 16). Nor could prazosin antagonize the clonidine effect in 16-day old pup (Figure 17).

The results support the hypothesis that functionally separate a_1- and a_2- adrenoceptors at about two weeks after birth emerge out of an initially undifferentiated a-adrenoceptor population.

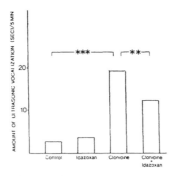

FIG. 4.16. The effects of the following treatments on the amount of ultrasonic vocalization emitted by rat pups, 18 days of age, exposed to an ambient temperature of 4° C. Upper figure: The effects of prazosin (0.5 mg/kg) or clonidine (0.1 mg/mg/kg) alone, or in combination. Lower figure: The effects of idazoxan (0.1 mg/kg) or clonidine (0.1 mg/mg/kg) alone, or in combination.
*** P<0.001; ** P<0.01.

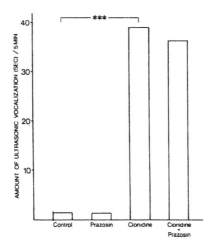

FIG. 4.17. The effect of prazosin (0.5 mg/kg) or clonidine (0.1 mg/mg/kg) alone, or in combination, on the amount of ultrasonic vocalization emitted by rat pups, 16 days of age, exposed to an ambient temperature of 4° C.
*** $P<0.001$.

Conclusion

In contrast to the results obtained in experiments with 6-OHDA, the results of the experiments with clonidine suggest that noradrenaline is involved in distress crying in rat pups, exerting an excitatory influence on this behavior.

GENERAL CONCLUSIONS

Ultrasonic vocalization in the rat pup may be viewed as an early fear/anxiety reaction to separation from the mother which may constitute a model for separation anxiety in other species, including man. The function of the reactions to separation is to attract the attention of the mother, forcing her to take care of her offspring and is, as such, vital for survival. This implies that the reactions to separation should be firmly based in the neonatal organization of the nervous system.

The present series of studies demonstrate, not unexpectedly, the involvement of several neurotransmittor systems in the regulation of this vital behavior. The results suggest that the noradrenaline system exerts an excitatory influence on distress crying, whereas an inhibitory function may be attributed to the GABA and dopamine systems. The serotonin system appears to have at least a permissive role for the manifestation of distress crying. The dynamic interplay between these systems and their dependence on various stimulus conditions awaits further research.

ACKNOWLEDGEMENTS

This study was supported by the Swedish Medical Research Council (4247 and 7202), the Swedish Council for Research in the Humanities and Social Sciences, the Bank of Sweden Tercentenery Foundation, Wilhelm and Martina Lundgren's Foundation, Claes Groschinsky's Memorial Foundation, Greta Jeansson's Foundation, and Torsten and Ragnar Söderberg's Foundation.

REFERENCES

Adolph, E.F. (1957). Ontogeny of physiological regulations in the rat. *Quart. Rev. Biol., 32,* 80-137.

Aghajanian, G.K., & Van der Maelen, C.P. (1982). a_2-adrenoceptor-mediated hyperpolarization of locus coeruleus neurons: intracellular studies in vivo. *Science, 215,* 1394-1396.

Allin, J.T., & Banks, E.M. (1971). Effects of temperature on ultrasound production by infant albino rats. *Dev. Psychobiol., 4,* 149-156.

Allin, J.T., & Banks, E,M. (1972). Functional aspects of ultrasound production by infant albino rats (Rattus Norvegicus). *Anim. Behav., 20,* 175-185.

Breese, G.R., Vogel, R.A., Mueller, R.A. (1978). Biochemical and behavioral alterations in developing rats treated with 5,7-dihydroxytryptamine. *J. Pharmacol. Exp. Ther., 205,* 587-595.

Broekkamp, C.L., & Jenck, F. (1989). The relationship between various animal models of anxiety, fear-related psychiatric symptoms and response to serotonergic drugs. In P. Bevan, A. Colls, and T. Archer (Eds.), *Behavioral Pharmacology of 5-HT* (pp. 321-335). New Jersey: Lawrence Erlbaum Assoc., Publishers.

Conely, L., & Bell, R.W. (1978). Neonatal ultrasounds elicited by odor cues. *Dev. Psychobiol., 11,* 193-197.

Doxey, J.C., Roach, A.G., & Smith, C.F.S. (1983). Studies on RX781094: a selective, potent and specific antagonist of a_2-adrenoceptors. *Br. J. Pharmacol., 78,* 489-505.

Engel, J., & Hård, E. (1987). Effects of diazepam, ethanol and Ro 15-1788 on ultrasonic vocalization, locomotor activity and body righting in the neonatal rat. *Alcohol Alcohol., Suppl. 1,* 709-712.

Gardner, C.R. (1985a). Inhibition of ultrasonic distress vocalizations in rat pups by chlordiazepoxide and diazepam. *Drug Dev. Res., 5,* 185-193.

Gardner, C.R. (1985b). Pharmacological studies of the role of serotonin in animal models of anxiety. In A.R. Green (Ed.), *Neuropharmacology of Serotonin* (pp. 281-325). Oxford: Oxford University Press.

Gold, M.S., Redmond, D.E., Jr., & Kleber, H.D. (1978). Clonidine blocks acute opiate withdrawal symptoms. *Lancet ii,* 929-930.

Hård, E., Ahlenius, S., & Engel, J. (1983). Effects of neonatal treatment with 5,7-dihydroxytryptamine or 6-hydroxydopamine on the ontogenetic development of the audiogenic reaction in the rat. *Psychopharmacology, 80,* 269-274.

Hård, E., & Engel, J. (1988). Effects of 8-OH-DPAT on ultrasonic vocalization and audiogenic immobility reaction in pre-weanling rats. *Neuropharmacology, 27,* 981-986.

Hård, E., Engel, J., & Lindh, A-S. (1988). Effect of clonidine vocalization in preweaning rats. *J. Neural Transm., 73,* 217-237.

Hård, E., Engel, J., Musi, B. (1982). The ontogeny of defensive reactions in the rat: influence of the monoamine transmission systems. *Scand. J. Psychol. (Suppl. 1),* 90-96.

Hård, E., & Larsson, K. (1975). Development of air righting in rats. *Behav. Evol., 11,* 159-164.

Hjort, S., Carlsson, A., Lindberg, P., Sanchez, D., Wikström, H., Arvidsson, L-E., Hacksell, U., & Nilsson, J.L. (1982). 8-hydroxy-2-(di-n-propylamino)tetralin, 8-OH-DPAT, a potent and selective simplified ergot congener with central 5-HT-receptor stimulating activity. *J. Neural Transm., 55,* 169-188.

Hoehn-Saric, R., Merchant, A.F., Keyser, M.L., & Smith, V.K. (1981). Effects of clonidine on anxiety disorders. *Arch. Gen. Psychiatry, 38,* 1278-1282.

Hofer, M.A. (1987). Shaping forces within early social relationships. In N.A. Krasnegor, E.M. Blass, M.A. Hofer, and W.P. Smotherman (Eds.), *Perinatal Development. A Psychobiological Perspective* (pp. 251-274). Academic Press, Inc.

Hofer, M.A., & Shair, H.N. (1987). Isolation distress in two-week-old rats: Influence of home cage, social companions, and prior experience with littermates. *Dev. Psychobiol., 20,* 465-476.

Hommer, D.W., Skolnick, P., & Paul, S.M. (1987) The benzodiazepine/GABA receptor complex and anxiety. In H.Y. Meltzer (Ed.), *Psychopharmacology: The Third Generation of Progress* (pp. 977-983). New York: Raven Press.

Hunkeler, W., Mohler, H., Pieri, L., Polc, P., Bonetti, E.P. et al. (1981). *Nature, 290,* 514-516.

Insel, T.R., Hill, J.L., & Mayor, R.B. (1986). Rat pup ultrasonic isolation calls: possible mediation by the benzodiazepine receptor complex. *Pharmacol. Biochem. Behav., 24,* 1263-1267.

Jans, J.E., & Leon, M. (1983). Determinants of mother-young contact in Norway rats. *Physiol. Behav., 30,* 919-935.

Lader, M. (1974). The peripheral and central role of the catecholamines in the mechanisms of anxiety. *Int. Pharmacopsychiatry, 9,* 125-137.

Levine, S. (1986). Psychobiological consequences of disruption in mother-infant relationships. In N. Krasnegor, E. Blass, M. Hofer, and W. Smotherman (Eds.), *Perinatal Behavioral Development* (pp. 359-376). New York: Academic Press.

Liebowitz, M.R., Fyer, A.J., McGarath, P., & Klein, D.F. (1981). Clonidine treatment of panic disorder. *Psychopharmacol. Bull., 17,* 122-123.

Lloyd, K.G., & Morselli, P.L. (1987). Psychopharmacology of GABAergic drugs. In H.Y. Meltzer (Ed.), *Psychopharmacology: The Third Generation of Progress* (pp. 183-195). New York: Raven Press.

Lyons, D.M., & Banks, E.M. (1982). Ultrasounds in neonatal rats: Novel, predator and conspecific odor cues. *Dev. Psychobiol., 15,* 455-460.

Menkes, D.B., Baraban, J.M., & Aghajanian, G.K. (1981). Prazosin selectively antagonizes neuronal responses mediated by a_1-adrenoceptors in brain. *Naunyn-Schmiedeberg's Arch. Pharmacol., 317,* 273-275.

Mos, J., & Olivier, B. (1989). Ultrasonic vocalizations by rat pups as an animal model for anxiolytic activity: effects of serotonergic drugs. In T. Archer, P. Bevan, and A. Cools (Eds.), *Behavioural Pharmacology of 5-HT* (pp. 361-366). New Jersey: Lawrence Erlbaum Assoc., Inc., Publishers.

Noirot, E. (1966). Ultrasons et comportements maternels chez les petits rongeurs. *Ann. Soc. Roy. Zool. Belg., 95,* 47-56.

Noirot, E. (1968). Ultrasounds in young rodents. II. Changes with age in albino rats. *Anim. Behav., 16,* 129-134.

Noirot, E. (1972). Ultrasounds and maternal behavior in small rodents. *Dev. Psychobiol., 5,* 371-387.

Nomura, Y., & Segawa, T. (1979). The effect of a-adrenoceptor antagonists and metiamide on clonidine-induced locomotor stimulation in the infant rat. *Br. J. Pharmacol., 66,* 531-535.

Nomura, Y., Oki, K., & Segawa, T. (1980). Pharmacological characterization of central a-adrenoceptors which mediate clonidine-induced locomotor hypoactivity in the developing rat. *Naunyn-Schmiedeberg's Arch. Pharmacol., 311,* 41-44.

Okon, E.E. (1970). The effect of environmental temperature on the production of ultrasounds in non-handled albino mouse pups. *J. Zool. (London), 162,* 71-83.

Okon, E.E. (1971). The temperature relation of vocalization in infant golden hamsters and Wistar rats. *J. Zool. (London), 164,* 227-237.

Redmond, D.E., Jr. (1977). Alterations in the function of the nucleus locus coeruleus: a possible model for studies of anxiety. In I. Hanin and E. Usdin (Eds.), *Animal Models in Psychiatry and Neurology* (pp. 293-305). New York: Pergamon.

Sales, G.D., Cagiano, R., De Salvia, A.M., Colonna, M., Racagni, G., & Cuomo, V. (1986). Ultrasonic vocalization in rodents: biological aspects and effects of benzodiazepines in some experimental situations. In G. Racagni and A.O. Donoso (Eds.), *GABA and Endocrine Function. Adv. Biochem. Psychopharmac., 42,* (pp. 87-92.

Shirley, M., & Poyntz, L. (1941). Influence of separation from the mother on children's emotional responses. *J. Psychol., 12,* 251-282.

Smith, R.D., Cooper, B.R., & Breese, G.R. (1973). Growth and behavioral changes in developing rats treated intracisternally with 6-hydroxydopamine: Evidence for involvement of brain dopamine. *J. Pharmacol. Exp. Ther., 185,* 609-619.

Soubrié, P. (1986) Reconciling the role of central serotonin neurons in human and animal behavior. Behav. Brain Sci., 9, 319-364.

Svensson, T.H., Bunney, B.S., Aghajanian, G.K. (1975). Inhibition of both noradrenergic and serotonergic neurons in brain by the a-adrenergic agonist clonidine. *Brain Res., 92,* 291-306.

Wålinder, J., Balldin, J., Bokström, K., Karlsson, I., & Lundström, B. (1981). Clonidine suppression of the alcohol withdrawal syndrome. *Drug Alcohol Depend., 8,* 345-348.

5 Neuroendocrine Integration of Social Behavior in Male Songbirds

Cheryl F. Harding
Hunter College, CUNY and
American Museum of Natural History
New York, NY

ABSTRACT

This paper reviews studies examining the specificity of hormone-sensitive neural mechanisms controlling male social behavior in zebra finches and red-winged blackbirds. As in other species, androgen metabolism proved to be an important step in activating behavior. Despite great disparities in the reproductive strategies and behavior of these two species, the specificity of the neural mechanisms controlling behavior were quite similar. With one exception, only hormone treatments which provided both androgenic and estrogenic metabolites consistently restored normal levels of courtship displays, copulatory and aggressive behaviors. Estrogens alone were able to reinstate normal aggressive behavior in finches. The hormonal control of singing behavior varied, depending on the social context in which the song was used. The metabolism of androgens to estrogens proved to be an obligatory step in activating behavior, and as expected, estrogen as well as androgen receptors were found throughout the vocal control system and the hypothalamus. Androgens and estrogens profound effects on cholinergic function in the syrinx and syringeal nerve and catecholamine levels and turnover in hypothalamic and vocal control nuclei. It appears that these neurotransmitter systems may be involved in mediating hormone-induced changes in behavior.

Neuroendocrine integration of social behavior in male songbirds

In most male vertebrates, the repertoire of social behaviors associated with reproduction, including heightened intermale aggression, courting of females, and copulation, is dependent on gonadal steroids. The relationship between endocrine status and social behavior is complex, but in most cases, castration reduced both the quantity and quality of male social behavior (Leshner, 1978). Since treatment with testosterone (T) usually reversed the effects of castration and T appeared to be the primary androgen secreted by the testes in most species, it was assumed for many years that T was *the* hormone which caused these changes in male behavior. But, over the years, evidence accumulated suggesting that I served primarily as a prehormone. Although hormone-dependent tissues concentrate T, they often metabolize it to other hormones, and it was suggested that these metabolites, not T itself, activate behavior. Target tissues may also release metabolites back into circulation. The two metabolic pathways most often hypothesized to be involved in activating behavior are 5alpha-reduction, through with T is converted to other androgens, such as 5alpha-dihydrotestosterone (DHT), and aromatization, through which T may be converted to a variety of estrogens. These are opposing metabolic processes in that once a hormones has been 5alpha-reduced, it cannot be aromatized and vice versa (Mainwaring, 1977).

The 5alpha-reduction pathway proved to be important in many male sexual accessory tissues, and studies have documented the activity of DHT in such diverse tissues as ventral prostrate in a variety of mammals, the comb of chickens, and the vocal organ of songbirds (Lieberburg & Nottebohm, 1979; Martini, 1982). In contrast, brain tissue contains both 5alpha-reducing and aromatizing enzymes. Comparative studies indicate that both aromatizing and 5alpha.reducing enzymes are present in the brains of representatives of every major vertebrate group, suggesting that the ability to metabolize androgens such as T is a primitive character of the brain which been widely conserved (Martini, 1982). Data from these species have been studied demonstrate that both of these metabolic pathways are involved in the activation of male social behavior. In a variety of species as diverse as Japanese quail and red deer, particular behaviors, most often copulatory or high-intensity aggressive behaviors, can be activated by very low doses of estrogen, but treatment with nonaromatizable androgens, such as DHT, also reinstates some level of sexual and/or aggressive behavior in castrated males, though nonaromatizable androgens are rarely as effective as T (Harding, 1986).

Such data have lead to heated controversy over the hormonal specificity of the neural mechanisms regulating male social behavior. For the most part, this controversy has focused on the relative efficacies of androgens versus estrogens in activating behavior, ignoring the fact that concurrent treatment with a combination of 5alpha-reduced androgen and estrogen is more effective than treatment with either hormone alone and that treatment with aromatizable androgens such as T or androstenedione (AE) which can be converted to a variety of estrogenic and androgenic metabolites leads to consistent restoration of behavior in castrated males in all species tested (see Harding, 1986). Knut Larsson (Larsson, Södersten, Beyer, Morali, & Perez-Palacios, 1976) was among the first to suggest that both androgenic and estrogenic metabolites act at the neural level to induce normal patterns of male social behavior, and the best data now available suggest that this is true (Martini, 1982; Harding, 1986). The more interesting question is not whether androgens or estrogens are more effective in activating male social behavior in a given species, but how androgens and estrogens normally interact in regulating male behavior. One also wonders why such interactive hormonal control of behavior evolved. What is the advantage of controlling behavior through T's metabolites rather than through the actions of T itself?

A particularly striking aspect of androgen metabolism is its dramatic variation from one target tissue to another. Interspecies differences are the rule, rather than the exception (Mainwaring, 1977). This has made it very difficult to develop a general model of how androgens modulate behavior. Research on the specificity of the hormone-sensitive neural mechanisms regulating male behavior is further complicated by the imperfect specificity of steroid hormone receptors, the multiple sites of hormone action in activating a single behavior, multiple sites of hormone metabolism, and the inherent limitations of the castration-replacement therapy paradigm (see Harding, 1986 for a more complete discussion of these issues). Greater attention to these issues would go a long way towards resolving many of the conflicting reports published in the past 25 years.

My introduction to some of these issues came in a graduate course taught by Knut Larsson at the Institute of Animal Behavior, Rutgers University. Intrigued by the problems of a) which specific hormone(s) act in the brain to modulate male behavior and b) determining the intervening processes through which hormones elicit their effects on behavior, I began research in this area 11 years ago. My research has focused on songbirds, a previously unstudied group, which offers significant advantages for research in this area. Songbirds have a unique system of interconnected brain nuclei which control their vocal behavior. This recently-evolved neural system is found only in songbirds and is the only clearly delineated, hormone-sensitive neural system with a clearly delineated behavioral function--it controls vocalizations, and to the best of our knowledge, only vocalizations. Thus, the vocal control system provides an excellent model system for examining the physiological mechanisms underlying hormonal modulation of behavior, since any hormone-induced changes found in these areas are probably involved in the modulation of singing behavior. It is much more difficult to study these mechanisms in such classic hormone target tissues as the hypothalamus, because of the wealth of hormone-sensitive behaviors and physiological processes modulated by such areas.

Castration hormone replacement studies

Which steroids activate behavior in songbirds?

Before we could study the neural mechanisms controlling male behavior, we had to determine which hormones they responded to. Two species of songbirds with very different reproductive strategies were chosen for this research, allowing examination of the relative stability of endocrine control mechanisms following divergent evolution within songbirds. Male red-winged blackbirds are usually polygynous, and, outside the breeding season, flock and migrate separately from females. There is little behavioral overlap between male and female redwings. During the breeding season, male redwings spend most of their time on territorial defense, while females build nests, incubate eggs, and feed the young. In contrast, zebra finches form strong pair bonds and reportedly pair for life. Pairs stay together even when not breeding. Males are strongly involved in every aspect of offspring care, from nestbuilding, to incubation and feeding the young. If the hormonal mechanisms controlling reproductive behavior differ among various songbird species, it seemed likely that they would differ between these two species which showed such great differences in behavior. In particular, it seemed likely that finches might be more estrogen dependent, because male finches engaged in behaviors similar to those of females and that male redwings might more androgen dependent, because males engaged in completely different behaviors from females.

We examined the abilities of six hormone treatments to elicit behavior in castrated finches and redwings compared to two control groups, sham-treated castrates and intact males (see Harding, 1983; Harding, Sheridan, & Walters, 1983; Harding, Walters, Collado, & Sheridan, 1988, for complete details). The hormones tested were 1) two nonaromatizable androgens, DHT and androsterone (AN), which can only be metabolized to other androgens; 2) the estrogen, estradiol (e), which only provides estrogenic metabolites; 3) two aromatizable androgens, T and AE, which can be metabolized to a variety of other androgens and estrogens; and 4) a combination of E + DHT which should be equivalent to treatment with an aromatizable androgen..

Although there were some differences between the abilities of the various hormones to activate behavior in the two species. I will first highlight the similarities. In both species, singing behavior was clearly activated by a combination of androgenic and estrogenic metabolites. Only treatment aromatizable androgens or the combination of E + DHT restored singing behavior to the levels shown by intact males. These treatments not only restored singing, they also restored the species-typical visual displays which often accompany singing behavior. During high-intensity courtship displays, a male finch pivots and dances towards the female as he sings. Only treatment with T or AE activated normal levels of the dance display. Male redwings have striking red epaulets which contrast with their black plumage. The epaulets are normally covered by other wing feathers, but males can expose them at will. Singing males often give a species-typical song-spread display, spreading the wings to varying degrees and exposing the epaulets as they sing. Epaulet display plays an important role in territorial defense; males with blackened epaulets are rarely able to successfully defend territories (see review in Searcy & Yasukawa, 1983). Males whose hormone treatments provided both androgenic and estrogenic metabolites showed the highest levels of song-spread displays and were more likely than other birds to display their epaulets during entire observation periods. In redwings, only treatment providing both androgenic and estrogenic metabolites restored other vocalizations which are important during the reproductive cycle, the ti-ti-ti call given during precopulatory displays and the growl vocalization used during nest site demonstration displays and intense aggressive interactions.

Because they are domesticated, finches showed more complete reproductive behavior in the laboratory than the feral redwings. So for finches, we were able to demonstrate that like courtship singing, pair bonding and copulatory behavior were also activated only by treatments providing both estrogenic and androgenic metabolites. Females solicited males given these treatments significantly more often than other males. These data are not surprising, since we have evidence that female finches choose mates primarily on the basis of their courtship songs (Sheridan & Harding, unpublished observations). Although male redwings did not copulate in the laboratory, only males whose treatments provided both androgens and estrogens showed the species-specific precopulatory display.

The greatest difference between the two species was the control of high-intensity aggressive behaviors. In male finches, these behaviors could be elicited by any treatment providing estrogen; androgen availability was not important. But in redwings, high levels of these behaviors were shown only by males whose treatments provided both androgenic and estrogenic metabolites.

Thus, despite broad differences in reproductive strategies, hormonal control of reproductive behavior in the two species is quite similar as summarized in Table 1.

Table 5.1

Summary of the Hormone Treatments which Restore Behavior in two Species of Songbirds		
BEHAVIOR	REDWINGS	FINCHES
Singing	A + E	A + E
Calls	A + E	NI*
Visual courtship displays	A + E	A + E[a]
Mounting and copulation	NI	A + E
Pairbonding	NI	A + E
Aggressive behaviors	A + E	E

Note. A + E: provides androgenic and estrogenic metabolites, T, AE, E + DHT; E: provides estrogenic metabolites, E, T, AE, E + DHT; NI: not investigated
a: in this specific case, E + DHT was ineffective.

Treatments providing androgenic metabolites alone did not reinstate normal levels of hormone-dependent behavior on either species. Combined stimulation by androgenic and estrogenic metabolites was necessary to activate all hormone-dependent behavior patterns in both species, with the single exception of high intensity aggressive behavior in finches. The profound synergism between androgen and estrogen was clearly seen when one compared the data from males treated with E alone, DHT alone, or the combination of E + DHT. When given alone, neither hormone increased behavior above the level shown by sham-treated castrates, but when the two were combined, there was a clear synergism, increasing behavior to levels shown by intact males.

These data agree with those from many other avian and mammalian species showing that treatments which provide both classes of metabolites are significantly more effective in stimulating the full repertoire of male behavior (see Harding, 1986). In these songbirds, most individual behaviors clearly require stimulation by both androgens and estrogens, whereas in other species, individual behaviors can sometimes be activated at reasonably high levels by androgens or estrogens alone, but it takes the combination to restore the full repertoire of behaviors.

While most of the data obtained in these studies were fairly straightforward, there were some findings which cannot easily be explained on the basis of our current knowledge. For example, the most effective hormone treatment clearly differed between the two species, being in T in redwings and AE in finches. There were clear qualitative as well as quantitative differences in the behaviors of males receiving these treatments, which allowed observers blind to the animals' treatment to easily pick out finches receiving AE or redwings receiving T. Differences in the relative abilities of various hormones in the same class to activate the same behaviors in different species are common, but there is no compelling hypothesis to explain why this should occur. Many such findings have been attributed to methodological differences between studies, but in this particular case, the same methodology and design were used in the redwing and finch studies. Since T may easily be metabolized to AE and vice versa, these differences in behavioral efficacy must be attributable to differences in metabolic fate or in the affinity of neural receptors for particular metabolites.

The same type of effect can be seen if one examines the ability of different hormones within the same class to activate several related behaviors in the same species. Figure 1A illustrates the ability of the various hormones tested to restore courtship singing.

One can see that only T, AE, and E + DHT significantly increased levels of courtship singing above that shown by sham-treated castrates and that there is a clear synergism when E and DHT are given concurrently. AE was clearly the most effective treatment, increasing courtship displays 50% above the mean level shown by intact males. If one examines the ability of these same treatments to reinstate high-intensity displays (i.e., only those courtship songs accompanied by the dance display, Figure 1B), one sees a somewhat different story. The two major disparities are the abilities of AN and E + DHT to activate these displays. Although AN males sang few courtship songs (Figure 1A), 63% of their songs involved high-intensity displays, bringing their level above that shown by intact males (Figure 1B). E + DHT, on the other hand, did a good job of restoring normal levels of courtship songs, but these males showed 45% fewer high-intensity displays than intact males. If T and A activate behavior through their androgenic and estrogenic metabolites, why was E + DHT so much less effective at restoring high-intensity displays than either of these two treatments? And since all of our other data suggest that both androgenic and estrogenic metabolites are necessary to restore patterns of sexual behavior in this species, how did a treatment which is believed to supply only androgenic metabolites, AN, restore normal levels of high-intensity courtship displays? We thought that these disparities might be explained by the affinities of neural tissue for the various metabolites, but as discussed below, this does not seem to be the case.

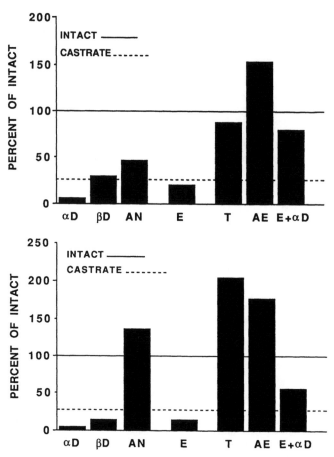

Fig. 5.1. The ability of 7 hormone treatments to reinstate courtship displays (A) and high-intensity courtship displays (B) in castrated finches compared to levels shown by sham-treated castrates and intact males. (aD = 5alpha-dihydrotestosterone [DHT], b-D = 5b-DHT, AN = androsterone, E = estradiol, T = testosterone, AE = androstenedione, E + aD = estradiol + 5alpha-DHT)

Another difference between the two species is more easily explained. In both species, castration caused a profound decrease in singing behavior. In redwings, castration totally eliminated singing behavior, while in finches castrates sang 75% less than intact males. The degree to which singing behavior is dependent on endocrine status in the two species is probably explicable in terms of their normal breeding strategies. Redwings live in seasonal environments and show a strong annual cycle in reproductive physiology and breeding behavior which is cued by changes in photoperiod. Zebra finches, in contrast, are opportunistic breeders. Rainfall which is important in initiating their reproductive cycles occurs unpredictably. Several workers have suggested that the strong, life-long pair bonds shown by these birds are an adaptation to this unpredictable environment. Because birds already have mates, they are poised to begin breeding whenever rainfall occurs. Since singing behavior is crucial to pair bond initiation and maintenance, it has probably become somewhat independent of endocrine status so that basal levels of singing may be maintained even when birds are not actively breeding and testes size is minimal. Since redwings do not maintain territories or pair bonds outside the breeding season, their behavior may be more sensitive to seasonal changes in hormone titers.

Inhibiting estrogen formation inhibits behavior

These behavioral studies demonstrated that only treatments capable of providing both androgenic and estrogenic metabolites (i.e., T, AE, E + DHT) reinstated normal levels of male sexual behavior. The most parsimonious explanation of these data is that aromatizable androgens (T, AE) are metabolized to provide androgenic and estrogenic metabolites which then activate behavior. To determine if conversion to estrogenic metabolites is necessary for aromatizable androgens to stimulate behavior, we determined if AE could activate behavior when its conversion to estrogenic metabolites was blocked through use of the aromatization inhibitor 1,4,6-androstatriene-3,17-dione (ATD, Walters & Harding, 1988). This study clearly demonstrated that the ability of AE to activate a variety of sexual and aggressive behaviors depends on its conversion to estrogenic metabolites. Castrated finches treated with AE + ATD showed significantly fewer courtship displays, high-intensity courtship displays, and less nest-building and aggressive behavior. These effects were specific to the inhibition of aromatization by ATD, since castrates treated with AE + AT + E (i.e., restoring estrogen after ATD blocked its formation) showed the same high levels of behavior shown by males treated with AE alone. These data strongly suggest that the conversion of androgens to estrogens is an obligatory step in activating behavior in normal males. Differences in efficacy between various aromatizable androgens may be related to their ability to be converted to estrogens. Data available to date suggest that, in contrast, conversion of androgens to 5alpha-reduced metabolites is not critical to their ability to activate behavior in finches.

Hormonal modulation of behavior in finches

One issue which particularly intrigued me was whether the hormonal control of a particular behavior is invariant or if it depends on the context in which the behavior is used. For example, is singing during courtship displays activated by the same hormone treatments as singing which occurs during aggressive interactions? To examine this issue, we observed finches in a variety of social situations. The hormonal control of singing behavior proved dependent on social context.

Male zebra finches sing relatively simple, stereotyped songs, allowing listeners to identify the singer as a male zebra finch, but songs also vary sufficiently to allow individual identification (Immelmann, 1968; Sossinka & Böhner, 1980). Two types of song are generally recognized in the literature. Songs directed towards females are typically labeled courtship songs. Males also sing without directing the song towards any bird in particular; these have been labeled undirected songs. Such undirected songs may occur when males are isolated from other birds or in the midst of a crowded aviary. Males also direct songs towards other males, but such songs have not received much attention, probably because they occur relatively infrequently. Songs sung in the three contexts sound similar, and one can easily learn to identify particular males. Sossinka and Böhner (1980) showed that there are subtle differences between courtship and undirected songs. Courtship songs have more introductory elements, are sung more rapidly, and are less variable than undirected songs. Both song types were previously shown to be hormone dependent. Castration reduced the frequencies of both courtship and undirected singing, and T-therapy reinstated normal behavior (Pröve, 1974). One study (Pröve, 1978) found that the frequencies of both song types were positively correlated with circulating T levels, but that undirected songs could be reinstated with half the dose of T necessary to restore normal levels of courtship singing.

We found that the hormonal control of singing depended on whether the song was directed at a female, a male, or no bird in particular (Walters, Collado, & Harding, submitted for publication). When ATD was used to block formation of estrogens, it significantly reduced 1) the percentage of birds which directed songs to other birds, both males and females, and 2) the total number of directed songs sung. This treatment had no effect on the number of undirected sons sung by the same males. ATD treatment also altered the motor patterning of female-directed songs, 1) lowering the speed at which they were sung and 2) reducing the number of songs sung per bout. It did not affect these measures in undirected or male-directed songs sung by the same birds.

Thus, hormonal status differentially affects songs sung in these three contexts. Estrogen is important in motivating males to sing to other birds, both males and females, but there was no indication that estrogen affects the frequency of undirected songs as androgen does. Estrogen also has important effects on the motor patterning of songs directed towards females, causing them to be sung faster and in longer bouts, while not affecting these measures in undirected or male-directed songs. We have previously shown that estrogens are important in activating successful nest site selection and defense by male finches (Harding, 1983). Thus, the estrogenic effects on female-directed songs might allow female finches to assess the estrogen status of courting males and predict their later nesting behavior.

Neural androgen and estrogen receptors

The effects of steroid hormones on behavior are typically mediated through interactions with intracellular receptors in neural areas involved in the control of the behavior in question. I decided to measure androgen and estrogen receptors in finch hypothalamus and VCS to determine if receptor characteristics would shed further light on differences in the relative abilities of the various steroids to elicit behavior as well as demonstrating where androgens and estrogens could act in the brain. We did not know whether all hormone-sensitive areas would contain both types of receptors, or if some areas would be androgen dependent and others estrogen dependent. Previous research had shown that all VCS nuclei examined, with the exception of area X, showed heavy labeling following an injection of tritiated T (Arnold, Nottebohm, & Pfaff, 1976; Arnold & Saltiel, 1979), suggesting that T acted throughout the VCS to stimulate singing behavior. The high degree of labeling seen in the telencephalic vocal control areas is unique; comparably recently-evolved areas in the brains of other animals do not show evidence of such high concentrations of steroid receptors (Arnold et al., 1976). Further research was necessary, however, since this autoradiographic work demonstrated where T or its metabolites were bound, but not which steroids were involved. Further autoradiographic work with tritiated DHT demonstrated similar heavy labeling throughout the hypothalamus and VCS, indicating widespread distribution of androgen receptors (Arnold, 1979). However, autoradiographic studies (Arnold, 1979; Nordeen, Nordeen, & Arnold, 1987), cytosol receptor assays (Siegel, Akutagawa, Fox, Konishi, & Politch, 1986), and immunocytochemical techniques (Gahr, Flügge, & Güttinger, 1987) failed to find any evidence of significant levels of estrogen receptors within the VCS, with the exception of one nucleus. Yet our data indicated that estrogens play a major role in stimulating singing behavior, suggesting that there should either be significant concentrations of estrogen receptors within the VCS, or there should be a major input to the VCS from an estrogen-sensitive area. To date, no one has published evidence of the latter.

We used *in vitro* binding assays to quantify unoccupied androgen (Harding, Walters, & Parsons, 1984) and estrogen receptors (Walters, McEwen, & Harding, 1988) in microdissected hypothalamic and vocal control nuclei from castrated finches. Initially, we examined the specificity of these receptors, using competition assays to determine the abilities of various hormones to block binding by the tritiated standard. These studies showed that the androgen receptor showed much higher affinities for T and DHT than for AE, AN, and a variety of other androgens. Thus, the greater abilities of AE and AN to stimulate behavior compared to the actions of T and DHT, respectively, could not be attributed to differences in receptor affinity, because the differences were in precisely the wrong direction. Similarly, while estrogen receptors showed much lower affinity for all the androgens tested compared to estradiol, estrone, moxestrol, diethylstilbestrol, and estriol, they also showed higher affinities for T and DHT than for the other androgens. Thus, the greater efficacy of AN in inducing behavior compared to DHT could not be explained through crosstalk with estrogen receptors.

As expected, the hypothalamic areas examined showed higher levels of both androgen and estrogen receptors than the vocal control nuclei. But counter to receptor measurements made by other techniques, binding assays documented the presence of both estrogen and androgen receptors in all vocal control areas examined, except area X. Area X showed low levels of estrogen binding, but we did not look for androgen binding in this area, and no other study has found evidence of androgen binding in X.

All brain areas contained higher levels of androgen than estrogen receptors. This is rather unusual, since hypothalamic levels of these two receptors are usually quite similar. This led me to question whether castration removed all endogenous estrogens, or if endogenous hormone was binding to receptors and lowering the number of receptors free to bind to tritiated estradiol in our assay. So the assays were repeated in castrated males treated with ATD to block estrogen formation. Sure enough, ATD treatment significantly increased estrogen receptor levels in three of five vocal control nuclei examined. The differential effects of ATD treatment on receptor levels in different nuclei suggest that there may be regional differences in receptor regulation. Such differences have been documented in rat brain (Lieberburg, MacClusky, & McEwen, 1980). Additional data from my lab (Walters, Gonzalez, & Harding, 1988) suggest that our measurements still underestimate the number of estrogen receptors in the finch brain because of competition from a non-estrogenic steroid, presumably of adrenal origin, which is able to bind to estrogen receptors, decreasing the number of receptors measured. Regardless of what the relative concentrations of androgen and estrogen receptors in each nucleus prove to be, this work demonstrated the presence of significant quantities of both receptors in all nuclei examined, except X. Thus, both androgens and estrogens may modulate singing behavior at almost every level of the VCS.

The avian vocal organ, the syrinx, is also hormone dependent, but unlike the VCS, the syrinx contains only androgen receptors. Thus, the hormonal specificity of the neural and muscular tissue controlling singing differ. This finding is, however, not unusual. Hormonal activation of a particular behavior pattern often involves the actions of multiple metabolites at multiple sites. In males of many species, sexually-dimorphic structures in the periphery, like scent glands, are best characterized as androgen dependent, while the brain areas which control the behavioral use of these structures are primarily estrogen dependent or dependent on the combined actions of estrogens and androgens (see Harding, 1986).

Hormone effects on neurotransmitter systems

Having documented the widespread distribution of androgen and estrogen receptors in hypothalamic and vocal control nuclei, the next step was to determine how these receptors affect brain function in these areas. Hormone-receptor complexes act within the cell nucleus to increase transcriptional activation of protein synthesis and/or the production of new types of protein in the cell (Yamamoto, 1985). Rather than focus on this step per se, I chose to concentrate on hormone-induced changes in the synthesis and degradation of neurotransmitters thought to be involved in the activation of hormone-sensitive behaviors. In particular, my lab is focusing on hormonal modulation of cholinergic and monoaminergic function in the VCS, always using classic hypothalamic target areas as reference points. Although these are not only neurotransmitter systems found in the VCS, I chose to work with them because 1) the VCS contains high levels of acetylcholine, and the monoamines, norepinephrine (NE), dopamine (DA), and serotonin (5-HT), 2) these neurotransmitters have previously been implicated in the regulation of hormone-dependent male social behaviors in other species, and 3) a variety of sensitive biochemical techniques are available to quantify neurotransmitter function in microdissected nuclei.

Hormonal modulation of cholinergic function

Acetylcholine obviously plays an important role in singing behavior, since it is the primary neurotransmitter in the neuromuscular junctions of the syrinx (Luine, Harding, & Bleisch, 1980). It has also been identified as a putative neurotransmitter in the VCS (Ryan & Arnold, 1981). We began examining hormonal modulation of cholinergic function in the syrinx. The finch syrinx is quite sensitive to circulating gonadal steroid levels. Castration reduced syrinx weight 30%, the activity of choline acetyltransferase, the enzyme involved in acetylcholine synthesis 40%, the activity of acetylcholinesterase, the enzyme involved in acetylcholine degradation 50%, and acetylcholine receptors 40% (Luine et al., 1980; Bleisch, Luine, & Nottebohm, 1984). Castration-induced changes in acetylcholinesterase were reversed by treatment with a variety of androgens (Luine, Harding, & Bleisch, 1983). As one would expect from the lack of estrogen receptors in the syrinx, these effects were purely androgen dependent -- aromatizable androgens were no more effective than 5alpha-reduced androgens, and E + DHT was no more effective than treatment with DHT alone. As mentioned before, this androgen dependence is typical of peripheral target tissues in male vertebrates. However, this pure androgen dependence is limited to the syrinx. The syringeal nerve is sensitive to estrogen as well as androgen. We are now examining the effects of androgens and estrogens on cholinergic activity in the VCS and predict that both androgens and estrogens will be involved in the regulation of cholinergic function in the brain.

Hormonal modulation of monoaminergic function

There is a large literature documenting steroid-monoamine interactions in the central nervous system. An increasing number of studies have found that monoaminergic cells often contain androgen or estrogen receptors, and steroids have been shown to modulate every aspect of monoamine function from syntheses, to release, reuptake, and degradation (see references in Barclay & Harding, 1988). Other studies document the involvement of monoamines in mediating the effects of changing steroid levels on reproductive behavior in a variety of species. NE and DA, in particular, have been shown to modulate male sexual behavior (Crowley & Zemlan, 1981; Meyerson, 1984). There is also evidence that NE is involved in the regulation of courtship vocalizations in ring doves (Barclay, Johnson, & Cheng, 1985). We therefore decided to investigate whether monoamines were involved in mediating steroid-induced changes in singing behavior.

Our first study (Barclay & Harding, 1988), found that endocrine status had profound effects on DA levels and turnover in 25% of the hypothalamic and vocal control nuclei examined and similarly significant effects on NE levels and turnover in 44% of the areas. In these areas, castration altered catecholamine (CA) function and treatment with AE, the hormone which activates the highest levels of singing behavior in this species, restored function to that typical of intact males. The degree of steroidal modulation of CA levels and turnover found suggested that NE and DA may be involved in mediating neural responses to changing endocrine status in finches.

More recently, we determined the relative contributions of androgenic and estrogenic stimulation to these effects. Since courtship singing requires stimulation by both androgens and estrogens, this study (Barclay & Harding, submitted for publication) examined whether estrogen alone, androgen alone or combined treatment with estrogen + androgen exerted differential effects on CA levels and turnover in brain areas involved in the control of singing behavior. A few additional brain areas were examined in this study, and even more profound effects were found (see Table 2). Hormone treatment affected NE function in 78% of the comparisons, and DA function in 50%.

Table 5.2

Type of Metabolite Necessary to Restore Normal Catecholamine Function in Various Hypothalamic, Auditory, and Vocal Control Nuclei					
		Dopamine		**Norepinephrine**	
Brain Area	**Change[a]**	**Levels**	**Turnover**	**Levels**	**Turnover**
Hypothalamic					
POA	+		E	E	E
IN	+	A		A+E	A+E
PVM	-		A	E	E
Auditory					
L	-		A+E		
Vocal Control					
X	-	A	A	E	E
MAN	-		E		E
NIf	+		A+E		A+E
HVc	0				
RA	+		E	E	E
DM	+			E	E

Note. Brain Areas: POA--nucleus preopticus anterior, IN--nucleus infundibularis, PVM--nucleus paraventricularis magnocellularis, L--Field L, X--area X, MAN--nucleus magnocellularis neostriatum anterior, NIf--nucleus interfacialis, HVc--hyperstriatum ventrale pars caudale, RA--nucleus robustus archistriatalis, DM--dorsomedial portion of nucleus intercollicularis.
a: Metabolites listed increase (+), decrease (-), or have no effect (0) on catecholamine function.

Interestingly, the great majority of hormone effects did not require stimulation by both androgens and estrogens. Estrogen alone was sufficient to restore normal CA function in 57% of the hormone-sensitive areas, androgen alone was sufficient in another 22%, and the combination was only required in 22% of the hormone-sensitive areas examined. As found in the previous study, whether steroids up or down regulate CA function appears to depend on the brain area examined. In two hypothalamic areas and 3 vocal control nuclei involved in producing vocalizations, effective hormone treatment always up- regulated CA function, while in a third hypothalamic nucleus and three areas involved in song learning, effective hormone treatment always down-regulated CA function.

All modulation of NE function required estrogenic stimulation. 86% of the changes in NE function could be stimulated by estrogen alone. The remainder required combined stimulation by estrogen + androgen. DA function, in contrast, proved more androgen sensitive. 56% of the changes in DA function were elicited by treatment with androgen alone, 22% by combined treatment, and the remainder by estrogen alone. Thus, although both androgenic and estrogenic stimulation are required to restore singing behavior, in most cases, stimulation by either androgen or estrogen is sufficient to restore normal CA function in a particular brain area. As one can see in Table 2, normal NE and DA function in the same area were often modulated differently. This is not particularly surprising, since NE and DA inputs to the various nuclei are themselves coming from different brain regions which might be expected to have different hormone sensitivities.

The role of norepinephrine in activation courtship singing

These data on androgenic and estrogenic modulation of CA function in the VCS strongly suggest that the CAs may be involved mediating hormone-dependent changes in singing behavior. We now have pilot data implicating NE in mediating courtship singing. We used the neurotoxin N-(2-chlorethyl)-N-ethyl-2-bromobenzylamine (DSP-4) to lower noradrenergic function. DSP-4 is a potent neurotoxin which appears to selectively destroy noradrenergic axon terminals from the locus coeruleus, leaving noradrenergic function in hypothalamic areas largely unaffected (Fritschy & Grzanna, 1989). As expected, treating finches with DSP-4 lowered NE levels in the vocal control system without affecting NE levels in hypothalamic areas or DA or 5-HT at any sites. A dose of DSP-4 which reduced NE levels in MAN and X over 60% and in RA and DM about 45% caused a 35% reduction in courtship singing and accompanying displays without affecting copulatory behavior. Birds with greater reductions in NE levels in RA and DM showed greater reductions in behavior. These data suggest that the hormone-induced increases in noradrenergic function in RA and DM, nuclei involved in the motor patterning of song, may be involved in mediating increases in courtship singing in finches.

FUTURE DIRECTIONS

The use of songbirds to investigate the question of hormonal specificity of the neural mechanisms controlling male social behaviors has proved quite useful. The exquisite hormone sensitivity of the VCS and its unique characteristics have allowed us to investigate the interactions of androgens and estrogens in modulating behavior in ways not feasible in other animals. Our research is currently moving along three tracks. First, now that we know that androgens and estrogens affect CA levels and turnover, we are determining how this is accomplished (e.g., through changes in enzyme syntheses or enzyme activity). Second, we are using the paradigm used to study steroid influences on CA function to study regulation of other important neurotransmitters in the VCS, such as acetylcholine and serotonin. Finally, and most importantly, we have begun neuropharmacological studies to determine what effects hormone-induced changes in neurotransmitter function have on behavior.

ACKNOWLEDGEMENTS

Preparation of this chapter was supported by NIMH RSDA Award MH00591. NIH Grant HD15191, and PSC CUNY Award 668237.

REFERENCES

Arnold, A.P. (1979). Hormone accumulation in the brain of the zebra finch after injection of various steroids and steroid competitors. *Soc. Neurosci. Abst., 5,* 437.

Arnold, A.P., Nottebohm, F., & Pfaff, D. (1976). Hormone concentrating cells in vocal control and other areas of the brain of the zebra finch (Poephila guttata). *J. Comp. Neurol., 165,* 487-512.

Arnold, A.P., & Saltiel, A. (1979). Sexual difference in pattern of hormone accumulation in the brain of a songbird. *Science, 205,* 702-705.

Barclay, S.R., & Harding, C.G. (1988). Androstenedione modulation of monoamine levels and turnover in hypothalamic and vocal control nuclei in the male zebra finch. *Brain Res., 459,* 333-343.

Barclay, S.R., & Harding, C.F. Differential regulation of monoamine levels and turnover rates by estrogen and/or androgen in hypothalamic and vocal control nuclei of male zebra finches, submitted.

Barclay, S.R., Johnson, & Cheng, M.F. (1985). Male courtship vocalization and the noradrenergic system. *Soc. Neurosci. Abst., 11,* 736.

Bleisch, W.V., Luine, V.N., & Nottebohm, F. (1984). Modification of synapses in androgen sensitive muscle. *J. Neurosci., 4,* 786-792.

Crowley, W.R., & Zemlan, F.D. (1981). The neurochemical control of mating behavior. In N.T. Adler (Ed.), *Neuroendrocrinology of Reproduction* (pp. 451-484). New York: Plenum Press.

Fritschy, J-M., & Grzanna, R. (1989). Immunohistochemical analysis of the neurotoxic effects of DSP-4 identifies two populations of noradrenergic axon terminals. *Neuroscience, 30,* 181-197.

Gahr, J.L., Flügge, G., & Güttinger, H.R. (1987). Immunocytochemical localization of estrogen-binding neurons in the songbird brain. *Brain Res., 402,* 173-177.

Harding, C.F. (1983). Hormonal specificity and activation of social behaviour in the male zebra finch. In J. Balthazart, E. Pröve, and R. Gilles (Eds.), *Hormones and Behaviour in Higher Vertebrates* (pp. 275-289). Berlin, FRG: Springer Verlag.

Harding, C.F. (1986). The role of androgen metabolism in the activation of male behavior. In B.R. Komisaruk, H.I. Siegel, M.F. Cheng, and H.H. Feder (Eds.), *Ann. N.Y. Acad, Sci., Vol. 74, Reproduction: A behavioral and neuroendocrine perspective* (pp. 371-378).

Harding, C.F., Sheridan, K., & Walters, M. (1983). Hormonal specificity and activation of sexual behavior in the male zebra finch. *Horm. Behav., 17,* 111-133.

Harding, C.F., Walters, M., Collado, D., & Sheridan, K. (1988). Hormonal specificity and activation of social behavior in male red-winged blackbirds. *Horm. Behav., 22,* 402-418.

Harding, C.F., Walters, M., & Parsons, B. (1984). Androgen receptor levels in hypothalamic and vocal control nuclei in the male zebra finch. *Brain Res., 306,* 333-339.

Immelmann, K. (1968). *Australian Finches in Bush and Aviary.* Sydney: Halstead.

Larsson, K., Södersten, P., Beyer, C., Morali, G., & Perez-Palacios, G. (1976). Effects of estrone, estradiol, and estriol combined with dihydrotestosterone on mounting and lordosis behavior in castrated male rats. *Horm. Behav., 7,* 379-390.

Leshner, A.I. (1978). *An Introduction to Behavioral Endocrinology.* New York: Oxford University Press.

Lieberburg, I., MacClusky, N., & McEwen, B.S. (1980). Cytoplasmic and nuclear estradiol-17b binding in male and female rat brain. *Brain Res., 193,* 487-503.

Lieberburg, I., & Nottebohm, F. (1979). High-affinity androgen binding proteins in syringeal tissues of song birds. *Gen. Comp. Endocrinol., 37,* 286-293.

Luine, V.N., Harding, C.F., & Bleisch, W.V. (1983). Specificity of gonadal hormone modulation of cholinergic enzymes in the avian syrinx. *Brain Res., 279,* 339-342.

Mainwaring, W.I.P. (1977). The Mechanisms of Action of Androgens. *Monographs of Endocrinology, Vol. 10.* New York: Springer Verlag.

Martini, L. (1982). The 5alpha-reduction of testosterone in the neuroendocrine structures. *Endocrine Rev., 3,* 1-24.

Meyerson, B.J. (1984). Hormone-dependent socio-sexual behaviors and neurotransmitters. In G.J. DeVries, J.C.P. DeBruin, H.B.M. Uylings, and M.A. Corner (Eds.), *Progress in Brain Research, Vol. 61* (pp. 271-281). Amsterdam: Elsevier.

Nordeen, K.W., Nordeen, E.J., & Arnold, A.P. (1987). Estrogen accumulation in zebra finch song control nuclei: Implications for sexual differentiation and adult activation of song behavior. *J. Neurobiol., 18,* 569-582.

Pröve, E. (1974). Der Einfluss von Kastration und Testosteronsubstitution auf das Sexualverhalten männlicher Zebrafinken (Taeniopygia guttata castanotis Gould) *J. Ornithologie, 115,* 338-347.

Pröve, E. (1978). Quantitative Untersuchungen zur Wechselbezeiehungen zwischen Balzaktivität und Testosterontitern bei männlichen Zebrafinken. *Z. Tierpsychologie, 48,* 47-67.

Ryan, S.M., & Arnold, A.P. (1981). Evidence for cholinergic participation in the control of bird song. *Gen. Comp. Neurol., 202,* 211-219.

Searcy, W.A., & Yasukawa, K. (1983). Sexual selection and red-winged blackbirds. *Am. Scientist, 71,* 166-174.

Siegel, L.I., Akutagawa, E., Fox, T.O., Konishi, M., & Politch, J.A. (1986). Androgen and estrogen receptors in the adult zebra finch brain. *J. Neurosci. Res., 16,* 617-628.

Sossinka, R., & Böhner, J. (1980). Song types in the zebra finch, poephila guttata castanotis. *Z. Tierpsychologie, 53,* 123-131.

Walters, M.J., & Harding, C.F. (1988). The effects of an aromatization inhibitor on the reproductive behavior of male zebra finch. *Horm. Behav., 22,* 207-218.

Walters, M.J., Collado, D., & Harding, C.F. Estrogenic control of male zebra finch song in different social contexts, submitted.

Walters, M.J., Gonzalez, A., & Harding, C.F. (1988). Stress, estrogen receptors and sexual behavior in male zebra finches. *Soc. Neurosci. Abst., 14,* 435.

Walters, M.J., McEwen, B.S., & Harding, C.F. (1988). Estrogen receptor levels in hypothalamic and vocal control nuclei in the male zebra finch. *Brain Res., 459,* 37-43.

Yamamoto, K.R. (1985). Steroid receptor regulated transcription of specific genes and gene networks. *Ann. Rev. Genetics, 19,* 209-252.

II Sexual Behavior I

Thorsten Klint
Pharmacia Leo Therapeutics
Malmö

This section comprises four chapters, all of them reviewing different aspects of sexual behavior. A common denominator is to get an understanding of the basic mechanisms of sexual behavior. An important aspect is to comprehend how these mechanisms are integrated into a functional context. Function can be interpreted at both an adaptive, evolutionary level and at a physiological mechanistic level. When studying sexual behavior at the evolutionary level, the functional significance of this behavior is obvious and the subsequent interpretation of the overt behavior is relatively easy. At this level we have to observe the timing and sequencing of behavior, e.g., between male and female rats in a sexual encounter. At another level we must investigate how hormonal mechanisms influence, e.g., sexual or courtship behavior and, yet at another level, we have to work out how, e.g., changes of mono- and catecholamines integrate with the function of hormones and/or directly influence overt behavior.

In Chapter 6, van de Poll discusses the physiological significance of how acute hormonal changes may act as direct cues in a drug-discrimination paradigm: in other words, how hormonal changes can be used by the animal as a conditioning stimuli in different situations. It is of importance that these changes also seem to have intrinsic rewarding effects. To understand the functional significance, van de Poll draws attention to the phenomenon of conditioning of sexual behavior. This, coupled with the observations that the mere exposure of male rats or doves to a stimulus female gives an immediate rise in testosterone in the blood plasma (Feder Storey, Goodwin, Reboulleau, & Silver, 1977; Purvis & Haynes, 1974), points to the possibility that these hormonal changes are part of a hormonal-behavioral feedback system. On an adaptive level, this system may act as a tuner to integrate behavioral-hormonal as well as motivational factors, e.g., in a courtship situation. It is tempting to fit these observations into the behavioral-hormonal interaction scheme proposed by Lehrman (1965) for the courtship sequence between a male and a female ring dove. Motivational and learning aspects seem to be of great importance in this situation to achieve a precise sequencing and synchronization between the mates. In this context, it is also of interest that courtship and sexual behavior change with experience (Cheng, Klint, & Johnson, 1986).

In Chapter 7, Meyerson points to the importance of hormonal-neurotransmitter interaction as well as interactions between different neurotransmitters and desribes, in particular, the functional significance of serotonergic mechanisms involved in the regulation of female rat lordosis behavior. There seems to be a consensus that an increased serotonergic acitivty, in general, inhibits the lordosis response (LR) but that the opposite effect, a stimulation of the LR by a decreased serotonergic activity, has not been as readily demonstrated. When discussing these inconsistencies, Meyerson brings up a number of important pitfalls when neuropharmacologists attempt to relate properties of drugs to behavioral effects. Among these, he discusses the problem of to what extent receptor affinity *in vitro* is comparable to a functional *in vivo* response, but also the possibility of an agent's affinity to multiple receptors. In a strict sense, drugs are always 'dirty'. He addresses the possibility that the agents may act as agonists at one site but at the same time as partial agonists or even antagonists at other sites. Of crucial importance is, of course, knowledge of the anatomy of the 5-HT receptor localization and pathways.

Södersten (Chapter 8) outlines a third line of research where he demonstrates the functional importance of testosterone metabolism in relation to sexual behavior in male rats. Södersten fits the pieces in an interesting jig-saw puzzle where he sorts out the effects of different metabolites and also their combined effects. Peripheral and central effects are separated and discussed in relation to the expression of sexual behavior. This field has flourished, since different androgen and estrogen metabolites, antagonists, and aromatase and reductase inhibitors have been available as research tools.

In the fourth and last chapter in this section (Chapter 9), Ahlenius, Hillegaart, and Larsson describe biochemical changes in selected brain regions as a consequence of sexual activity in male rats. An interesting scenario emerges where they find a selective increase in striatal monoamine synthesis, but only in sexually naive rats. An important distinction is suggested between the performance of overt sexual behavior and the underlying motivational factors in relation to changes in dopaminergic and serotonergic neurotransmission. A general point is that drugs can, of course, influence functional mechanisms at many different levels: consequently, the drug effect can either be direct or indirect. Thus, serotonergic drugs may influence sexual behavior either by a direct effect on serotonergic pathways or mediate their effect via gonadal hormone mechanisms.

When studying function, we always work with an underlying assumption that an animal behavior is determined by a set of causal factors. What we, in fact, are doing in our research are to isolate these factors and subsequently try to manipulate them in such a way that we can observe changes in the animal's behavior. All the chapters in this section present ways to isolate different factors, all of them of equal importance for an understanding of the basic mechansims of sexual behavior.

REFERENCES

Cheng, M-F., Klint, T., & Johnson, A.L. (1986). Breeding experience modulating androgen dependent courtship behavior in male ring doves (Streptopelia risoria). *Physiol. Behav., 36,* 625-630.

Feder, H.H., Storey, A., Goodwin, D., Reboulleau, C., & Silver, R. (1977). Testosterone and 5α-dihydrotestosterone levels in peripheral plasma of female and female ring doves (Streptopelia risoria) during the reproductive cycle. *Biol. Reprod., 16,* 666-677.

Lehrman, D.S. (1965). Interaction between internal and external environments in the regulation of the reprodutive cycle of the ring dove. In F.A. Beach (Ed.), *Sex and Behavior* (pp. 355-380). New York: Wiley.

Purvis, K., & Haynes, N.B. (1974). Short-term effects of copulation, human chorionic gonadotrophin injection and non-tactile association with a female on testosterone levels in the male rat. *J. Endocrinol., 60,* 429-439.

6 The Physiological Significance of Acute Activation of the Gonadal Hormonal Axis

N.E. van de Poll
Netherlands Institute for Brain Research
Amsterdam

INTRODUCTION

For years, behavioral research on the influence of gonadal hormones has been dominated by the now well-established organizing and activating effects. The former effects results in drastic and largely irreversible changes in the central neural substrate underlying classes of behavior, characteristically related to sex differences. Activating effects, in contrast, are not permanent, although chronic treatment is required to induce specific behaviors, especially in the male. From the perspective of behavioral change and adaptation, however, both these aspects of hormonal influences affect behavior in a relatively slow and protracted manner and only long-term psychological functions can logically be involved, such as adaptation to seasonal changes or social stratification.

Concentration upon these two prime constituents of hormonal influences distracted attention from a third category of hormonal effects: the "behavioral feedback function" of acute activation of the gonadal hormonal axis, as it can be related to specific social stimulation (Leshner, 1979). According to this principle, hormonal changes occurring as a direct consequence of psychological stimulation, affect the individual and change responding on subsequent similar occasions, thus introducing potential elements of behavioral adaptation and plasticity. The present article will focus attention on the behavioral significance of activation of the gonadal hormonal axis and present recent promising data, indeed suggesting that acute hormonal changes are more than merely functional epiphenomena of central activation.

Activation of the gonadal axis and behavioral change

Among the few early experiments on behavioral change in sexuality, those of Hård and Larsson (1969) established that sexual stimuli (exposure to copulating conspecifics) can facilitate subsequent sexual behavior in male rats. Such an excitatory influence was originally ascribed to a factor of general arousal, as mere handling was shown to have comparable effects. However, the finding that neutral stimuli associated with sexual arousal may greatly and specifically facilitate the male's sexual responding, drew attention to the relevance of mechanisms of conditioning in sexual behavior. In a series of experiments (Zamble, Mitchell, & Findley, 1986) it was shown that neutral stimuli preceding mating tests could facilitate behavior when these were conditioned to sexual arousal (the presentation of a receptive female), especially if such confrontations aroused and activated the males' sexual behavior without the subsequent resolution of ejaculation.

Remarkably little attention has thus far been paid to the significance of conditioning in sexual behavior. In parallel, but largely independent of the former experiments, effects of sexual stimulation upon the hormones of the gonadal axis were established. It is now well-established that males of many species, including the human, may show a characteristic central hormonal activation when sexually stimulated. Such changes are remarkably acute and specific, and appear to result from the earliest phase of copulation and prime stimulation, rather than from actual mating performance as such (Kamel, Wright, Mock, & Frankel, 1977). Like the behavioral responses, this hormonal activation is easily conditioned to neutral stimuli, as suggested by an animal experiment (Graham & Desjardins, 1980), but also by the often cited anecdote of the conditioning of fluctuations in beard-growth, published by a "Crusoë-Endocrinologist" on a British island (Anon., 1970). All together these data suggest a close and direct causal and functional correlation of psychological and neuroendocrine mechanisms in conditioning and adaptation.

Central activation of the gonadal axis as stimulus factor

If hormonal changes conditioned to specific environmental stimuli subsequently determine or modulate effects of environmental contingencies upon behavior, and indeed play a vital role in behavior change and adaptation, it is to be expected that such changes will easily generate internal stimulus characteristics. State-dependent learning and the technique to investigate stimulus properties of centrally acting drugs have now been known for many years and numerous experiments have shown that drug-induced internal alterations may act as a discriminative stimulus in learning. Hardly any experiment, however, has thus far established the relevance of endogenous stimuli, such as hormonal changes. In an early study on hormones as a factor in state-dependent learning (Stewart, Krebs, & Kaczender, 1967), rats were trained to escape from shock in a T-maze to one of the two arms, cued by a previous injection of "viadril", a hypnotic compound with chemical resemblance to progesterone (in males), or progesterone in a high dose (100 mg/kg, in females). Both sexes learned to escape from shock on the basis of the hormonal condition.

Recent results of an experiment using pentobarbital and progesterone, further corroborate these findings (Heinsbroek, van Haaren, Zantvoord, & van de Poll, 1987). Rats were trained to discriminate pentobarbital from saline in a two-lever drug-discrimination (DD) paradigm. Subsequently, generalization was established by injecting different doses of pentobarbital or progesterone. As can be seen in Fig. 1, ovariectomized female rats (in contrast to males), trained to discriminate 12 mg/kg pentobarbital from saline, generalized pentobarbital to test doses of progesterone dose-dependently, thus confirming the resemblance of these two compounds as internal stimuli, probably based upon common central sedative activity.

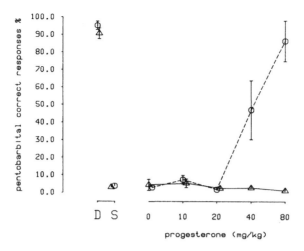

Fig. 6.1. Dose-response curve of progesterone for intact males (Δ) and overiectomized females (o) trained on disrimination of pentobarbital (12 mg/kg) from saline. Averaged percentage of pentobarbital correct responses are presented (+SE). D: baseline performance during pentobarbital training session; S: baseline performance during saline training sessions.

The results of a second experiment further indicate that hormonal change may directly act as a stimulus (unpublished). Male and female rats were trained in a two-lever DD-paradigm with 5 mg/kg Luteinizing Hormone Releasing Hormones (LHRH) as contrasted to saline, as stimulus. Characteristically in such a procedure, internal changes resulting from LHRH-treatment are the only cue complex for the animal to select the lever which is reinforced. Fig. 2 shows the typical learning curve of male rats which gradually but reliably learned to press the hormone- or saline-associated levers on the appropriate days. IP-injections of LHRH were given 45 minutes prior to the discrimination sessions, an interval of 15 min being ineffective. Females never learned to discriminate between LHRH and saline.

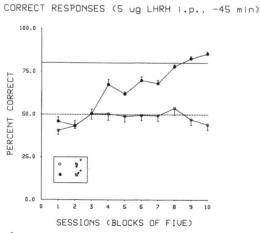

Fig. 6.2. Averaged percent correct responses per blocks of five daily sessions of male and female rats trained to discriminate between two levers associated with ip LHRH and saline treatment (presented in semi-random order). Males and females were gonadectomized and implanted with a silastic tube containing testosterone.

71

The results of these three experiments indicate that under certain conditions, rats can indeed be made to use hormonal changes as an internal discriminative stimulus. Self-evidently, these data do not provide any indication of the behavioral significance and the animals' readiness to use such cues. The next paragraph, however, provides further data and argues for an intrinsic significance of hormonal activation.

Central activation of the gonadal axis and affect

Now that hormonal activation has been shown to act as a stimulus and cue for behavioral adaptation, it is important to further study the possible intrinsic (unconditioned) quality such changes might have. Many drugs possessing stimulus properties have now been shown to have a positive or negative affective quality. Aversive and rewarding effects associated with treatment with a certain drug can be reliably established as approach toward or withdrawal from environmental stimuli in a place-preference (PP) paradigm. If animals are confronted with a distinct environment after treatment with a compound inducing positive or negative affect, a clear preference or aversion develops, which can be shown when the rat is given the choice between the drug-associated environment and the alternative, associated with the non-drug condition.

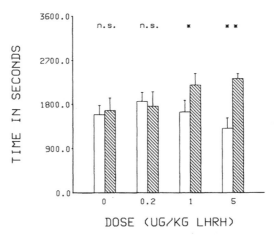

Fig. 6.3. Time in seconds, spend on the LHRH-associated side of a place-preference test cage before (blank bars) and after (hatched bars) association with LHRH treatment. Male rats were goandectomized and implanted with a silastic tubing containing testosterone. LHRH or saline was injected 15 min before the start of association sessions on 10 preceding days.

In our study of effects of the central activation of the gonadal axis, affective properties of ip injections with LHRH in male rats (gonadectomized, with a silastic implant containing testosterone) were established (De Beun et al., submitted; DeBeun, Geerts, Vreeburg, Slangen, & van de Poll, 1989). In Fig. 3, preference is clearly seen as it develops after treatment with different doses of LHRH, injected 15 minutes before the start of association sessions (30 min). The animals in such a paradigm are typically tested in a non-drug condition. Doses of LHRH in this and in the former DD-experiment, were chosen on the basis of separate determination of LHRH effects upon LH or testosterone release, as an index of biological activity.

Experiments with females under various hormonal conditions (ovariectomized with implants of testosterone or estradiol) never showed any significant conditioned preference effect for stimuli associated with LHRH-treatment, unexpectedly suggesting that the rewarding characteristics of central activation of the gonadal axis by LHRH are exclusively the males' domain of sexuality. As to the effects in males, an important question has to be answered: which aspects of hormonal activation actually trigger rewarding effects? A comparison of these data with those of the DD-paradigm indicates that the rewarding effect needs the development through a comparable interval, which strongly argues for a central mediation of the effect and, thus, for a significant involvement of such effects in actual behavioral adaptation.

Central activation of the gonadal axis and human sexuality

The present results suggest that the central activation of the gonadal axis by LHRH significantly contributes to behavioral change in sexual behavior, and stress the relevance of the hormones most directly involved in the animals' specific response to environmental stimulation. Behavioral effects of LHRH have now been observed for several years (Dornan & Malsbury, 1989). Peripherally or centrally administered LHRH was shown to facilitate lordosis responses and mounting in female and male rats, primed with low doses of gonadal hormones. In these experiments, LHRH had to be injected at an interval of at least two hours for a behavioral effect to develop. Together with the present data, which show a critical interval of less than one hour, these results suggest that the present stimulus and rewarding aspects, associated with LHRH-treatment, are different by nature. Thus, short-term activation of this hormonal axis by environmental contingencies may serve as a separate mechanism appropriate for behavioral adaptation. Interestingly, Zamble et al. (1986), conditioning a facilitation of sexual responding, showed that at least a period of 8 minutes had to elapse as an interval between the CS and the establishment of its facilitatory effect on sexual behavior, which strongly argues for a hormonal mediation of the conditioning effect.

There have been reports of activation of the gonadal axis by sexual stimulation in the human male, although failures to find such an effect can also be found in literature. Rowland, Heiman, Gladue, Hatch, Doering, and Weiler (1987), for instance, reported an increase of LH as a reaction to sexual arousal, which facilitated responsiveness to later stimulation. Direct effects of LHRH were established in a study in which sexual arousal and alertness were increased after infusion of this hormone (McAdoo, Doering, Kraemer, Dessert, Brodie, & Hamburg, 1978). These data, together with the animal data mentioned in this article, provide strong arguments to follow the line of experiments initiated by the simple experiments of Hård and Larsson, and to continue concentrating on the intricate interaction of a central hormonal activation and behavioral change.

SUMMARY

The phenomenon of a central activation of the gonadal hormonal axis by social stimuli and contingencies has been known for many years, although its behavioral significance is not clear as yet. The present article presents some recent data showing that such hormonal changes may act as internal discriminative stimuli and have intrinsic rewarding aspects in male rats. These results, together with data on the relevance of conditioning in sexuality, suggest that hormonal feedback mechanisms play an important role in behavioral change and adaptation.

REFERENCES

Anon. (1970). Effects of sexual activity on beard growth in man. *Nature, 226,* 869-870.

De Beun, R., Geerts, N.E., Jansen, E., Slangen, J.L., & van de Poll, N.E. Luteinizing hormone releasing hormone induced conditioned place preference in the male rat (submitted).

De Beun, R., Geerts, N.E., Vreeburg, J.T.M., Slangen, J.L., & van de Poll, N.E. (1989). Sex differences in luteinizing hormone releasing hormone-induced conditioned place preference in the rat. *Drug Dev. Research, 16,* 375-383.

Dornan, W.A., & Malsbury, C.W. (1989). Neuropeptides and male sexual behavior. *Neurosci. Biobehav. Rev., 13,* 1-15.

Graham, J.M., & Desjardins, C. (1980). Classical conditioning: induction of luteinizing hormone and testosterone secretion in anticipation of sexual activity. *Science, 210,* 1039-1041.

Hård, E., & Larsson, K. (1969). Effects of precoital exposure of male rats to copulating animals upon subsequent mating performances. *Animal Behav., 17,* 540-541.

Heinsbroek, R.P.W., van Haaren, F., Zantvoord, F., & van de Poll, N.E. (1987). Discriminative stimulus properties of pentobarbital and progesterone in male and female rats. *Pharmacol. Biochem. Behav., 28,* 371-374.

Kamel, F., Wright, W.W., Mock, E.J., & Frankel, A.I. (1977). The influence of mating and related stimuli on plasma levels of luteinizing hormone, follicle stimulating hormone, prolactine and testosterone in the male rat. *Endocrinology, 101,* 421-429.

Leshner, A.I. (1979). Kinds of hormonal effects on behavior: A new view. *Neurosci. Biobehav. Rev., 3,* 69-73.

McAdoo, B.C., Doering, C.H., Kraemer, H.C., Dessert, N., Brodie, H.K.H., & Hamburg, D.A. (1978). A study of the effects of gonadotrophin-releasing hormone on human mood and behavior. *Psychosom. Med., 40,* 199-209.

Rowland, D.L., Heiman, J.R., Gladue, J.P., Hatch, J.P., Doering, C.H., & Weiler, S.J. (1987). Endocrine, psychological and genital response to sexual arousal in men. *Psychoneuroendocrinology, 12,* 149-158.

Stewart, J., Krebs, W.H., & Kaczender, E. (1967). State-dependent learning produced with steroids. *Nature, 216,* 1223-1224.

Zamble, E., Mitchell, J.B., & Findley, H. (1986). Pavlovian conditioning of sexual arousal: parametric and background manipulations. *J. Exp. Psychology: Animal Behav. Proc., 12,* 403-411.

7 The Neuropharmacology of Lordosis Behavior: Serotonergic Mechanisms

Bengt J. Meyerson
University of Uppsala

INTRODUCTION

Over the last twenty-five years, the functional connections and interactions between neuronal populations involved in the lordotic behavior of the female rat have become better understood. Lordosis behavior has been a very useful model, both from a behavioral pharmacology point of view and for the understanding of the interaction between gonadal hormones and neurotransmitter function. However, in the light of the present rapid progress in various fields of neurobiology, it is also clear that we are still merely beginning to understand this fascinating area. At this point, it might be appropriate to call attention to some phases of the development of neuropharmacology which have been of significance for our present knowledge of the biochemical basis of lordosis behavior.

At the time when the first neuropharmacological studies of lordosis behavior were undertaken, drugs were used with a known action on the biosynthesis (precursors, inhibitors of the rat limiting enzyme involved in the biosynthesis), metabolism (enzyme-inhibitors like monoamine oxidase inhibitors) or reuptake of the monoamines (tri-cyclic antidepressants). The attempt was to relate transmitter levels to function (for reviews and references cited herein, see: Meyerson & Eliasson, 1977; Meyerson & Malmnäs, 1978; Meyerson, Malmnäs, & Everitt, 1984; Pfaff & Schwartz-Giblin, 1988). The next phase was influenced by the issue of feedback regulatory mechanisms, especially the discovery of presynaptic receptors with a functional relationship to the biosynthesis and release of the transmitter (autoreceptors) (Everitt & Fuxe, 1977, Meyerson & Sietniks, 1981). The discovery of the coexistence of several transmitters in the same neuron shook our neurobiological concepts, which were heavily based on the theory of Sir H.H. Dale (Dale & Feldberg, 1934), i.e., 'one neuron one transmitter'. Simultaneously, the number of putative transmitter candidates increased tremendously, especially when it became clear that neuropeptides could also participate in central nervous transmitter mechanisms.

Today, interest has focused on the receptor protein and subsequent ligand-receptor complex interaction with ion channels or a G-protein second messenger-mediated system.

The cloning and sequency of the receptor proteins gave an alternative basis for classification (Hartig, 1989). Hopefully, it will be possible to correlate the knowledge of the structure to the physiological function. So far, lordosis behavior has proved to be a very useful toll for gaining such knowledge.

Several types of transmitters (monoamines, ACH, GABA, neuropeptides) have been demonstrated to be involved in the central nervous regulation of the lordosis response (LR). However, the scope of the present paper is limited to the role played by serotonergic mechanisms. There is experimental support for the involvement of more than one population of serotonin (5-HT) receptors. Some recent data have been used as the basis for the hypothesis that serotonin is implicated in inhibitory as well as stimulatory mechanisms regulating lordosis behavior. The assumptions connected with this hypothesis will be discussed here, mainly against the following pharmacological background.

Pharmacological consideration

In the discussion and interpretation of results from neuropharmacological studies on the lordosis response (LR), the following points recur.

1. Hormonal factors: In estrogen alone-induced LR, the facilitatory effect may be indirect because of drug-induced adrenal progesterone release - especially when the onset of facilitation does not occur until hours after the drug administration. There is also increasing support for other mechanisms being involved in the estrogen alone-activated LR that are partly different from the mechanisms involved in the estrogen + progesterone-activated LR.

2. Time and dose relationship: The effects achieved after more than one hour can be due to drug metabolism with other properties than the original compound. Furthermore, for some compounds the effect on the LR appears within 10-15 min. Hours after the administration, low concentrations may still remain which act at different loci compared to the initial high concentration, e.g., a post-versus pre-synaptic action.

3. Selectivity: Especially with 5-HT antagonists, the compounds have affinity for more than one receptor. The relationship between an in vivo effect and the in vitro binding is a guide to receptor involvement, but hardly more. Some antagonists are also partial agonists.

Serotonin and lordosis inhibition

At an early stage of the psychopharmacological exploration of the LR, it was suggested that serotonergic pathways exert an inhibitory effect on the LR. Serotonergic pathways mediating inhibition of the LR have well-established experimental support. The evidence stems from inhibition of the LR after 5-hydroxytryptamine (5-HT) precursor treatment (5-HTP), treatment with 5-HT agonists (LSD, quipazine, alphamethyl-tryptamine), reuptake inhibitors (tricyclic antidepressants), 5-HT displacing agents (fenfluramine), and monoamine oxidase inhibitors (for ref., see Introduction). There are a number of agents which increase postsynaptic serotonergic activity that also inhibit the LR. Compounds that decrease serotonergic activity should subsequently facilitate the behavior. This has been much more difficult to show, however. Actually, most data which indicate facilitation are indirect (e.g., based on autoreceptor stimulation, prevention of an inhibitory effect) (Everitt & Fuxe, 1977; Sietnieks & Meyerson, 1983) or based on effects after the selection of the material into different levels of receptivity (Wilson & Hunter, 1985). Decreased biosynthesis by para-chlorophenylalanine (PCPA) does not have an unequivocal effect in this respect. In estrogen + progesterone-activated LR, several studies show that PCPA given 24 hours or more before testing inhibits the LR (Gorzalka & Whalen, 1975; Meyerson & Malmnäs, 1978; Wilson & Hunter, 1985). If PCPA is given simultaneously with or after progesterone, no inhibition but rather enhancement is seen (Ahlenius, Engel, Eriksson, Modigh, & Södersten, 1972; Everitt, Fuxe, Hökfelt, & Jonsson, 1975; Meyerson & Lewander, 1970). PCPA might induce adrenal progesterone release with a subsequent decreased response 24 hours later due to the dual effect of progesterone (see Morin, 1977).

76

Further evidence for serotonin in mediating inhibition of the LR is provided by the selective destruction of certain 5-HT pathways by the neurotoxin 5,7 dihydroxytryptamine (Everitt, Fuxe & Jonsson, 1975).

5-HT antagonists and lordosis behavior

The effects of 5-HT antagonists on the estrogen + progesterone-activated LR have been investigated in several laboratories. There are generations of 5-HT antagonists with increasing selectivity in the agents which have become available more recently. However, this is not to say that any antagonist possesses so selective effects that we can exclude other receptors or partial agonism (Colpaert & Janssen, 1983). The amount of the administered antagonist is apparently also crucial. The most frequently studied, but also rather non-selective agent, is methysergide. Both inhibitory and facilitatory effects have ben reported (Davis & Kohl, 1978; Foreman & Moss, 1978; Ward, Crowley, Zemlan, & Margules, 1975; Zemlan, Ward, Crowley, & Margules, 1973). If we stay with the effects on the estrogen + progesterone-activated LR and effects seen with 1 hour, it is mainly inhibition which is achieved. Considering that all compounds have affinity for several of the various subpopulations of 5-HT receptors, it might be advantageous to compare a series of agents which possess antagonistic properties to 5-HT receptors, with respect to their relationship to the 5-HT1 and 5-HT2 receptor populations.

Testing such a series of antagonists, Sietnieks (1985) found that some of these compounds inhibited the LR, but none had facilitatory effects. A dose-response relationship was obtained for each compound and it was found that the inhibition was achieved at relatively high doses. Cyproheptadine and pirenperone inhibited the LR. Both compounds have affinity for the 5-HT2 type of receptor (Leysen, Niemegeers, van Neuten, & Laduron, 1982; Martin & Sanders-Bush, 1982). However, altanserine which has a rather selective 5-HT2 affinity and cinnanserine (Martin & Sanders-Bush, 1982) also fail to inhibit the LR. Thus, no clear relationship was found between the inhibitory effect on the LR of these compounds and their 5-HT2 affinity in vitro. The inhibitory effects are more likely to be associated with other properties, such as 5-HT1 affinity, than with the 5-HT2 antagonistic ones. Comparisons of various 5-HT antagonists have also been done by Mendelson and Gorzalka (1986a,b). The most effective agent was metitepine, which was more effective than ketanserin and cyproheptadine, whereas pipamperone was reported not to give any significant inhibitory effects. Metitepine has affinity for the 5-HT2 sites but also has affinity for the 5-HT1 receptor (Martin & Sanders-Bush, 1982) using the classification of Peroutka and Snyder (1979). This is even true for e.g., methysergide which also exerted inhibition.

It should be considered that the antagonists may also act as partial agonists and/or may bind to receptors other than 5-HT receptors (Colpaert & Janssen, 1983; Janssen, 1983; Leysen et al., 1981;). The fact that some of these compounds inhibit the LR does not therefore exclude that the lordosis inhibition is mediated by serotonergic activity.

5-HT agonists and lordosis behavior

Several putative 5-HT agonist indolalkylamines (Eliasson & Meyerson, 1976; Espino, Sano & Wade, 1975; Everitt & Fuxe, 1977) were shown to inhibit the LR. More recently, Wilson and Hunter (1985) used the compounds (5-methoxy-3(1,2,3,6-tetrahydropyridin-4-yl1)1H indole (RU 24969) and the piperazine-containing agonist MK212 which is structurally related to another agonist, quipazine. All compounds except MK212 inhibit the LR. An important point here is the time-response relationship. The inhibitory effects of the agonists appear after about 10-15 min and the LR has usually returned to pretreatment levels after 60 min. MK212 treatment increased LR in ovariectomized and adrenalectomized estrogen-primed females 60 min after injection. There are no data available about any initial inhibitory effects (cf. quipazine, Table 1). Almost all other indole derivatives with 5-HT agonistic properties have been shown to facilitate the LR at low doses (Everitt & Fuxe, 1977; Sietnieks & Meyerson, 1983). It has been suggested that this is due to stimulation of presynaptic autoreceptors with subsequently decreased 5-HT release. Effects seen 1 hour after agonist administration might be caused by the low concentration of the compound remaining in the tissue. This may explain the stimulatory effect of MK212.

5-HT and stimulation of lordosis behavior

There seems to be general agreement that an inhibitory 5-HT system is involved in the regulation of lordosis behavior. This system may involve more than one receptor population and also encompass various areas of the CNS.

To explain the inhibitory effects of PCPA and 5-HT-antagonists, a hypothesis has been put forward involving dual serotonergic mechanisms (Hunter, Hole, & Wilson, 1985; Mendelson & Gorzalka, 1985; Wilson & Hunter, 1985), one stimulatory, the other inhibitory. This theory raises two questions. First, is there a stimulatory system and second, what kind of 5-HT receptors are involved in the two mechanisms. As to the last question, it is claimed that inhibition is mediated by the 5-HT1 and facilitation by the 5-HT2 receptors (Hunter et al., 1985; Mendelson & Gorzalka, 1985; Wilson & Hunter, 1985).

Regarding the effects of 5-HT antagonists, the lack of inhibition after certain 5-HT2 antagonists makes it equally relevant to relate the inhibitory effects to 5-HT1 mechanisms.

One possibility is that the inhibitory effect is coupled to the relative 5-HT2/5-HT1 affinity of the compound. Hypothetically, this means that blockade of 5-HT1-type receptors with autoreceptor properties located presynaptically leads to decreased inhibition of 5-HT release and subsequent increased stimulation of postsynaptic LR-inhibitory 5-HT2 receptors. Alternatively, the inhibitory effects of some of the antagonists may be due to each compound having a spectrum of receptor antagonistic and agonistic properties including receptors other than 5-HT, or differing bioavailabilities, or even, considering the time-effect relationship when inhibition does not appear until hours after injection, pharmacologically active metabolites. However, significant evidence for the inhibition of LR by activation of the 5-HT2 type of receptor comes from the inhibitory effects of lysergic acid diethylamide (LSD) (Eliasson & Meyerson, 1976). LSD has affinity for both 5-HT1 and 5-HT2 receptors (Peroutka & Snyder, 1979). It was demonstrated that LR-inhibitory effect of LSD is prevented by low doses of 5-HT antagonists (Sietnieks, 1985). The most effective were those which had a high affinity for the 5-HT2 but only low or almost no affinity for the 5-HT1 receptor type (altanserine, pirenpirone), whereas compounds with relatively high binding to 5-HT1 sites, like methysergide and mianserin were ineffective in blocking the LSD effect. It is clear from these data that the ability to prevent the LSD effect is related to a relatively high 5-HT2 versus low 5-HT1 receptor affinity.

We conclude that so far the overwhelming number of experiments carried out has led to inhibition of the LR. Facilitation of the response has been obtained only when very low doses of agonists (e.g., LSD) have been used. Such effects have been claimed to be due to decreased 5-HT release by means of stimulation of presynaptic autoreceptors (Everitt & Fuxe, 1977; Sietnieks & Meyerson, 1983).

As discussed above, some 5-HT antagonists decrease the LR. Let us assume this effect is due to inhibition at the presynaptic 5-HT1 receptors and subsequent decreased control of 5-HT release. In a study by Mendelson and Gorzalka (1986a) it was found that the inhibitory effect of certain 5-HT antagonists (ketanserin, pizotefin, cyproheptadine, metitepine) was reversed by the 5-HT2 agonist quipazine. The effects were studied 1 hr after administration. The initial (10-20 min after i.p. injection) effect of quipazine is a very clear-cut inhibition of the LR (Table 1). Regarding the initial inhibitory effect of quipazine, the reversal of the antagonist inhibition of the LR reported demonstrated 1 hr after injection, does not provide unequivocal support for the hypothesis of a 5-HT2 LR stimulatory mechanism. Quipazine has several effects both post- and pre-synaptically (Blier & de Montigny, 1983). The initial effect of quipazine rather indicates stimulation of postsynaptic 5-HT2 receptors with subsequent inhibition of the LR. At 1 hr after the quipazine injection, the main action of the remainder of the drug in the tissue may be to activate autoreceptors and, in so doing, to compete with a presynaptic effect of the antagonist. Other rebound phenomena are also available.

Table 7.1. The effect of quipazine and cyproheptadine on the lordosis behavior in ovariectomized estradiol benzoate + progesterone primed female rats.

TREATMENT	LORDOSIS			
Min after last injection	10		60	
L%	LQ	L%	LQ	
Quipazine	25**	17±9***	75	63±10
Cyproheptadine	54	35±6***	70	53±7**
Cyproheptadine & quipazine	75	61±11	80	68±8
Saline	75	72±8	80	80±6

Hormonal treatment: estradiol benzoate 10 µg/kg followed 48 hrs later by progesterone 0.4 mg/rat. The drugs were given 5 hrs after the progesterone injection: quipazine 3 mg/kg i.p., cyproheptadine 1 mg/kg i.p.
LQ=number of LR/number of mounts (6) x 100 ± s.e.m., L%=proportion of group responding. For laboratory conditions and testing procedure, see Sietnieks and Meyerson (1983).
N=12-24 per treatment. Statistical difference (Chi-square test or Mann-Whitney U test) between drug vs. saline treatment ** $p \leq 0.01$, *** $p \leq 0.001$.

5-HT1a agonists and LR

Compounds that are agonistic to the 5-HT1a receptor have evident and well-documented behavioral effects. The most used and specific compound is 8-hydroxy-2-(di-N-propyl-amino)tetralin (8-OH-DPAT) which facilitates elements of the copulatory behavior in the male rat, induces hyperphagia, and causes changes in motor performance (Ahlenius, Larsson, Svensson, Hjorth, Carlsson, Lindberg, et al., 1981; Dourish, Hutson, & Curson, 1985; Hjorth Carlsson, Lindberg, Sanchez, Wikström, Arvidsson, et al., 1982). The LR is inhibited (Table 2; Ahlenius, Fernandez-Guasti, Hjorth, & Larsson, 1986). There are biochemical and pharmacological data suggesting a presynaptic as well as a postsynaptic action (Hjorth et al., 1982). One way to analyze a pre- versus postsynaptic effect is to pretreat with a biosynthesis inhibitor. Then the effect of a presynaptic agonist on the autoreceptors should be negligible, whereas a postsynaptic agonist action should be increased. As can be seen from Table 2, the inhibitory effect of 8-OH-DPAT was, if anything, decreased. This indicates an autoreceptor action rather than a postsynaptic agonist effect. Furthermore, the effect is probably not exerted at those sites which change the locomotor performance (flat body posture, Tricklebank, Forler, & Fozard, 1984). Chronic treatment with 8-OH-DPAT leads to tolerance for the 8-OH-DPAT-induced effect on motor performance, but no tolerance was seen with the effect on copulatory behavior in the male or with the LR in the female rat (Johansson, Meyerson, & Höglund, submitted for publication). If the 8-OH-DPAT-induced inhibition of the LR is due to activation of autoreceptor mechanisms, then it suggests that 5-HT neurons mediate stimulation of the LR. The type of postsynaptic 5-HT receptor in this particular system is so far unknown.

Table 7.2

TREATMENT		LORDOSIS		
Min after last injection	15		120	
L%	LQ	L%	LQ	
8-OH-DPAT	20***	8±4***	100	80±6
PCPA	95	77± 6*	91	80±6
PCPA and 8-OH-DPAT	43**	29±8** °	87	58±5 °°
Saline	82	62±7	88	70±7

For hormonal treatment see legend to Table 1.
Doses and time of drug treatment in relation to the progesterone treatment: PCPA, 10 mg/kg i.p. at 0 hrs; 8-OH-DPAT, 50 µg/kg i.p. and saline , 02, ml at 6 hrs.
N = 17-23/group. Statistical differences (Chi-square or Mann-Whitney U test) between 8-OH-DPAT alone vs after PCPA treatment (°), drug vs salone (*), number of symbols p < 0.05 (1), 0.01 (2), 0.001 (3).

Progesterone enhances the effects of 5-HTP and 5-HT agonists

In a series of reports, we showed that the effect of LSD on the LR was dependent on progesterone treatment (Sietnieks & Meyerson, 1980, 1982, 1983). The inhibitory effect of the 5-HT precursor 5-HTP was also enhanced by progesterone (Sietnieks & Meyerson, 1982). Similar facilitatory effects by progesterone had earlier been shown by Espino et al. (1985) using alpha-methyltryptamine. An interaction between progesterone and 5-HT agonists is also suggested from the data by Mendelson and Gorzalka (1986c). Some putative anxiolytic drugs, buspirone, ipsapirone, and gepirone, which all have 5-HT1a receptor affinity, were shown to induce a slight facilitatory effect on the LR. Higher doses were inhibitory. The facilitatory effect was only seen in estradiol + progesterone treated females, whereas LR induced by estrogen alone showed only inhibition. It is not clear whether these effects are related to the 5-HT1a agonistic effect as an analogous effect was not seen after 8-OH-DPAT (Ahlenius et al., 1986). We have confirmed the latter data. In our study, a dose-related effect is obtained from 0.025 to 0.100 mg/kg i.p. The inhibitory effect of 0.05 mg/kg was not influenced by different doses of progesterone. Complete inhibition was obtained in females primed with estradiol benzoate alone (Meyerson, unpublished data).

CONCLUDING REMARKS

Obviously, serotonergic inhibition of the lordosis behavior exists. From the discussion above, it follows that the 5-HT2 type of receptor is the prime candidate for the LR inhibitory system. If it can be demonstrated that the inhibitory action of 8-OH-DPAT is achieved by presynaptic autoreceptor mechanisms with subsequently decreased 5-HT release, this would suggest that a 5-HT stimulatory mechanism also comes into the picture. Which type of 5-HT receptor might be associated with that system has to await proof that stimulatory mechanisms actually exist. In interpreting the data above, one can get the feeling that the serotonergic system involved in the LR has been looked upon as detached from its context. Of course, the serotonergic mechanisms are part of a neuronal network which comprises cholinergic, dopaminergic, and peptidergic transmitter mechanisms (see Pfaff & Schwartz-Giblin, 1988). The possibility of ever understanding the neurobiological basis of the LR may seem remote, and impression compounded by the existence of some kind of cooperative mechanism between different kinds of transmitters at the same synapse. However, the problem may not be in understanding the neurological basis of the LR as such, but rather in using the LR as a functional model in order to study the cooperative mechanisms between various components in a neuronal network which has hormone-dependent elements. The effect of progesterone on serotonergic mechanisms is one of these interesting relationships, which is not only of basic neurobiological interest but also has clinical importance with the implication of serotonergic mechanisms in various emotional states and the mood-changes associated with the menstrual cycle. It is of obvious importance to investigate the cooperative mechanisms which are involved in the interneuronal process of communication. However, it may also be of importance to establish the set-point of sensitivity of the system. By this I mean that the neuronal network, which is activated for the manifestation of a particular behavior, has a certain level of readiness to respond. At a cell biological level, this might mean the number of receptors, or the proportions of various receptors at the postsynaptic membrane, or autoreceptor mechanisms for the feedback control of the synaptic activity, etc. Such setting or quantitative imprinting of the system might be established perinatally. The extent to which the not yet fully-mature neuronal network is influenced by various trophic factors could be one factor involved in such chemical imprinting. Besides the polypeptide nerve growth factors, such agents might be neuropeptides which participate as neurotransmitters or modulators in the manifestation of the behavior. Several laboratories have initiated research on the developmental effects of neuropeptides. Based on the hypothesis just described we have started a series of experiments, some of which also include the effects on the LR of neonatal exposure to neuropeptide agonists and antagonists (Meyerson & Höglund, 1989). Although the investigation is at an early phase, the data are so far consistent with the view that early stimulation of neuropeptide receptors has consequences for the readiness to respond to these particular neuropeptides in the mature individual with behavioral consequences. There is also evidence that early manipulation of serotonergic mechanisms influences adult sexual behavior (Jarzab & Döhler, 1984). The study of the cooperative mechanisms between the various transmitter systems at the perinatal period also seems a relevant line of research.

REFERENCES

Ahlenius, S., Engel, J., Eriksson, H., Modigh, K., & Södersten, P. (1972). Importance of central monoamines in the mediation of lordosis behavior in ovariectomized rats treated with estrogen and inhibitors of monoamine sythesis. *J. Neur. Transm., 33,* 247-255.

Ahlenius, S., Fernandez-Guasti, A., Hjorth, S., & Larsson, K. (1986). Suppression of lordosis behavior by the putative 5-HT receptor agonist 8-OH-DPAT in the rat. *Eur. J. Pharmacol., 124,* 361-363.

Ahlenius, S., Larsson, K., Svensson, L., Hjorth, S., Carlsson, A., Lindberg, P., Wikström, H., Sanchez, D., Arvidsson, L-E., Hacksell, U., & Nilsson, J.L.G. (1981). Effects of a new type of 5-HT receptor agonist on male rat sexual behavior. *Pharmacol. Biochem. Behav., 15,* 785-792.

Blier, P., & de Montigny, C. (1983). Effects of quipazine on pre- and postsynaptic serotonin receptors: single cell studies in the rat CNS. *Neuropharmacology, 4*, 495-499.

Colpaert, F.C., & Janssen, P.A.J. (1983). The head-twitch response to intraperitoneal injection of 5-hydroxytryptophan in the rat: antagonist effects of purported 5-hydroxytryptamine and of pirenperone, and LSD antagonist. *Neuropharmacology, 22*, 993-1000.

Dale, H.H., & Feldberg, W. (1934). Chemical transmitter of vagus effects to stomach. *J. Physiol., 81*, 329.

Davis, G.A., & Kohl, R.L. (1978). Biphasic effects of the antiserotonergic methysergide on lordosis in rats. *Pharmacol. Biochem. Behav., 9*, 487-491.

Dourish, C.T., Hutson, P-H., & Curzon, G. (1985). Low doses of the putative serotonin agonist 8-hydroxy-2-(di-N-propylamino)tetralin (8-OH-DPAT) elicit feeding in the rat. *Psychopharmacology, 86*, 197-204.

Eliasson, M., & Meyerson, B.M. (1976). Comparison of the action of lysergic acid diethylamide and apomorphine on the copulatory response in the female rat. Psychopharmacology, 49, 301-306.

Espino, C., Sano, M., & Wade, G.N. (1975). Alpha-methyltryptamine blocks facilitation of lordosis by progesterone in spayed, estrogen-primed rats. *Pharmacol. Biochem. Behav., 3*, 557-559.

Everitt, B.J., & Fuxe, K. (1977). Serotonin and sexual behaviour in female rats. Effects of hallucinogenic indoleakylamines and phenylethylamines. *Neurosci. Lett., 4*, 215-220.

Everitt, B.J., Fuxe, K., Hökfelt, T., & Jonsson, G. (1975). Role of monoamines in the control by hormones of sexual receptivity in the female rat. *J. Comp. Physiol. Psychol., 89*, 556-572.

Everitt, B.J., Fuxe, K., & Jonsson, G. (1975). The effects of 5,7-dihydroxytryptamine lesions of ascending 5-hydroxytryptamine pathways on the sexual and aggressive behaviour of female rats. *J. Pharmacol. (Paris), 6*, 25-32.

Foreman, M.M., & Moss, R.L. (1978). Role of hypothalamic serotonergic receptors in the control of lordosis behavior in the female rat. *Horm. Behav., 10*, 97-106.

Gorzalka, B.B., & Whalen, R.E. (1975). Inhibition not facilitation of sexual behavior by PCPA. *Pharmacol. Biochem. Behav., 3*, 511-513.

Hartig, P.R. (1989). Molecular biology of 5-HT receptors. *Trends Pharmacol. Sci., 10*, 64-69.

Hjorth, S., Carlsson, A., Lindberg, P., Sanchez, D., Wikström, H., Arvidsson, L-E., Hacksell, U., & Nilsson, J.L.G. (1982). 8-Hydroxy-2-(di-n-propylamino)tetralin, 8-OH-DPAT, a potent and selective simplified ergot congener with central 5-HT-receptor stimulating activity. *J. Neural Transm., 55*, 169-182.

Hunter, A.J., Hole, D.R., & Wilson, C.A. (1985). Studies into the dual effects of serotonergic pharmacological agents on female sexual behavior in the rat: Preliminary evidence that endogenous 5HT is stimulatory. *Pharmacol. Biochem. Behav., 22*, 5-13.

Janssen, P.A.J. (1983). 5-HT2 receptor blockade to study serotonin-induced pathology. *Trends Pharmacol. Sci., 4*, 198-206.

Jarzab, B., & Döhler, K.D. (1984). Serotonergic influences on sexual differentiation of the rat brain. In G.J. De Vries et al. (Eds.), *Progress in Brain Research, Vol. 6* (pp. 119-125). Amsterdam: Elsevier.

Leysen, J.E., Awouter, F., Kemis, L., Laduron, P.M., Vandenverk, J., & Janssen, P.A.J. (1981). Receptor binding profile of R 41468, a novel antagonist at 5-HT2 receptors. *Life Sci., 28*, 1015-1022.

Leysen, J.E., Neimegeers, C.J.E., Van Nueten, J.M., & Laduron, P.M. (1982). (3H) Ketanserin (R 41468), a selective 3H-ligand for serotonin2 receptor binding sites. *Molec. Pharmacol., 21*, 301-304.

Martin, L.L., & Sanders-Bush, E. (1982). Comparison of the characteristics of 5-HT1 and 5-HT 2 binding sites with those of serotonin autoreceptors which modulate serotonin release. *Naunyn-Schmiedebergs Arch. Pharmacol., 321*, 165-170.

Mendelson, S.D., & Gorzalka, B.B. (1985). A facilitatory role for serotonin in the sexual behavior of the female rat. *Pharmacol. Biochem. Behav., 22,* 1025-1033.

Mendelson, S.D., & Gorzalka, B.B. (1986a). Methysergide inhibits and facilitates lordosis behavior in the female rat in a time-dependent manner. *Neuropharmacology, 25,* 749-755.

Mendelson, S.D., & Gorzalka, B.B. (1986b). Serotonin type 2 antagonists inhibit lordosis behavior in the female rat: Reversal with quipazine. *Life Sci., 38,* 33-39.

Mendelson, S.D., & Gorzalka, B.B. (1986c). Effects of 5-HT1a selective anxiolytics of lordosis behavior: interactions with progesterone. *Eur. J. Pharmacol., 132,* 323-326.

Meyerson, B.J., & Lewander, T. (1970). Serotonin synthesis inhibition and estrous behaviour in female rats. *Life Sci., 9,* 661-671.

Meyerson, B.J., & Eliasson, M. (1977). Pharmacological and hormonal control of reproductive behavior. In L.L. Iversen, S.D. Iversen, and S.H. Snyder (Eds.), *Handbook of Psychopharmacology, Vol. 8: Drugs, Neurotransmitters, and Behavior* (pp. 159-232). New York and London: Plenum Press.

Meyerson, B.J., & Malmnäs, C.O. (1978). Brain monoamines and sexual behaviour. In J.B. Hutchison (Ed.), *Biological Determinants of Sexual Behaviour* (pp. 521-554). New York: Wiley.

Meyerson, B.J., & Sietnieks, A. (1981). Progesterone, monoamines and sexual behavior. In K. Fuxe et al. (Eds.), *Steroid Hormone Regulation of the Brain* (pp. 293-299). Oxford and New York: Pergamon Press.

Meyerson, B.J., Malmnäs, C.O., & Everitt, B. (1984). Neuropharmacology, neurotransmitters and sexual behaviour in mammals. In N. Adler, D. Pfaff, and R.W. Goy (Eds.), *Handbook of Behavioural Neurobiology, Vol. 7: Reproduction* (pp. 495-536). New York: Plenum.

Meyerson, B.J., & Höglund, U. (1989). The behavioral effects of neonatal exposure to peptides and peptide antagonists. In J. Balthazart (Ed.), *Hormones, Brain and Behaviour.* Karger (in press).

Morin, L.P. (1977). Progesterone: Inhibition of rodent sexual behavior. *Physiol. Behav., 18,* 701-715.

Peroutka, S.J., & Snyder, S.H. (1979). Multiple serotonin receptors: differential binding of (3H)-5-hydroxytryptamine, (3H)-lysergic acid diethylamide and (3H)-spiro-peridol. *Molec. Pharmacol., 16,* 687-699.

Pfaff, D.W., Schwartz-Giblin, S. (1988). Cellular mechanisms of female reproductive behaviors. In E. Knobil and J. Neil (Eds.), *The Physiology of Reproduction* (pp. 1487-1568). New York: Raven Press.

Sietnieks, A., & Meyerson, B.J. (1980). Enhancement by progesterone on lysergic acid diethylamide inhibition of the copulatory response in the female rat. *Eur. J. Pharmacol., 63,* 57-64.

Sietnieks, A., & Meyerson, B.J. (1982). Enhancement by progesterone of 5-hydroxytryptophan inhibition of the copulatory response in the female rat. *Neuroendocrinology, 35,* 321-326.

Sietnieks, A., & Meyerson, B.J. (1983). Progesterone enhancement of lysergic acid diethylamide and levo-5-hydroxytryptophan stimulation of the copulatory response in the female rat. *Neuroendocrinology, 36,* 462-467.

Sietnieks, A. (1985). Involvement of 5-HT2 receptors in the LSD- and L-5-HTP-induced suppression of lordotic behavior in the female rat. *J. Neural Transm., 61,* 65-80.

Tricklebank, M.D., Forler, C., & Fozard, J.R. (1984). The involvement of subtypes of the 5-HT1 receptor and of catecholaminergic systems in the behavioural response to 8-hydroxy-2-(di-n-propylamino)tetralin in the rat. *Eur, J. Pharmacol., 106,* 271.

Ward, I.L., Crowley, W.R., Zemlan, F.P., & Margules, D.L. (1975). Monoaminergic mediation of female sexual behavior. *J. Comp. Physiol. Psychol., 88,* 53-61.

Wilson, C.A., & Hunter, A.J. (1985). Progesterone stimulates sexual behavior in female rats by increasing 5-HT activity on 5-HT2 receptors. *Brain Res., 333,* 223-229.

Zemlan, F.P., Ward, I.L., Crowley, W.R., & Margules, D.L. (1973). Activation of lordotic responses in female rats by suppression of serotonergic activity. *Science, 179,* 1010-1011.

8 Testosterone Metabolism and Sexual Behavior in Male Rats

Per Södersten
Karolinska Institute, Huddinge

INTRODUCTION

I was fortunate in that one of the first experiments Knut Larsson encouraged me to participate in turned out to be of great importance for our subsequent analysis of the role of testosterone metabolites in the control of sexual behavior in male rats. The experiment concerned the role of reduced sensory feedback to the penis in the display of sexual behavior in male rats. The control of sexual behavior by sensory stimuli was a main interest of Knut's at the time; one of his first experiments on the role of sensory stimulation of the genitals used the application of a local anesthetic to the glans penis of the rat and was performed in collaboration with Dr. Sven Carlsson (Carlsson & Larsson, 1964). In the subsequent experiment Knut performed, and I assisted in, the penile nerve was sectioned and the behavioral effects were subsequently investigated (Larsson & Södersten, 1973). Both of these maneuvers had similar effects; they prevented ejaculation in the rats but apparently did not affect the motivation to copulate because the rats continued to mount at very high frequencies.

Not only was this a good start for me, but Knut also suggested that I visit the United States to get acquainted with American behavioral endocrinology. He generously arranged for me to spend some time in the laboratory of Dr. Howard Moltz at the University of Chicago. I benefited tremendously from my stay in Chicago and in particular I appreciated, intellectually and otherwise, the company of Drs. Michael Leon and Klaus Miczek, as well as that of Dr. Michael Numan.

During my stay in Chicago, I investigated the comparative actions of testosterone and estradiol on the sexual behavior of castrated male rats (Södersten, 1973). A popular notion at the time was that the effects of estradiol on male rat sexual behavior, which were considered weak (Davidson, 1969), reflected "non-specificity" and which normally were believed to be specifically sensitive to testosterone (Davidson & Bloch, 1969; Pfaff, 1970). I found strong effects of the estrogen treatment on the sexual behavior of the castrated rats

(Södersten, 1973) and the further development of this observation will be outlined in this chapter.

Effects of estradiol on sexual behavior

Daily treatment with high doses of estradiol benzoate (EB, 100 µg) of sexually experienced male rats induced ejaculations in all rats and the effect was comparable to that of treatment with the same dose of testosterone propionate (TP) with the exception that the EB-treated rats tended to show more mounts and intromissions before ejaculation (Södersten, 1973). This tendency was more pronounced in a group of prepuberally-castrated rats treated with 50 µg EB/day. Although these rats showed more mounts and as many intromissions as normally preceded ejaculation in intact rats, only one ejaculated, and did so only after a very high number of mounts and intromissions (Södersten, 1983). The behavior of these rats was reminiscent of that of the rats with reduced sensory feedback to the penis (Carlsson & Larsson, 1964; Larsson & Södersten, 1973). It was long known that androgen treatment of castrated male rats restores not only their sexual behavior, but also the morphology of the cornified papillae or spines on the glans penis, and that these effects occur in close parallel (Beach & Levinson, 1950). The penile spines are considered to house the receptors which are stimulated during copulation and they are believed to be a requisite for normal sexual behavior (Beach & Levinson, 1950). Inspection of glans penis of the EB-treated rats revealed the absence of any stimulatory effect on penile morphology (Södersten, 1973). The conclusion, therefore, was that estradiol exerts strong central neural effects, as revealed by behavioral responses, but no peripheral "androgenic" effects.

The "aromatization-5α-reduction" hypothesis

Testosterone is extensively metabolized in the brain. Since this field was recently and admirably reviewed by Balthazart (1989), there is no need for an additional review here. Suffice it to say testosterone is aromatized to estradiol and 5α- and 5ß-reduced to 5α- and 5ß-dihydrotestosterone (5α-DHT and 5ß-DHT) in the brain. The DHTs cannot be aromatized. Early experiments (e.g., Feder, 1971) showed that while 5α-DHT has pronounced stimulatory effects on penile morphology, it fails to stimulate sexual behavior in castrated male rats. The "aromatization hypothesis" - testosterone must be aromatized to estradiol in the brain to stimulate sexual behavior - was therefore advanced (see Balthazart, 1989). However, since estradiol treatment has no effect on peripheral androgen sensitive structures (see above), the aromatization hypothesis had to be modified to what one perhaps might term the "aromatization-5α-reduction hypothesis": testosterone must be aromatized in the brain and 5α-reduced in the penis for proper control of sexual behavior (Balthazart, 1989). Three reports supported the implication of this hypothesis, i.e., that combined treatment with EB and 5α-DHT is highly effective in stimulating sexual behavior in castrated rats (Baum & Vreeburg, 1973; Feder, Naftolin, & Ryan, 1974; Larsson, Södersten, & Beyer, 1973).

Estradiol - 5α-dihydrotestosterone interaction in the brain

Perhaps the most interesting finding on the interactive effects of combined EB-5α-DHT treatment of castrated rats was that very small amounts of EB, in fact as little as 0.05 µg/day, were sufficient to induce the behavior if combined with 5α-DHT in 60% of the rats (Larsson et al., 1973; Figure 1). Thus, while neither EB nor 5α-DHT exerted behavioral effects by themselves, they were highly effective in stimulating the behavior if given in combination. This suggested an estradiol-androgen interaction in the brain in the control of sexual (see Södersten & Gustafsson, 1980a). A series of experiments was carried out to explore the nature of this interaction.

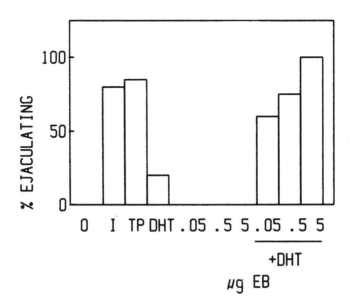

FIG. 8.1. Percent of intact (I) or castrated male rats ejaculating after treatment with vehicle (0), testosterone propionate (TP, 1 mg/day), 5α-dihydrotestosterone (DHT, 1 mg/day) or various doses of estradiol benzoate (EB) alone or in combination with 5α-dihydrotestosterone (DHT, 1 mg/day).

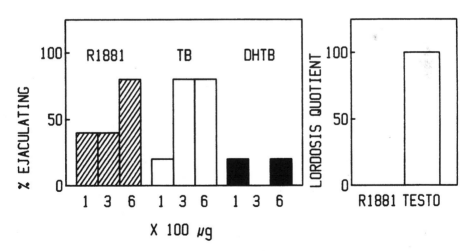

FIG. 8.2. Percent of castrated male rats ejaculating after treatment with methyltrienolone (R1881), testosterone benzoate (TB) or 5α-dihydrotestosterone benzoate (DHTB). Failure of R1881 to stimulate sexual behavior in ovariectomized rats is shown to the right.

Androgenic activation of sexual behavior

5α-DHT is rapidly and reversibly metabolized to 3α-androstanediol (3α-ADIOL) and irreversibly to 3ß-androstanediol (3ß-ADIOL), weak androgens which are considered devoid of androgenic activity (see Balthazart, 1989). It appeared, therefore, that a possible mechanism of action of estradiol in the control of sexual behavior might be that estradiol normally acts by preventing or modifying androgen metabolism such that the androgens are permitted to bind to neural androgen receptors and activate the behavior (Södersten & Gustafsson, 1980a). As a test of this hypothesis we administered methyltrienolone (R1881) to castrated rats. R1881 was found to be as effective as testosterone in stimulating male sexual behavior and, as expected, 5α-DHT caused no significant stimulation (Södersten & Gustafsson, 1980b; Figure 2). The stimulatory effect of R1881 on male sexual behavior was replicated in an independent series of experiments in which we also found that R1881 failed to stimulate female sexual behavior (Mode, Gustafsson, Södersten, & Eneroth, 1985; Figure 2), a known estradiol-dependent response, perhaps even a "behavioral estradiol bioassay" (Blaustein & Olster, 1989). These results show that androgenic stimulation, in the absence of any behavioral signs of estradiol stimulation, is highly potent in activating male sexual behavior in castrated rats, and argue against the suggestion that R1881 activates male sexual behavior only in high doses and by an action that is mediated via estrogen receptors (see Balthazart, 1989).

Interference with aromatization and 5α-reduction effects on testosterone-activated sexual behavior

These results opened the possibility that interference with aromatization, long known to disrupt the stimulatory effect of testosterone on sexual behavior in castrated male rats (see Balthazart, 1989), had its behavioral effect, not by preventing the formation of an estrogenic metabolite required for display of the behavior, but by shunting androgen metabolism via the 5α- and 5ß-reduction pathways so that only behaviorally inert androgens were formed (Södersten et al., 1986). In line with this hypothesis we found that sexual behavior, which had been blocked in castrated testosterone-treated rats by administration of the aromatization inhibitor 1,4,6-androstatriene-3,17-dione (ATD) could be reactivated by treatment with 5a-reductase inhibitor 17ß-N,N-diethylcarbamoyl-4-aza-5α-androstan-3-one (4MA) or with a low dose of estradiol, by itself behaviorally inert (Södersten et al., 1986; Figure 3).

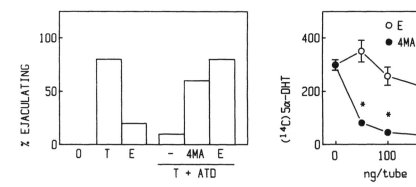

FIG. 8.3. Percent of castrated male rats ejaculating after treatment with vehicle (0), 30 mm long testosterone-filled implants (T) or implants filled with 50 μg/ml estradiol in oil (E). T-treated rats were treated with the aromatization inhibitor ATD in combination with the 5a-reductase inhibitor 4MA or E implants. The comparative action of 4MA and E to inhibit the formation of 5a-dihydrotestosterone (5α-DHT, ng/g tissue per h) by hypothalamic tissue in vitro is shown to the right.

In ovariectomized female rats ATD prevented testosterone from stimulating female sexual behavior and this inhibition could not be overcome by treatment with 4MA but could be overcome by by treatment with estradiol. These results offer additional support to the notion that androgenic stimulation, in the absence of estrogenic stimulation, can activate male sexual behavior.

However, although 4MA exerted a powerful inhibitory effect on the formation of 5α-DHT in the brain (Fig. 3), estradiol did not. The possibility that our failure to detect antagonistic effects of estradiol on neural 5α-reduction was merely a methodological artifact, e.g., due to an inappropriate method of sampling, a relevant neuronal population (Södersten et al., 1986) should perhaps be considered in view of the potent inhibitory effects of estradiol on 5α-reduction in peripheral androgen-sensitive organs (Danzo & Eller, 1984).

3α-Androstanediol: A "weak androgen" that is not weak

Rather than being a methodological artifact, the failure to prevent the formation of 5α-DHT from testosterone by reasonable doses of estradiol (see above) might be a true falsification of our original hypothesis that this estrogen acts as a neural 5α-reductase inhibitor (Södersten et al., 1986). Perhaps then estradiol acts by preventing the formation of the ADIOLs from 5α-DHT. This hypothesIs was tested by comparing the behavioral effects of co-administration of estradiol and the ADIOLs with those of combined 5α-DHT - estradiol administration in castrated male rats (Södersten, Eneroth, & Hansson, 1988). Somewhat surprisingly, 3α-ADIOL, but not 3β-ADIOL, was found as effective as 5α-DHT in activating the sexual behavior of castrated rats if given in combination with a subthreshold dose of estradiol (Södersten et al., 1988; Figure 4). These results show that an androgen metabolite, traditionally considered "weak" is a highly potent stimulator of sexual behavior if given with estradiol.

These behavioral data argue against our hypothesis that estradiol controls sexual behavior in male rats by preventing rapid androgen metabolism. The hypothesis was further weakened by biochemical data showing no effects of in vivo treatment with behaviorally effective doses of estradiol on the formation of ADIOLs from 5α-DHT by hypothalamic tissue (Södersten et al., 1988; Figure 4).

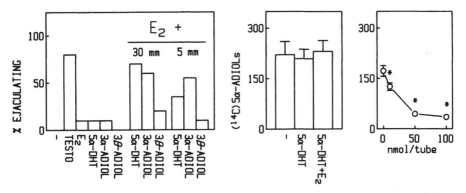

FIG. 8.4. Percent of castrated male rats ejaculating after treatment with vehicle (-) 30 mm long testosterone-filled implants (TESTO), implants filled with 50 µg/ml estradiol in oil (E2) or 30 mm long implants filled with 5α-dihydrotestosterone (5α-DHT) 3α-androstanediol (3α-ADIOL) or 3β-androstanediol (3β-ADIOL). Rats were treated with implants filled with 5α-DHT, 3α-ADIOL or 3β-ADIOL of various lengths in combination with E2 filled implants. The middle panel shows the failure of treatment with 5α-DHT+E2 in vivo to affect the formation of 5α-ADIOLSs (nmol/g tissue per h) by hypothalamic tissue in vitro, and the right panel shows inhibition of 5α-ADIOL formation by E2 in vitro.

Possible mechanisms of action of estradiol

If estradiol does not control male sexual behavior by modifying androgen metabolism in the brain, in what way, then, does estradiol participate in the neuroendocrinology of this behavior? It is interesting that brain aromatase activity is stimulated by testosterone (Roselli, Ellinwood, & Resko, 1984; Roselli, Horton, & Resko, 1985, 1987), that testosterone exerts this effect via neural androgen receptors (Roselli et al., 1984), and that estradiol increases the number of androgen receptors in the preoptic area of the brain (Handa, Roselli,, Horton, & Resko, 1987). Possibly, therefore, testosterone induced brain aromatase and the estradiol thus formed might interact with an androgen, e.g., 5α-DHT or 3α-ADIOL, by facilitating the binding of the androgen to a neural androgen receptor, thereby inducing the behavior. Since binding of androgens to their neural receptors would reduce the amount of the androgen which might serve as a substrate for metabolism to inactive androgens, this hypothesis is really a modification of the idea that estradiol controls sexual behavior by modifying androgen metabolism. In this context, it is noteworthy that it was recently suggested that testosterone acts primarily via androgen receptors in the control of sexual behavior (McGinnis & Dreifus, 1989).

Role of androgen inactivation

In what is generally regarded as the first experiment in endocrinology, Berthold (1849) reported activation of sexual behavior in capons by treatments with testicular extracts. Since then, virtually thousands of experiments have explored the mechanisms whereby testicular secretions activate the behavior (see Balthazart, 1989). Less attention has been paid to the possibility that the brain might normally need mechanisms for inactivating the endocrine secretion of the testes. Testosterone is secreted episodically from the rat testes and can increase up to tenfold in the serum within a brief period of time (Södersten, Eneroth, & Petterson, 1983). Since there is no barrier that might prevent testosterone from entering the brain from the blood (Partridge, 1983), and since there is no sex steroid binding protein in rat plasma (Renoir, Mercier-Bodard, & Baulieu, 1980), the rat brain is likely to be exposed to excessive amounts of testosterone each time a pulse of testosterone is secreted by the testes. The rat nervous system might require protection from these high testosterone levels. Inactivation by 5α-reductase and 3β-hydroxysteroid dehydrogenase might serve such a function.

SUMMARY

The interactive effects of estradiol and testosterone metabolites on sexual behavior in male rats are discussed. It is possible that estradiol participates in the control of the behavior by inhibiting or modifying androgen metabolism although the nature of this interaction remains to be clarified because testosterone metabolites such as 5α-dihydrotestosterone and 3α-androstanediol are equally effective in stimulating sexual behavior in castrated rats if combined with a low dose of estradiol. Alternatively, estradiol might affect male sexual behavior by facilitating androgen binding to neural androgen receptors. In addition to participating in behavioral activation, androgen metabolism may be important for androgen inactivation in the brain.

ACKNOWLEDGEMENTS

This review is dedicated Professor Knut Larsson for his inspiration and support. The work is supported by the Swedish MRC (7516).

REFERENCES

Balthazart, J. (1989). Steroid metabolism and the activation of social behavior. In J. Balthazart (Ed.), *Advances in Comparative & Environmental Physiology, Vol. 3. Molecular and Cellular Basis of Social Behavior in Vertebrates* (pp. 105-159). Berlin: Springer-Verlag.

Baum, M.J., & Vreeburg, J.T.M. (1973). Copulation in castrated male rats following combined treatment with estradiol and dihydrotestosterone. *Science, 182,* 283-285.

Beach, F.A., & Levinson, G. (1950). Effects of androgen on the glans penis and mating behavior of castrated rats. *J. Exp. Zool., 114,* 159-171.

Berthold, A.A. (1849). Transplantation der Hoden. *Arch. Anat. Physiol. Wiss. Med., 16,* 42-55.

Blaustein, J.D., & Olster, D.M. (1989). Gonadal steroid hormone receptors and social behaviors. In J. Balthazart (Ed.), *Advances in Comparative & Environmental Physiology, Vol. 3. Molecular and Cellular Basis of Social Behavior in Vertebrates* (pp. 31-104). Berlin: Springer-Verlag.

Carlsson, S.G., & Larsson, K. (1964). Mating in male rats after local anesthetization of the glans penis. *Z. Tierpsychologie, 21,* 854-856.

Danzo, B.J., & Eller, B.C. (1984). The effects of various steroids on testosterone metabolism by the sexually mature rabbit epididymis. *Steroids, 44,* 435-46.

Davidson, J.M. (1969). Effects of estrogen on the sexual behavior of male rats. *Endocrinology, 84,* 1365-1372.

Davidson, J.M., & Bloch, G.N. (1969). Neuroendocrine aspects of male reproduction. *Biol. Reprod., Suppl. 1,* 67-92.

Feder, H.J. (1971). The comparative action of testosterone propionate and 5α-androstan-17β-3-one propionate on the reproductive behaviour, physiology and morphology of male rats. *J. Endocrinol., 51,* 242-252.

Feder, H.H., Naftolin, F., & Ryan, K.J. (1974). Male and female sexual responses in male rats given estradiol benzoate and 5α-androstan-17β-3-one propionate. *Endocrinology, 94,* 136-141.

Handa, R.J., Roselli, C.E., Horton, L., & Resko, J.A. (1987). The quantitative distribution of cytosolic androgen receptors in microdissected areas of the male rat brain: effects of estrogen treatment. *Endocrinology, 121,* 233-240.

Larsson, K., & Södersten, P. (1973). Mating in male rats after section of the dorsal penile nerve. *Physiol. Behav., 10,* 567-571.

Larsson, K., Södersten, P., & Beyer, C. (1973). Sexual behavior in male rats treated with estrogen in combination with dihydrotestosterone. *Horm. Behav., 4,* 289-299.

McGinnis, M.Y., & Dreifus, R.M. (1989). Evidence for a role of testosterone-androgen receptor interactions in mediating masculine sexual behavior in male rats. *Endocrinology, 124,* 618-626.

Mode, A., Gustafsson, J-Å., Södersten, P., Eneroth, P. (1985). Sex differences in behavioural androgen sensitivity: possible role of androgen metabolism. *J. Endocrinol., 100,* 245-248.

Partridge, W.M. (1981). Transport of protein-bound hormones into tissues in vivo. *Endocr. Rev., 2,* 103-135.

Pfaff, D.W. (1970). Nature of sex hormone effects on rat sex behavior: Specificity of effects and individual patterns of response. *J. Comp. Physiol. Psychol., 73,* 348-358.

Renoir, J-M., Mercier-Bodard, C., & Baulieu, E-E. (1980). Hormonal and physiological aspects of the phylogeny of sex steroid binding plasma proteins. *Proc. Natl. Acad. Sci. U.S.A., 77,* 4578-4582.

Roselli, C.E., Ellinwood, W.E., & Resko, J.A. (1984). Regulation of brain aromatase activity in rats. *Endocrinology, 114,* 192-200.

Roselli, C.E., Horton, L.E., & Resko, J.A. (1985). Distribution and regulation of aromatase activity in the rat hypothalamus and limbic system. *Endocrinology, 117,* 2471-2477.

Roselli, C.E., Morton, L.E., & Resko, J.A. (1987). Timecourse and steroid specificity of aromatase induction in rat hypothalamus-preoptic area. *Biol. Reprod., 37,* 628-633.

Södersten, P. (1973). Estrogen-activated sexual behavior in male rats. *Horm. Behav., 4,* 247-256.

Södersten, P., Eneroth, P., & Hansson, T. (1988). Oestradiol synergizes with 5a-dihydrotestosterone or 3α- but not 3β-androstanediol in inducing sexual behaviour in castrated rats. *J. Endocrinol., 199,* 461-465.

Södersten, P., Eneroth, P., Hansson, T., Mode, A., Johansson, D., Näslund, B., Liang, T., & Gustafsson, J-Å. (1986). Activation of sexual behaviour in castrated rats: the role of oestradiol. *J. Endocrinol., 111,* 455-462.

Södersten, P., Eneroth, P., & Pettersson, A. (1983). Episodic secretion of luteinizing hormone and androgen in male rats. *J. Endocrinol., 97,* 145-153.

Södersten, P., & Gustafsson, J-Å. (1980a). A way in which estradiol might play a role in the sexual behavior of male rats. *Horm. Behav., 14,* 271-274.

Södersten, P., & Gustafsson, J-Å. (1980b). Activation of sexual behavior in castrated rats with the synthetic androgen 17n-hydroxy-17α-methyl-estra-4,9,11-trien-3-one (R1881). *J. Endocrinol., 87,* 279-283.

9 Motivation and Performance: Region-Selective Changes in Forebrain Monoamine Synthesis Due to Sexual Activity in Rats

Sven Ahlenius
Astra Research Center, Södertälje and University of Göteborg

Viveka Hillegaart
University of Göteborg

Knut Larsson
University of Göteborg

INTRODUCTION

It is well known that monoaminergic drugs can affect the performance of male rat sexual behavior (see Bitran & Hull, 1987). For example, stimulation of brain dopaminergic neurotransmission by the administration of apomorphine or d-amphetamine facilitates certain aspects of the sexual behavior, whereas dopamine (DA) antagonists inhibits the behavior (e.g., Ahlenius, Heiman, & Larsson, 1979; Butcher, Butcher, & Larsson, 1969). With the notable exception of the 5-hydroxytryptamine (5-HT) agonist 8-hydroxy-2-(di-n-propyl-amino) tetralin (8-OH-DPAT), the situation is the opposite when activating or inhibiting brain 5-HT neurotransmission (see Ahlenius & Larsson, 1989). Together, these observations indicate an important role for brain monoaminergic neurotransmission in the mediation of rat masculine sexual behavior. Needless to say, the performance can concomitantly be influenced by drug effects at many levels of the nervous system, from supra-spinal to spinal, and also by effects more directly on sensory afferents and motor efferents in the peripheral nervous system.

Monoaminergic drugs also affect important mechanisms of gonadal hormone regulation in the hypothalamus (e.g., Meites & Sonntag, 1981) and, secondary to this, influence sexual functions. Most drug effects, however, have an onset and duration that is incompatible with an involvement of the hypothalamic-gonadal axes, as evidenced by the observation that castration or androgen treatment does not affect the behavior immediately or in a dose-related fashion (see Beach & Fowler, 1959; Bloch & Davidson, 1968). Thus, although it is a distinct possibility that monoaminergic mechanisms in the hypothalamus may directly influence sexual behavior beyond hormonal control, it is equally important to look at brain areas outside the hypothalamus for effects of monoaminergic drugs on male rat sexual behavior.

A distinction has been made between appetitive and consummatory components of the male rat sexual behavior (see Beach, 1956; Sachs, 1978). The appetitive or motivational aspect of the sexual behavior, in all probability, is subserved by neuronal mechanisms separate from those regulating the consummatory or sensory-motor aspect. In the human context, this distinction between motivation and performance is well illustrated by Shakespeare in his play Macbeth (Act II, Scene 3), where he wrote of alcohol that "it provokes the desire but it takes away the performance". The series of experiments to be presented here were designed to study possible brain regions of particular importance for effects of monoaminergic drugs on sexual motivation, as distinct from effects on sexual performance, in the male rat. In contrast to the Porter's remark in Macbeth, however, we can only record the motivation or "desire" of the rat by observing its performance, and our rationale and procedures require some further explanations.

Brain monoaminergic pathways occupy strategic positions in sensory-motor regulation, as well as in motivational aspects of behavior. This applies to the limbic forebrain, which is of great importance for motivation, and to the neostriatum, which is of critical importance for extrapyramidal motor functions. Experimental studies have revealed separate dopaminergic projections to these two areas, namely the meso-limbic DA pathway, and the nigro-striatal (DA) pathway (Andén, Dahlström, Fuxe, Larsson, Olson, & Ungerstedt, 1966; Ungerstedt, 1971), and this separation of afferent supply lends further support for the functional separation of the target areas. Furthermore, the serotonergic innervation of the basal forebrain and the striatum is provided by separate projections from the dorsal and median raphe nuclei of the lower brain stem. Thus, it can be hy-pothesized that behavioral events related to the dynamics of monoaminergic neurotransmission in the limbic forebrain reflect brain mechanisms of motivation, whereas concomitant or specific changes in the neostriatum might reflect motor activation. This assumption is validated by comparative studies, including clinical observations (see Alheid & Heimer, 1988). Consideration of homologue brain areas is an important aspect of animal experimentation on integrative brain functions, and has to be observed when attempting to correlate regional brain activity with behavioral events. Unfortunately, even carefully designed experiments on behavioral phenomena or functions, defined in the human context, may have an entirely different meaning from the rat's vantage point of view (Fox, 1968). Conversely, region-selective changes in brain neurotransmission, related to specific behavioral activities, may help us define aspects of animal behavior of importance in the formulation of animal models related to clinical issues.

In the first series of experiments presented here, we examined the rate of monoamine synthesis separately in the ventral and the dorsal striatum during sexual interactions. In addition, we estimated the monoamine synthesis in two areas known to be rich in sex steroid receptors, namely, the septum and the anterior hypothalamus (see Stumpf & Sar, 1981). Furthermore, the animals had various amounts of sexual experience, from minimal or none, to sexual experience provided by repeated pre-tests. Finally, the interaction with receptive, as well as non-receptive, females was studied. The results of this first series of experiments indicated a selective increase in striatal monoamine neurotransmission as a result of sexual interaction in sexually naive animals, but not in sexually experienced rats. This prompted us to perform a second series of experiments where we examined, in more detail, regional forebrain monoamine synthesis in animals with minimal sexual experience.

Experimental procedures

There is reason to suppose that the nerve impulse flow is linked to availability and synthesis of the neurotransmitter, and, therefore, estimation of the monoamine synthesis at the rate-limiting step, tyrosine and tryptophan hydroxylation, should indicate functional requirements of the monoaminergic projections (see Carlsson, 1974). The accumulation of the monoamine precursors DOPA and 5-HTP, following inhibition of cerebral aromatic amino acid decarboxylase, can be used for this purpose, although the method primarily has been used to study receptor-mediated changes in brain monoamine synthesis (Carlsson, Davis, Kehr, Lindqvist, & Atack, 1972; Andén, 1980). In the present series of experiments, adult male Wistar rats were treated with the decarboxylase inhibitor 3-hydroxy-benzylhydrazine 2HCl (NSD-1015), 100 mg kg^{-1} IP 5 min before they were presented with a receptive female brought into estrus by sequential treatment with estradiol benzoate (12.5 µg animal^{-1} IM - 54h) and progesterone (0.5 mg animal^{-1} IM - 6h). The males were allowed from 10 to 35 min of sexual interactions before decapitation and subsequent biochemical analysis of regional brain DOPA and 5-HTP accumulation. Brain dissections were performed as previously described (Ahlenius, Carlsson, Hillegaart, Hjorth, & Larsson, 1987; Hillegaart, Hjorth, & Ahlenius, 1990). The levels of DOPA and 5-HTP were determined by HPLC with electrochemical detection according to standard procedures (Felice, Felice, & Kissinger, 1978; Magnusson, Nilsson, & Westerlund, 1980). During sexual interaction the various components of the male rat sexual behavior were recorded as detailed elsewhere (e.g., Ahlenius & Larsson, 1989). The treadmill consisted of a rotating drum (8 rpm, ∅=166 mm). Animals walking on the top of the drum moved at a speed of 4 m min^{-1}. For further details, see Ahlenius and Hillegaart (1986).

The interpretation of the outcome of a study of this kind is complicated by a number of factors, the most obvious of which include: degree of sexual experience of the male, effects of the social interactions, and psychomotor activation. In an attempt to control these factors, the following groups were used:

[A] Naive male + receptive female (Experimental group)
[B] Naive male + non-receptive female (Social interaction controls)
[C] Experienced male + receptive male (Sexual experience controls)
[D] Effects of 10-30 min treadmill locomotion on regional brain monoamine synthesis (Psychomotor activation controls)

Effects of NSD-1015 on the male rat sexual behavior

All animals in the present series of experiments were, as part of the experimental procedures, given the decarboxylase inhibitor NSD-1015. As shown in Fig. 1, the compound by itself did not affect the sexual behavior and all animals (n=10) initiated copulation. Thus, the procedure is ideal for the present purpose to examine regional brain monoamine synthesis in awake, freely moving animals.

Effects of rat sexual and social interactions on striatal monoamine synthesis

The original design was to allow 10 min of sexual interactions between a large group of sexually naive male rats, treated with NSD-1015 and a receptive female. It was expected that the heterogeneity in this group of naive males would ensure that in the 10 min period preceding decapitation, there would be great individual variability in the behavior displayed up to this point.

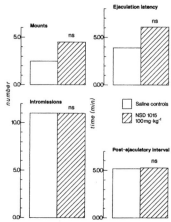

Fig. 9.1. Effects of NSD-1015 on male rat sexual behavior. NSD-1015, or the saline vehicle, were administered IP 30 min before observations started. The results are presented as medians. The animals (n=10) served as their own controls in a change-over design. Statistical analysis by means of the Wilcoxon matched-pairs signed-ranks test (from Ahlenius et al., 1987).
ns $p > 0.05$

Table 9.1. Correlation between the rate of intromission (per min) preceding ejaculation and forebrain DOPA and 5-HTP accumulation in the rat.

The table shows the correlation coefficients (r) based on the performance of 30 sexually naive animals. The statistical significance of the correlation was evaluated by t- tests (Guilford, 1956).

	DOPA	5-HTP
Dorsal striatum	-0.14[ns]	-0.21[ns]
Ventral striatum	-0.26[ns]	-0.18[ns]
Septum	-0.24[ns]	-0.21[ns]
Anterior hypothalamus	-0.18[ns]	-0.14[ns]

ns $p > 0.05$

Much to our surprise, there was no correlation between the DOPA or the 5-HTP accumulation during the preceding 10 min and the behavior of the animals. Thus, there was no statistically significant correlation between the rate of intromissions preceding ejaculation and the striatal DOPA or 5-HTP accumulation (Table 1). Separating out 8 particularly sexually active individuals from the remaining 22 animals did not indicate any statistically significant differences in striatal DOPA or 5-HTP accumulation between these two groups (Fig. 2). Consequently, these animals (n=30) were treated as one group in the following experiments (open bars in Fig. 3).

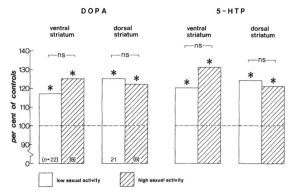

Fig. 9.2. Effects of intensity of sexual interaction on striatal monoamine synthesis in the male rat. A group of sexually naive animals (cf. "naive male + receptive female" group in Fig. 3) was split up into a particularly active subgroup (2 ejaculations within 10 min), as compared to the remaining, sexually less inactive animals (1 ejaculation within 10 min). Statistical comparisons with NSD-1015 controls and between the two subgroups by means of the Student's t-test (from Ahlenius et al., 1987).
ns $p > 0.05$ * $p < 0.05$

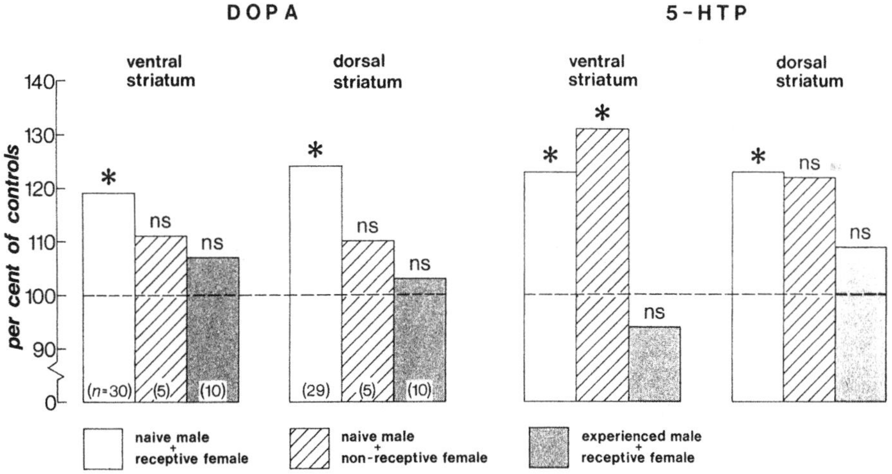

Fig. 9.3. Effects of sexual and social interactions on striatal monoamine synthesis in male rats. The animals were given NSD-1015, 100 mg kg^{-1} IP, 5 min before 10 min of interactions, as indicated in the figure. Statistical comparisons with NSD-1015-treated controls by means of the Student's t-test (from Ahlenius et al., 1987, and unpublished observations).
ns $p > 0.05$ * $p < 0.05$

The results presented in Fig. 3 show that the encounter with a receptive female brought about an increase in DOPA and 5-HTP accumulation in the striatum of the naive male rats. There was no difference between the dorsal and ventral striatum in this respect. The increase in DOPA accumulation in the dorsal striatum is possibly due to psychomotor activation since prolonged treadmill locomotion (25 min) brought about a selective increase in dorsal striatal DOPA accumulation (Fig. 4). It should be noted, however, that ten min of treadmill locomotion was not sufficient to produce any statistically significant changes in striatal monoamine synthesis (data not shown).

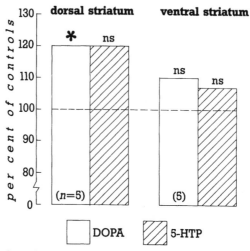

Fig. 9.4. Effects of treadmill locomotion on striatal monoamine synthesis in the male rat. The animals were administered NSD-1015, 100 mg kg^{-1} IP, 5 min before 25 min of treadmill locomotion at a speed of 4 m min^{-1}. The animals walked on a rotating drum, 8 rpm, with o.d. 166 mm. Statistical comparisons with NSD-1015 treated controls by means of the Student's t-test.
ns $p > 0.05$

The "social" interaction between a naive male and a non-receptive female produced a statistically significant increase in 5-HTP accumulation in the ventral striatum only (Fig. 3). In contrast, male rats with sexual experience obtained in at least four pre-tests, did not display any changes in striatal monoamine synthesis following sexual interactions with a receptive female. This finding indicates that the performance as such was of little importance for the increase in striatal monoamine synthesis in naive animals as compared to other aspects of the behavior. The most obvious factor is strong motivation in a novel, challenging situation. This observation also provides further support for the notion, put forward above, that the increase in striatal monoamine synthesis in naive male rats was due to other factors than psychomotor activation during these brief, 10 min sessions.

Together, the above observations suggest that the exposure of a sexually naive male rat with a receptive female produces an increase in striatal monoamine synthesis, that is not seen in sexually experienced males. The social interaction, as studied by naive male-non-receptive female interaction, produced a selective increase in ventral striatal 5-HT synthesis. Furthermore, the effect brought about in naive males shows a considerable degree of region-selectivity, since no effects were observed in the septum or in the anterior hypothalamus. Thus, in a subsequent experiment, to be described below, we further examined region-selective changes in the monoamine synthesis in the striatum and some adjacent areas in sexually naive male rats.

Effects of sexual interaction on forebrain monoamine synthesis in sexually naive male rats

The final series of experiments differed in some important respects from the experiments described above. First, the animals were allowed to ejaculate once in a single, brief pre-test, 2 weeks before the experiments. This was done in order to increase the likelihood that the limited number of rats used in this study would initiate copulation. Still, this experience was so limited in comparison with the "experienced" animals shown in Fig. 3, tested repeatedly, the last time 2-3 days before experiments, that they will be referred to as "naive". Secondly, the animals were decapitated following 15, 25, or 35 min of sexual interactions and comparisons were made with the DOPA and 5-HTP accumulation in home-cage controls at the corresponding time intervals.

In agreement with the results presented in Fig. 3 above, there was a statistically significant increase in striatal DOPA and 5-HTP accumulation (Figs. 6 & 7). A close inspection provides suggestive evidence that only the DOPA accumulation was increased in the sensory-motor, dorso-lateral part of the neostriatum (see Kelley, Domesick, & Nauta, 1982). Furthermore, the medial limbic forebrain, including the accumbens, the ventro-medial neostriatum, and, in this experiment, also the septal area appeared consistently activated as evidenced by the DOPA and 5-HTP accumulation. In addition, the DOPA, but not the 5-HTP, accumulation was increased in the neocortex and in the amygdala. Since the neocortex contains very little DA as compared to noradrenaline (NA), this observation suggests an important role for cortical noradrenergic mechanisms in this situation. Parts of the amygdala are strongly related to the medial limbic forebrain, rich in DA and 5-HT nerve terminals (see Alheid & Heimer, 1988). The fact that the DOPA, but not the 5-HTP, accumulation was increased in the amygdala, however, could also be related to an increased demand on cortical noradrenergic, in addition to limbic dopaminergic, neurotransmission. The increase in pallidal 5-HTP accumulation was unexpected and the significance of this finding must await information from continued studies.

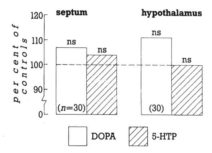

Fig. 9.5. Effects of sexual interactions on septal and hypothalamic monoamine synthesis in naive male rats. The animals were given NSD-1015, 100 mg kg[-1] IP, 5 min before 10 min of sexual interactions with a receptive female. Statistical comparisons with NSD-1015-treated controls by means of the Student's t-test (from Ahlenius et al., 1987).
ns $p > 0.05$

Motivation vs. performance

As mentioned above, the first series of experiments summarized in this chapter were originally designed to study brain monoaminergic mechanisms specifically related to the performance of sexual behavior. Our results, however, suggested something different: 1) in contrast to the increased striatal monoamine synthesis observed in sexually naive rats, there were no changes in the striatum of sexually experienced animals, and 2) the changes obtained in the naive animals were not directly related to the preceding behavioral activities. In fact, the confrontation of a naive male with a non-receptive female brought about an increase in ventral striatal 5-HT synthesis, not seen in the sexually experienced rats. Together these observations indicate that aspects of the situation, other than the specific sexual activity, like motivation or orienting behavior, dominated. Needless to say, reproductive behavior, of less importance for the individual than the species, strongly affects relations to other individuals in a number of situations. It is thus to be expected that the initial encounter of a sexually naive rat with a receptive female is an emotionally challenging situation. The process of arousing action, sustaining the activity in progress, and regulating the pattern of activity, offered as a definition of motivation (Young, 1961), are probable factors of great importance for the increase in striatal monoamine synthesis in the naive males of the present studies. On this assumption, the second part of our study provided details on forebrain monoamine synthesis as a result of sexual interactions in sexually naive male rats. Thus, it appears that the medial basal forebrain monoamine synthesis (accumbens, ventro-medial neostriatum, and the septum) is particularly involved. The fact that no changes were observed in the septum in the initial study (Fig. 5) is probably due to the fact the dissections were less restricted in comparison with the latter experiments (Figs. 6 & 7). In addition, noradrenergic mechanisms in the frontal cortex appear to be involved. The increase in DOPA accumulation in the amygdala is interesting but continued studies are necessary to determine the relative contributions of striatal dopaminergic mechanisms, related to the extended amygdala, as described by Alheid and Heimer (1988), and cortical noradrenergic mechanisms. The different portions of the amygdala were not separated in the present experiments.

It has been shown that, within the striatum, DA synthesis in the dorsal portion is preferentially increased by the motor activities associated with operant bar pressing in a Skinner-box (Heffner & Seiden, 1980; Seiden & Heffner, 1984). In general agreement with these observations, we observed an increase in dorsal, but not ventral, striatal DOPA accumulation in animals walking on a treadmill for 25 min (Ahlenius et al., 1987). It is doubtful, however, if these findings are relevant in the present context. The motor activity associated with sexual behavior was not sufficient to bring about an increased DA synthesis in the striatum of experienced male rats. Thus, it is possible that the increase in dorsal striatal DA synthesis in the two series of experiments described here, also is part of the motivational state of the animals.

SUMMARY AND CONCLUSIONS

The accumulation of DOPA and 5-HTP, following inhibition of cerebral aromatic amino acid decarboxylase, provides a good estimate of brain monoamine synthesis at the rate-limiting tyrosine and tryptophan hydroxylase steps. Treatment of male rats with the decarboxylase inhibitor NSD-1015, in a dose which completely inhibits cerebral decarboxylase (100 mg kg[-1] IP), did not by itself affect the performance of sexual behavior; this method, thus, eminently lends itself to studies on concomitant changes in brain monoamine synthesis.

The procedure to measure the accumulation of DOPA and 5-HTP, following decarboxylase inhibition was used to examine possible region-selective changes in brain monoamine synthesis as a result of sexual behavior in male rats. It was found that the striatal monoamine synthesis was increased in sexually naive, but not in sexually experienced, rats presented with a receptive female 10 min before decapitation and subsequent biochemical experiments. Furthermore, within the group of sexually naive rats, there was no correlation between the striatal monoamine synthesis, as observed here, and the intensity of sexual performance. These results suggest that motivational factors, rather than factors directly related to the performance of sexual behavior of such, were responsible for the increased striatal monoamine synthesis. This conclusion was partially supported by the observation that also the interaction between a naive male and a non-receptive female produced some changes in the monoamine synthesis, although less pronounced and limited to the 5-HT synthesis in the ventral striatum. Changes in striatal monoamine synthesis due to motor

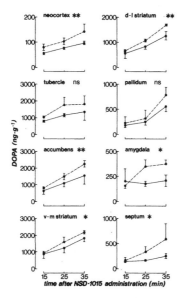

Fig. 9.6. Effects of sexual interaction on regional forebrain catecholamine synthesis in male rats. NSD-1015, 100 kg⁻¹ IP, was administered 5 min before presenting the male with a receptive female. The rats were decapitated at the time intervals indicated in the figure by the *broken line.* The DOPA accumulation in time-matched NSD-1015 controls is shown by the *solid line.* Shown are mean ± SD, based on observations of 3 animals per group. Statistical evaluation by means of 2-way ANOVA. A statistical significant difference between experimental animals and controls in the different brain areas, as indicated in the figure (Ahlenius et al., in preparation).

ns $p > 0.05$ * $p < 0.05$ ** $p < 0.01$

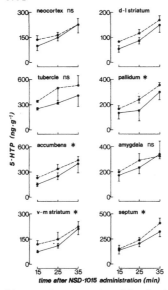

Fig. 9.7. Effects of sexual interaction on regional forebrain serotonin synthesis in male rats. For details, see legend to Fig. 5.

activation by the sexual behavior is not probable since no changes were found in sexually experienced males, but also since treadmill locomotion for 10-25 min produced no or only minor effects on the striatal monoamine synthesis. A closer examination of regional changes in forebrain monoamine synthesis as a result of sexual motivation in naive male rats disclosed further interesting details. Dopaminergic and serotonergic neurotransmission in the medial basal forebrain, including the nucleus accumbens, the ventro-medial part of the neostriatum and the septum, appear to be critical. In addition, there was evidence for the involvement of noradrenergic mechanisms in the frontal cortex. It is suggested that these regional changes in forebrain monoamine synthesis are related to sexual arousal and motivation. This also implies that the present results may be of limited relevance for understanding the effects of monoaminergic drugs on consummatory aspects of male rats sexual behavior.

REFERENCES

Ahlenius, S., Carlsson, A., Hillegaart, V., Hjorth, S., & Larsson, K. (1987). Region-selective activation of brain monoamine synthesis by sexual activity in the male rat. *Eur. J. Pharmacol., 144*, 77-82.

Ahlenius, S., Heimann, M., & Larsson, K. (1979). Prolongation of the ejaculation latency in the male rat by thioridazine and chlorimipramine. *Psychopharmacology, 65*, 137-140.

Ahlenius, S., & Hillegaart, V. (1986). Involvement of extrapyramidal motor mechanisms in the suppression of locomotor activity by antipsychotic drugs: A comparison between the effects produced by pre- and post-synaptic inhibition of dopaminergic neuro-transmission. *Pharmacol. Biochem. Behav., 24*, 1409-1415.

Ahlenius, S., & Larsson, K. (1989). New aspects on the serotonergic modulation of male rat sexual behaviour. In P. Bevan, A.R. Cools, and T. Archer (Eds.), *Behavioural Pharmacology of 5-HT* (pp. 35-53). Hillsdale, New Jersey: Lawrence Erlbaum Assoc., Inc., Publ.

Alheid, G.F., & Heimer, L. (1988). New perspectives in basal forebrain organization of special relevance for neuropsychiatric disorders: The striatopallidal, amygdaloid, and corticopetal components of substantia innominata. *Neuroscience, 27*, 1-39.

Andén, N-E. (1980). Regulation of monoamine synthesis and utilization by receptors. In L. Szekeres (Ed.), *Handbook of Experimental Pharmacology* (pp. 429-462). Berlin-Heidelberg: Springer-Verlag.

Andén, N-E., Dahlström, A., Fuxe, K., Larsson, K., Olson, L., & Ungerstedt, U. (1966). Ascending monoamine neurons to the telencephalon and diencephalon. *Acta Physiol. Scand., 67*, 313-326.

Beach, F.A. (1956). Characteristics of masculine "sex drive". In M.R. Jones (Ed.), *The Nebraska Symposium on Motivation* (pp. 1-32). Lincoln: Univ. of Nebraska Press.

Beach, F.A., & Fowler, H. (1959). Individual differences in the response of male rats to androgens. *J. Comp. Physiol. Psychol., 52*, 50-52.

Bitran, D., & Hull, E.M. (1987). Pharmacological analysis of male rat sexual behavior. *Neurosci. Biobehav. Rev., 11*, 365-389.

Bloch, G.J., & Davidson, J.M. (1968). Effects of adrenalectomy and experience on postcastration sex behavior in the male rat. *Physiol. Behav., 3*, 461-465.

Butcher, L.L., Butcher, S.G., & Larsson, K. (1969). Effects of apomorphine, (+)amphe-tamine, and nialamide on tetrabenazine-induced suppression of sexual behavior in the male rat. *Eur. J. Pharmacol., 7*, 283-288.

Carlsson, A. (1974). The *in vivo* estimation of rates of tryptophan and tyrosine hydroxyla-tion: Effects of alterations in enzyme environment and neuronal activity. In *Ciba Foundation Symposium 22 (New Series): Aromatic Amino Acids in the Brain* (pp. 117-134). Amsterdam: North Holland.

Carlsson, A., Davis, J.N., Kehr, W., Lindqvist, M., & Atack, C.V. (1972). Simultaneous measurement of tyrosine and tryptophan hydroxylase activities in brain *in vivo* using an inhibitor of the aromatic amino acid decarboxylase. *Naunyn-Schmiedeberg's Arch. Pharmacol., 275,* 153-168.

Felice, L.J., Felice, J.D., & Kissinger, P.T. (1978). Determination of catecholamines in rat brain parts by reverse-phase ion-pair liquid chromatography. *J. Neurochem., 31,* 1461-1465.

Fox, M.W. (1968). Ethology: An overview. In M.W. Fox (Ed.), *Abnormal Behavior in Animals* (pp. 21-43). Philadelphia: W.B. Saunders Co.

Guilford, J.P. (1956). *Fundamental Statistics in Psychology and Education.* New York: McGraw-Hill.

Heffner, T.G., & Seiden, L.S. (1980). Synthesis of ^{3}H-catecholamines from ^{3}H-tyrosine in brain regions of rats during performance of operant behavior. *Brain Res., 183,* 403-419.

Hillegaart, V., Hjorth, S., & Ahlenius, S. (1990). Effects of 5-HT and 8-OH-DPAT on forebrain monoamine synthesis after local application into the median and dorsal raphe nuclei of the rat. *J. Neural Transm.,* in press.

Kelley, A.E., Domesick, V.B., Nauta, W.J.H. (1982). The amygdalostriatal projection in the rat—an anatomical study by anterograde and retrograde tracing methods. *Neuroscience, 7,* 615-630.

Magnusson, O., Nilsson, L.B., & Westerlund, D. (1980). Simultaneous determination of dopamine, DOPAC and homovanillic acid. Direct injection of supernatants from brain tissue homogenates in a liquid chromatography-electrochemical detection system. *J. Chromatogr., 221,* 237-247.

Meites, J., Sonntag, W.E. (1981). Hypothalamic hypphyiotropic hormones and neurotransmitter regulation: Current views. *Ann. Rev. Pharmacol. Toxicol., 21,* 295-322.

Sachs, B.D. (1978). Conceptual and neural mechanisms of masculine copulatory behaviour. In T.E. McGill, D.A. Dewsbury, and B.D. Sachs (Eds.), *Sex and Behavior: Status and Prospectus* (pp. 267-295). New York: Plenum Press.

Seiden, L.S., & Heffner, T.G. (1984). Alteration of brain catecholamine metabolism by environmental and behavioral events: an explanation of drug-behavior interactions. In E. Usdin (Ed.), *Catecholamines: Neuropharmacology and Central Nervous System— Theoretical Aspects* (pp. 275-284). New York: A.R. Liss.

Stumpf, W.E., & Sar, M. (1981). Steroid hormone sites of action in the brain. *Wenner-Gren Cent. Int. Symp. Series, 34,* 41-50.

Ungerstedt, U. (1971). Stereotaxic mapping of the monoamine pathway in the rat brain. *Acta Physiol. Scand., S367,* 1-48.

Young, P.T. (1961). *Motivation and Emotion: A Survey of the Determinants of Human and Animal Activity.* New York: Wiley.

III Sexual Behavior II

A. Fernández-Guasti, O. Picazo,
B. Roldán-Roldán, and A. Saldivar
División de Investigaciones en Neurosciencias
Instituto Mexicano de Psiquiatría and
Centro Interdisciplinario de Ciencias de la Salud

INTRODUCTION

Anxiety is a pathological state, within the neuroses, characterized by feelings of uneasiness, apprehension, restlessness, irritability, sometimes accompanied by gastrointestinal distress, muscle tension, dizziness, or insomnia (cf. Treit, 1985). While the pharmacological treatment of anxiety is well studied (Greenblatt & Shader, 1984; Haefely, 1985), the physiological basis of this alteration remains unknown. Indeed, various particular behavioral states in preclinical tests selectively sensitive to anxiolytic drugs, are often considered to reveal a state of anxiety (Treit, 1985).

The influence that enhanced levels of anxiety and treatment with anxiolytics may exert on human sexual behavior is reviewed elsewhere (Norton & Jehu, 1984; Shader & Elkins, 1980). Additionally, the relationships between anxiety and the various phases of the human menstrual cycle have ben studied (cf. Hallman, 1987). It is worth noting that the alterations that anxiety causes on human sexual responses are further studied more than the relationships between anxiety and sexual behavior in animals. The reason for this difference could be based on the fact that enhanced anxiety levels disrupt the normal execution of human sexual responses. Moreover, high anxiety levels are considered one of the main features of the premenstrual syndrome in women.

The following section will summarize the possible relationships between anxiety and sexual behavior in male rats. These relationships have been studied from two different viewpoints, one considering the effects of modifying anxiety on the execution of sexual behavior, and the other analyzing the effects of copulation on the anxiety state. In the final section we studied whether the anxiety levels changed with the various phases of the rat estrous cycle. The relationships between anxiety and sexual state have been studied using behavioral and pharmacological approaches. The reason for including a pharmacological method is based on the extensive pharmacological knowledge of anxiety. The common male sexual behavior parameters were registered (cf. Larsson, 1956). On the basis of its advantages over other tests, the conditioned defensive burying behavior paradigm was used to determine the anxiety levels. In this paradigm, decreases in burying behavior time are interpreted as antianxiety actions (cf. Treit, 1985).

Influences of anxiety on male sexual behavior

Empirical and clinical data indicate that anxiety produces a biphasic effect on sexual behavior. Thus, moderate anxiety levels may facilitate ejaculation (Norton & Jehu, 1984), while extreme anxiety results in a complete inhibition of sexual behavior and impotence (Norton & Jehu, 1984; Shader & Elkins, 1980). In animal research similar alterations in mating behavior induced by anxiety have been observed. Beach and Fowler (1959), reported that "situational anxiety" facilitates rat sexual behavior by reducing the number of intromissions needed for ejaculation. On the other hand, copulation can be inhibited by pairing different aversive stimuli with the receptive female (McDonnell, Kenney, Menckley, & Garcia, 1985; Peters, 1983). Furthermore, the inhibitory action of the aversive conditioning may be prevented by treatment with diazepam (McDonnell et al., 1985). In our laboratory, we have demonstrated that the pharmacologically-induced anxiety also produced this biphasic action on copulation. The administration of low doses of the anxiogenic drug beta carboline ZK 39106, reduced the number of intromissions preceding ejaculation. This effect is abolished by the injection of the Ro 15-1788, a central benzodiazepine receptor antagonist. Conversely, high doses of ZK 39106 inhibited sexual activity by decreasing the proportion of copulating animals (Fernández-Guasti, Roldán-Roldán, & Saldívar, 1990).

Recently, we have extended the study of the facilitatory influence of anxiety on sexual behavior by using a very interesting model, namely, the enforced interval of copulation (EIC). This model consists in the interruption of mating by the female withdrawal during a fixed interval every time the male achieves an intromission. This manipulation results in a reduction of the number of intromissions accompanied by a behavioral anxiety-like repertoire (Larsson, 1956; 1959). In an extensive pharmacological characterization of this phenomenon, we have observed that diazepam blocks the reduction in the number of intromissions preceding ejaculation during the EIC. Moreover, Ro 15-1788 prevented the diazepam effect (Fig. 1).

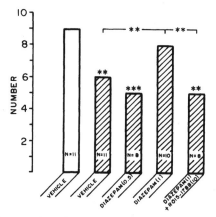

FIGURE III.1. Median number of intromissions preceding ejaculation during *ad libitum* (open bar) and the enforced interval (shadowed bard) conditions. Statistical analysis was made by means of the Mann-Whitney U-test. ** $p < 0.02$; *** $p < 0.002$.

This action was shared by other benzodiazepine and non-benzodiazepine anxiolytics (Roldán-Roldán & Fernández-Guasti, 1989; Fernández-Guasti, Roldán-Roldán, & Larsson, unpublished data), suggesting that the facilitation of sexual behavior induced by the EIC is an anxiety dependent process. Other authors have proposed that the excitatory action of the EIC could be due to an emotional or frustration-like effect (Bermant, 1964; Gerall, 1958);

however, an alternative explanation has been suggested, i.e., the modification in the timing of the interintromission interval results in an increase of the excitation produced by each intromission (Hård & Larsson, 1970; Larsson, 1959; 1961). Interestingly, none of these interpretations are mutually exclusive. In humans, differences between the influence of general and sexual (coital associated) anxiety on sexual dysfunctions have been recognized (Norton & Jehu, 1984). It appears that the facilitation produced by the EIC may correspond to the second type, since anxiety is elicited during copulation without any previous conditioning or pairing with an aversive stimulus. However, further investigation on this putative model of sexual anxiety is necessary in order to confirm this hypothesis.

Influences of male sexual behavior on anxiety

The neurochemical bases of anxiety are not well established (Gray, 1982); however, several lines of evidence strongly suggest that the GABAergic system is involved in this process (cf. Haefely, 1985). Thus, the stimulation of the GABAergic transmission (by GABA itself, GABA-A agonists or benzodiazepines) produces antianxiety effects (Graeff, Brandao, Audi, & Schutz, 1986; Haefely, 1985).

Recently, it has been reported that the GABAergic transmission plays an important inhibitory role in the neural control of masculine sexual behavior (Fernández-Guasti, Larsson, & Beyer, 1986). Thus, the administration of GABA agonists produced an inhibition of copulation while the injection of GABA antagonists resulted in a drastic shortening of the postejaculatory interval (PEI). In addition to the above mentioned pharmacological evidence, biochemical data have reported increases in the cerebrospinal fluid GABA levels after ejaculation (Qureshi & Södersten, 1986).

Since an increase in the GABAergic transmission has been proposed to occur during the PEI and the stimulation of this neurotransmitter system produces antianxiety responses, in the following series of experiments we proposed that: (1) low levels of anxiety could be found during the PEI and (2) the reduction in anxiety during the PEI could be modified by drugs acting on the GABAergic system.

In the first experiment we analyzed the anxiety levels along various phases of copulation in the male rat. A decrease in anxiety was found after the first and second ejaculations. Comparatively higher levels of anxiety (similar to those observed in the control group) were observed during the execution of the first series of copulation (Fernández-Guasti, Roldán-Roldán, & Saldívar, 1989). The second experiment performed revealed that administration of the GABA antagonist, bicuculline, and of the benzodiazepine blocker, Ro 15-1788, completely reversed the reduction in anxiety found after ejaculation. These drugs by themselves, at the doses tested, were unable to modify the anxiety levels (Fernández-Guasti & Saldívar, submitted). Additionally, the administration of either the GABA agonist, muscimol or diazepam immediately after ejaculation, resulted in a potentiation of the reduction in anxiety found during the PEI (Fernández-Guasti & Saldívar, submitted).

The series of experiments presented indicate that changes in anxiety occur during the different phases of copulation in male rats and that these changes are mediated via the GABAergic system. Since other neurotransmitters have been proposed to participate in the regulation of anxiety (cf. Johnston & File, 1986; Gray, 1982) and in the mediation of masculine sexual behavior (Ahlenius, Carlsson, Hillegaart, Hjorth, & Larsson, 1987), further experiments considering their involvement in the reduction in anxiety during the PEI should be performed.

Changes in anxiety along the estrous cycle

The study of gender differences in the neurotransmitter systems involved in the control of anxiety, GABA (Haefely, 1985), and serotonin (Johnston & File, 1986), remains incomplete. The main reason for the lack of knowledge of the steroid hormonal influences on these systems is primarily due to the fact that most of the anxiety studies are performed in male individuals. However, there is evidence which shows interesting differences in the activity of the serotonergic (5-HT) and the GABAergic pathways between males and females. For example, it has been established that the 5-HT turnover rate is greater in females than in males (Rosecrans, 1970). Additionally, the females' threshold to display the serotonergic syndrome is lower when compared with males (Biegon, Segal, & Samuel, 1979). Similarly, females are more sensitive to the convulsant action of GABAergic antagonists than males (Pericic, Manev, & Geber, 1986).

The differences in these neurotransmitter systems are not gender restricted but also occur along the various phases of the estrous cycle (Biegon, Bercovitz, & Samuel, 1980), probably influenced by sex steroid secretions. Thus, Biegon and McEwen (1982) have demonstrated that estrogen administration exerts a biphasic effect on 5-HT receptor density, reporting an initial acute reduction followed by a drastic increase. Other authors have shown that estrogen and progesterone increase the 3H muscimol binding sites in various regions of the rat brain (Beyer, present volume; Maggi & Pérez, 1984). These findings suggest an important hormonal influence on the number of receptors in the neurotransmitter systems that regulate anxiety.

On the bases of these data and of clinical reports (Hallman, 1987) we analyzed the levels of anxiety along the various phases of the rat estrous cycle. The results of this experiment are shown in Figure 2. During proestrus, the anxiety levels, measured in the burying behavior test, were significantly lower when compared to other phases of the cycle. The levels of anxiety shown by rats in the metestrous and diestrous phases were similar to those observed in ovariectomized females.

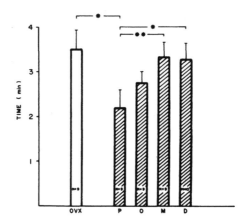

FIGURE III.2. Cumulative burying behavior during the rat estrous cycle. OVX = ovariectomized, P = proestrus, O = estrus, M = metaestrus, D = diestrus. Figure shows m ± S.E. Statistical evaluation was made by means of the Kruskal-Wallis one way ANOVA (H = 9.0524, G.L. = 4, P < 0.05) followed by the Mann-Whitney U-test for comparisons with OVX control animals. * p < 0.05; ** p < 0.02.

Previous reports have shown decreases in emotionality during proestrus (Burke & Broadhurst, 1966). Moreover, Rodríguez-Sierra, Howard, Pollard, and Hendricks (1984) showed that estrogen and progesterone in combination reduce anxiety in a similar manner than chlordiazepoxide. Our findings suggest that the reduction in anxiety observed during proestrus is due to the action of the main steroid hormones secreted during this period (17-beta-estradiol and progesterone). However, the mechanisms through which the steroid hormones modify the anxiety levels need further exploration.

ACKNOWLEDGEMENTS

Authors wish to thank Professor Knut Larsson for encouraging and fruitful discussions and Professor Trevor Archer for organizing the International Symposium in Biological Psychology and the present publication. The "Consejo Nacional de Ciencia y Tecnologia, Mexico" supported the AFG visit to Sweden and the GRR fellowship.

REFERENCES

Ahlenius, S., Carlsson, A., Hillegaart, V., Hjorth, S., & Larsson, K. (1987). Region selective activation of brain monoamine synthesis by sexual activity in the male rat. *Eur. J. Pharmacol., 144*, 77-82.

Beach, F., & Fowler, H. (1959). Effects of "situational anxiety" on sexual behavior in male rats. *J. Comp. Physiol. Psychol., 52*, 245-248.

Bermant, G. (1964). Effects of single and multiple enforced intercopulatory intervals on the sexual behavior of male rats. *J. Comp. Physiol. Psychol., 57*, 398-403.

Biegon, A., Segal, M., & Samuel, D. (1979). Sex differences in behavioural and thermal responses to pargyline and tryptophan. *Psychopharmacology, 61*, 77-80.

Biegon, A., Bercovitz, H., & Samuel, D. (1980). Serotonin receptor concentration during the estrous cycle of the rat. *Brain Res., 187*, 221-225.

Biegon, A., & McEwen, B.S. (1982). Modulation of estradiol of serotonin receptors in brain. *J. Neurosci., 2*, 199-205.

Burke, A.W., & Broadhurst, P.L. (1966). Behavioural correlates of the estrous cycle in the rat. *Nature, 209*, 223-224.

Fernández-Guasti, A., Larsson, K., & Beyer, C. (1986). GABAergic control of masculine sexual behaviour. *Pharmacol. Biochem. Behav., 24*, 1065-1070.

Fernández-Guasti, A., Roldán-Roldán, G., & Saldívar, A. (1989). Reduction in anxiety after ejaculation in the rat. *Behav. Brain Res., 32*, 23-29.

Fernández-Guasti, A., Roldán-Roldán, G., & Saldívar, A. (1990). Pharmacological manipulation of anxiety and male rat sexual behaviour. *Pharmacol. Biochem. Behav., 35*, 263-267.

Gerall, A. (1958). Effect of interruption of copulation on male guinea pig sexual behavior. *Psychol. Rep., 4*, 215-221.

Graeff, F.J., Brandao, M.L., Audi, E.A., & Schutz, M.T.B. (1986). Modulation of the brain aversive system by GABAergic and serotonergic mechanisms. *Behav. Brain Res., 21*, 65-72.

Gray, J.A. (1982). *The Neuropsychology of Anxiety: An Inquiry into the Function of the Septo-Hippocampal System.* Oxford: University Press.

Greenblatt, D.J., & Shader, R.I. (1984). *Benzodiazepam in Clinical Practice.* New York: Raven Press.

Haefely, W. (1985). Tranquilizers. In D.G. Grahame-Smith (Ed.), *Psychopharmacology 2, Part 1: Preclinical Psychopharmacology* (pp. 92-181). Amsterdam: Elsevier Science Publisher.

Hallman, J. (1987). *The Premenstrual Syndrome: Epidemiological, Biochemical and Pharmacological Studies.* Ph.D.. Thesis, Uppsala University.

Hård, E., & Larsson, K. (1970). Effects of delaying intromissions on the male rat's mating behavior. *J. Comp. Physiol. Psychol., 70*, 413-416.

Johnston, A., & File, S. (1986). 5-HT and anxiety: promises and pitfalls. *Pharmacol. Biochem. Behav., 24*, 1467-1470.

Larsson, K. (1956). *Conditioning and Sexual Behavior in the Male Albino Rat. Acta Psychologica Gothoburgensia I,* J. Elmgren (Ed.), Stockholm: Almqvist and Wiksell.

Larsson, K. (1959). The effect of restraint upon copulatory behaviour in the rat. *Anim. Behav., 7*, 23-25.

Larsson, K. (1961). The importance of time for the intromission frequency in the male rat mating behaviour. *Scand. J. Psychol., 2*, 149-152.

McDonnell, S.M., Kenney, RM., Menckley, P.E.., & Garcia, M.C. (1985). Conditioned suppression of sexual behavior in stallions and reversal with diazepam. *Physiol. Behav., 34,* 951-956.

Maggi, A., & Pérez, J. (1984). Progesterone and estrogens in rat brain: modulation of GABA (gamma-amino butyric acid) receptor activity. *Eur. J. Pharmacol., 103,* 165-168.

Norton, G.R., & Jehu, D. (1984). The role of anxiety in sexual dysfunctions: a review. *Arch. Sex. Behav., 13,* 165-183.

Pericic, D., Manev, H., & Geber, J. (1986). Sex related differences in the response of mice, rat and cats to administration of picrotoxin. *Life Sci., 38,* 905-913.

Peters, R.H. (1983). Learned aversions to copulatory behaviors in male rats. *Behav. Neurosci.,, 97,* 140-145.

Qureshi, G.A., & Södersten, P. (1986). Sexual activity alters the concentration of aminoacids in the cerebrospinal fluid of male rats. *Neurosci. Lett., 70,* 374-378.

Rodriguez-Sierra, J.F., Howard, J.L., Pollard, G.T., & Hendricks, S.E. (1984). Effect of ovarian hormones on conflict behavior. *Psychoneuroendocr., 9,* 293-300.

Roldán-Roldán, G., & Fernández-Guasti, A. (1989). Is the enforced interval of copulation a model of sexual anxiety? *21st Conference on Reproductive Behavior, Saratoga Springs, New York.*

Rosecrans, J.A. (1970). Differences in brain area 5-hydroxytryptamine turnover and rearing behaviour in rats and mice of both sexes. *Eur. J. Pharmacol., 9,* 379-382.

Shader, R.I., & Elkins, R. (1980). The effects of antianxiety and antipsychotic drugs on sexual behaviour. *Mod. Prob. Pharmacopsychiatry,* 15, 91-110.

Treit, D. (1985). Animal models for the study of antianxiety agents: a review. *Neurosci. Biobehav. Rev., 9,* 203-222.

10 Neuroendocrine and Psychological Mechanisms Underlying Masculine Sexual Behavior

Barry J. Everitt
University of Cambridge

INTRODUCTION

In his thesis (1956), Knut Larsson described the results of his detailed studies of the integrated pattern of sexual behavior in male rats and, by employing an operant procedure in which males responded on a lever to gain access to a sexually receptive female, other aspects of sexual behavior not readily appreciated by simply observing the unconditioned copulatory response sequence. Having demonstrated that a variety of behavioral methods might profitably be employed to dissect out the psychological processes underlying sexual behavior, he went on to investigate neuroendocrine substrates of sexual behavior - research which he continues even in "retirement". In 1966, Larsson, together with Lennart Heimer, published a fundamentally important paper describing the effects of lesions to various regions of the hypothalamus on the display of sexual behavior by male rats (Heimer & Larsson, 1966/67). In the same year, Julian Davidson (1966a) published an equally important paper on the effects on sexual behavior of implanting testosterone into the hypothalamus of castrated male rats. These reports describing the impairment and restitution, respectively, of sexual behavior following selective neuroendocrine manipulations of a discrete region of the hypothalamus firmly established the still widely held view that testosterone acting in the medial preoptic-anterior hypothalamic area (mPOA/AHA) is the necessary and sufficient event underlying the expression of the complete, integrated pattern of sexual behavior in the male rat and, indeed, other mammals. In this brief review, I will discuss the evidence that the mPOA/AHA is particularly important for the display of the copulatory behaviors of mounting, intromitting and ejaculating, but is not obviously a part of the neural mechanisms underlying appetitive aspects of sexual behavior, including reward in a sexual context. These aspects of sexual behavior, which have been studied using a variety of behavioral paradigms, appear to depend much more on dopamine-dependent processes in the ventral striatum and their interactions with components of the telencephalic limbic system, particularly the basolateral amygdala.

Effects of lesions of the medial preoptic anterior hypothalamic area

Heimer and Larsson (1966/67) first reported that substantial electrolytic lesions of the entire mPOA/AHA abolished copulatory behavior in male rats, but that partial lesions of the area, or lesions of the anterolateral and dorsomedial hypothalamus, or the medial and lateral septum were without significant effect. They concluded that neurons essential for the display of sexual behavior must be distributed over the entire preoptic-anterior hypothalamic continuum. The essential feature of these observations, that copulation is irreversibly prevented by inclusive lesions of the mPOA/AHA, has been replicated many times. However, interest has focused more recently on the behavioral specificity of mPOA/AHA lesions, in particular the selective sparing of elements of the male rat's sexual response repertoire. Indeed, Heimer and Larsson (1966/67) themselves reported that when mPOA/AHA-lesioned males were presented with a stimulus female, they approached, pursued and nuzzled her, engaged in genital investigation and displayed a curious form of climbing behavior in which they approached a female, often from the side rather than the rear, and proceeded to climb over her.

Almost twenty years later, Stefan Hansen, a former student of Larsson, re-investigated the behavioral consequences of the mPOA/AHA lesions to great effect. By showing that excitotoxic lesions of the mPOA/AHA induced by infusing ibotenic acid bilaterally also abolished copulatory acts in male rats, Hansen et al. (1982) demonstrated that it is indeed damage to intrinsic mPOA/AHA neurons, rather than pathways passing through the area, which is necessary to induce the behavioral impairment, since this type of lesion spares axons of passage while destroying neuronal cell bodies in the infused area. This was an important observation given the earlier, major re-appraisals of the lateral hypothalamic syndrome by Ungerstedt (1971), who showed that the marked alterations of feeding, drinking, and motor activity which followed electrolytic lateral hypothalamic lesions were largely due to inadvertant destruction of dopaminergic axons running *en passant* through the medial forebrain bundle to the neostriatum.

In a subsequent study of the effects of ibotenate-induced lesions of the mPOA/AHA, however, Hansen and Drake af Hagelsrum (1984) emphasized that investigative and other appetitive sexual responses, which included abortive mounting attempts, were spared following the lesions and, further, that such animals showed evidence of the thwarting of sexually motivated response tendencies. Thus, a wide range of so-called displacement behaviors such as hindlimb-to-flank scratching and increased levels of autogrooming occurred while marked increases in drinking emerged during the sexual behavior observation, even though the animals were not water deprived. The importance of these and our own observations (Everitt & Stacey, 1987) is that they indicate that the deficit in sexual behavior following mPOA/AHA lesions is one of *performance* of the reflexive acts of forepaw palpation following an adequate mount with clasping, pelvic thrusting, intromission, and ejaculation. Lesioned males clearly remain interested in estrous females, show locomotor excitement in their presence which may be channeled into apparently purposeless motor acts, and, in short, show every indication of being sexually aroused in an appropriate sexual context.

Observation of the effects of mPOA/AHA lesions in a variety of species, mainly by Hart and his collaborators, has confirmed the apparently selective disruption of copulatory behavior while sparing anticipatory or preparatory sexual responses (Hart, 1974; 1986; Hart, Haugen, & Peterson, 1973; Hart & Voith, 1978). For example, in male goats in which copulation and ejaculation had been abolished by mPOA/AHA lesions, courtship responses of babbling-snorting and pawing-stamping and even mounting persisted at preoperative levels, although the mounts were often inappropriately directed (Hart, 1986). Flehmen, self-enurination, and penis licking, which males readily display during sexual encounters were similarly unaffected by mPOA/AHA lesions and were sometimes increased (Hart, 1986).

Perhaps an even more explicit demonstration of the rather specific disruption of copulatory reflexes which are elicited following direct, usually tactile, contact with females while other sexual responses are spared comes from studies of the effects of mPOA/AHA lesions in male rhesus monkeys (Slimp, Hart, & Goy, 1978). As in non-primate male mammals, mPOA/AHA lesions severely impaired copulation when males were placed with females. However, in their home cages, males were seen to engage in bouts of masturbation and, from the evidence of seminal plugs in the drop pans under the cages, frequencies of ejaculation which were not different to controls. This strongly suggests that mPOA/AHA-lesioned males may be sexually aroused and display sexual acts resulting in ejaculation, but are unable to emit copulatory responses during encounters with females. It also suggests that the failure of ejaculation after mPOA/AHA lesions is not a direct consequence of damaging a hypothalamic "ejaculatory mechanism" or the capacity for erection, but indirectly the result of sub-effective levels of penile stimulation as a result of the failure of pelvic thrusting and intromission.

Effects of testosterone in the medial preoptic-anterior hypothalamic area

Since the majority of these appetitive sexual responses, as well as copulatory responses disappear with varying latency following castration (e.g., Bermant & Davidson, 1974; Beach, 1967; Davidson, 1966b; Hansen & Drake af Hagelsrum, 1984; Everitt & Stacey, 1987; Sachs & Meisel, 1988), it seems reasonable to suggest that testosterone or its metabolites acting within the mPOA/AHA cannot underlie the complete pattern of appetitive and investigatory acts which, together with copulatory responses, characterize the masculine sexual behavior pattern. A direct test of this view would require the detailed observation of the nature of restored versus unaffected sexual responses following the discrete implantation of testosterone into the mPOA/AHA in castrated males. But this does not appear to have been done in a systemically quantitative way, although in Davidson's study (1966a) some clues are to be found. Thus, males with testosterone implants in the mPOA/AHA, but not elsewhere, although generally showing reduced intromission latencies and ejaculatory patterns on some tests when compared with untreated castrates, clearly did not show the consistent levels of behavior seen in sham-operated controls. Indeed, complete behavioral restoration only occurred in two cases while analysis showed that the restored behavior "was not completely normal" (Davidson, 1966a, p. 794) in many cases. Of course, part of the explanation must involve the maintenance of the genital periphery, since its requirements in terms of 5-alpha dihydrotestosterone are well known. But this does not seem to be the whole story, especially when taken in conjunction with the rather specific behavioral consequences of removing the primary sex steroid target, the mPOA/AHA. Detailed re-examination of the effects of implanting testosterone in the mPOA/AHA, as well as in other limbic target sites—notably the medial amygdala, septum, and bed nucleus of the stria terminalis—would appear to be timely and worthwhile, especially if appetitive sexual responses are also measured using appropriate behavioral paradigms (see below).

Neurochemical manipulations of the medial preoptic-anterior hypothalamic area

In some instances, the putative hypothalamic locus of action of systemically administered drugs which often have marked effects on masculine sexual behavior has been studied. In general, parameters of copulatory behavior only have been quantified, although latencies to mount and intromit have frequently been viewed as measuring arousal or motivational, rather than performance, aspects of sexual behavior (Sachs & Barfield, 1976). The dangers of too narrow an interpretation of the latter have been pointed out elsewhere (e.g., Sachs, 1978; Sachs & Meisel, 1988; Everitt, 1990a) and do not need to be labored here other than to state that prolonged latencies may simply reflect an impaired *ability* to mount and intromit, while reduced latencies may reflect an enhanced erectile capacity (e.g., as occurs after intrathecal dopaminergic agonists, Pehek, Thompson, Eaton, Bazzett, & Hull, 1990) rather than indicating any *direct* effect on arousal/motivational processes *per se*. With these provisos in mind, it appears, for example, that dopamine receptor agonists infused into the mPOA/AHA rather selectively enhance copulatory parameters such as intromission rates and efficiency , while not affecting latencies to mount and intromit (Bitran & Hull, 1987; Bitran, Hull, Holmes, & Lokkingland, 1988; Hull, Bitran, Pehek, Warner, Band, & Holmes, 1986). Dopaminergic receptor antagonists, or presynaptic doses of the agonists tend to have opposite effects (Bitran & Hull, 1987; Bitran et al., 1988; Pehek et al., 1988). Thus, the marked changes in more appetitive aspects of masculine sexual behavior which follow systemic administration of these drugs (Bitran & Hull, 1987; Pfaus & Phillips, 1989; Everitt, 1990a) are not a feature of the intra-hypothalamic route of administration, even though they are seen to occur following intracerebroventricular (icv) infusion (Bitran & Hull, 1987; Bitran et al., 1988). Clearly, other sites of action of dopaminergic drugs must mediate these effects on non-copulatory measures of sexual activity.

Similarly, β-endorphin dose-dependently disrupted mounting, intromitting and ejaculation when infused into the mPOA/AHA, but had no effect on genital investigation and pursuit of receptive females, yet increased frequencies of hindlimb scratching and other purposeless acts (Hughes, Everitt, & Herbert, 1987; 1988) - rather as seen above to follow mPOA/AHA lesions (see also Hansen & Drake af Hagelsrum, 1984; Everitt & Stacey, 1987). The opiate receptor antagonist naloxone infused into the mPOA/AHA, on the other hand, markedly enhanced copulatory performance and also reduced postejaculatory refractory periods such that males ejaculated much more frequently during a standard fifteen minute observation period (Hughes, Herbert, & Everitt, 1990; Everitt, 1990a). However, there are more subtle features of this inhibitory effect of β-endorphin infused into the mPOA/AHA, since it seems not to be an all-or-none inhibitor of copulatory acts in the sense that lesions of the area are. Thus, the peptide has no inhibitory effect if infused into the mPOA/AHA following a single intromission, unless the male is presented with a novel female, that is, one with which he has not interacted sexually prior to the infusion (Stavy & Herbert, 1989). These observations either indicate that when a male has begun the copulatory sequence, the neural mechanisms underlying the rhythmic sequence of mounting and intromitting are insensitive to the peptide, or that the mPOA/AHA has orchestrated these mechanisms through interaction with a "downstream" site and is no longer involved with the response generating mechanism or, as Stavy and Herbert (1989) suggested, that this area is more concerned with the transition between pre-copulatory and copulatory acts; once this transition has been made following an intromission, then β-endorphin is ineffective. The latter view is a particularly interesting one since it argues that, following an intromission, a male rat experiences a change in central state which is specific to interaction with the female with which he has intromitted, since if she is replaced, he becomes sensitive again to the inhibitory effects of the opioid peptide.

Summary

Constraints of space do not permit en exhaustive appraisal of the consequences of all intrahypothalamic neuroendocrine manipulations in terms of the precise nature of the induced changes in sexual behavior—indeed, the necessary measures of a wide range of sexual responses are frequently not reported in the literature, perhaps because they do not change, perhaps because they were not recorded. Instead, I will summarize the above restricted review of the effects of lesion, androgenic and neurochemical manipulations of the mPOA/AHA by suggesting the the majority of data indicate this crucial region is more concerned with performance aspects of masculine sexual behavior, with what might be called "consummatory competence" (Everitt, Fray, Kostarczyk, Tayor, & Stacey, 1987; Everitt & Stacey, 1987), rather than with the rich array of appetitive or pre-copulatory responses which serve to bring a male into consummatory contact with a female such that proximal contact cues, which are largely somatosensory, come to control the reflexive acts of vigorous mounting with palpation, pelvic thrusting, and intromission (see Beach, 1956; Everitt, 1990a).

Studies of appetitive and reward-related aspects of sexual behavior following hypothalamic lesions and neuroendocrine treatments

Male rats will learn to respond on a lever to earn presentations of an estrous female with which they can copulate. To achieve reasonably high response rates, a second-order schedule of sexual reinforcement has been developed whereby a male's responding is maintained by the intermittent presentation of a previously arbitrary stimulus (usually a light) which has gained motivational significance, in fact conditioned reinforcing properties, through its prior association with sexual interaction with a female in heat (see Everitt & Stacey, 1987; Everitt, 1990a for further details). Using such a procedure, a male's appetitive sexual responses can be assessed relatively independently of his ability to execute the reflexive, species-specific copulatory pattern. Thus, this procedure provides one means of assessing the neuroendocrine and neuroanatomical determinants of appetitive acts which reflect arousal in a sexual context; in other words, those aspects of behavior which appear not to be affected by lesions and other treatments of the mPOA/AHA.

Lesions of the mPOA/AHA of male rats (Everitt & Stacey, 1987) were without effect on instrumental behavior to earn a receptive female, even though males were unable to copulate with females when they entered the operant chamber in which testing occurred. Comparable lesions of the mPOA/AHA in male rhesus monkeys were also without effect on lever pressing for access to a female (Slimp et al., 1978). Similarly, bilateral infusions of β-endorphin into the mPOA/AHA also had no effect on instrumental responding under the second-order schedule of sexual reinforcement, even though the same infusions prevented mounting, intromitting, and ejaculation (Hughes et al., 1987). Thus, manipulations of the mPOA/AHA which prevent copulation spare appetitive elements of sexual behavior in male rats. Castration of male rats both impaired instrumental behavior and copulatory responses, although this operant procedure proved to be less than ideal for studying the slow consequences of hormone withdrawal (Everitt & Stacey, 1987). However, the second-order paradigm proved to be exquisitely sensitive to the motivational effects of testosterone replacement to long term castrates, since it was shown that instrumental responding returned to control and pre-castration levels within seven days of administration and at a time when only a small proportion of males intromitted efficiently (Everitt & Stacey, 1987). The observation that systemic testosterone treatment affects instrumental, appetitive sexual acts whereas mPOA/AHA lesions or β-endorphin infusions do not, again suggests that important actions of the steroid occur outside this particular target area.

Place preference conditioned by sexual interaction with a receptive female provides another means of assessing appetitive aspects of sexual behavior, in this case conditioned approach to an environment previously associated with intromission and ejaculation (see Ágmo & Berenfeld, 1990; Edwards & Einhorn, 1986; Everitt, 1990a; Mehrara & Baum, 1990; Miller & Baum, 1987). Thus, it provides another means of assessing the neuroendocrine mechanisms underlying appetitive, especially reward-related aspects of sexual behavior which, like the instrumental procedure, does not rely on the ability to copulate in the test phase, since it is conducted in the absence of estrous females. In the way that we have used the procedure, the test of conditioned place preference (CPP) is always followed by a partner preference test, in which a male's choice between the estrous and anestrous females he interacted with in the conditioning phase is assessed.

Lesions of the mPOA/AHA had no *direct* effect on a CPP conditioned by sexual interaction with an estrous female, though it did decline after several tests in which males were unable to copulate, i.e., the CPP declined presumably because attempts to copulate were no longer reinforcing (Everitt, 1990a; Hughes et al., 1990). Indeed, the prevention of intromission in normal males resulted in a decrease in CPP over a similar time course. Surprisingly, mPOA/AHA lesions had no effect on a male's preference for a receptive over an unreceptive female, even though he was unable to copulate with her (Everitt, 1990a). However, while infusing β-endorphin into the mPOA/AHA did not affect CPP, it did promptly diminish a male's preference for a receptive female, so indicating that use of a variety of procedures to study the behavioral effects of neuroendocrine treatments does indeed reveal the operation of different processes underlying components of the sexual response repertoire (Everitt, 1990a). Systemic administration of the opiate receptor antagonist naloxone, in addition to decreasing instrumental behavior under the second-order schedule, also abolished a CPP (Everitt, 1990a; Hughes et al., 1990; Miller & Baum, 1987; Mehrara & Baum, 1990), yet the same drug infused into the mPOA/AHA had no effect on CPP, nor on instrumental behavior, yet it actually *increased* copulatory performance (see above). Thus, the rather selective disruption by systemic naloxone of a CPP and instrumental behavior for a sexual reward would appear to be mediated by mechanisms outside the mPOA/AHA. The opiate receptor-rich ventral midbrain and nucleus accumbens are strong candidates, given the literature on opiate and dopaminergic mechanisms mediating drug award, intracerebral self-stimulation and other more natural rewards (Phillips & Fibiger, 1985; Phillips, Pfaus, Blaha, & Fibiger, 1989; Wise, 1989a,b).

While the CPP procedure also reveals, then, that mPOA/AHA manipulations spare appetitive and reward-related sexual responses, it also reveals the more pervasive effects of castration. Intriguingly, a previously established CPP is promptly abolished by castration before males have experienced any failure of copulatory ability and, in any case, at a time when the majority of males are still able to mount, intromit and even ejaculate (Everitt, 1990a; Hughes et al., 1990). These observations clearly indicate that aspects of sexual reward are testosterone-dependent and, since mPOA/AHA lesions do not affect CPP, that a site outside this area must mediate such effects of the steroid.

Summary

Taken together, the data reviewed above strongly suggest that the special and unique importance of the mPOA/AHA in the display of sexual behavior in male rats and, probably, other male mammals, is most evident in terms of the control of copulatory reflexes. Thus, the mPOA/AHA appears to be one component of a final common motor outflow from this hormone-dependent region which allows the integrated expression of species-specific sequences of mounting, palpation, pelvic thrusting, and intromission. Appetitive acts such as instrumental responding, locomotor excitement in the presence of estrous females as well as reward-related processes measured in a CPP paradigm, all seem relatively immune to manipulations of the mPOA/AHA, but are nonetheless profoundly affected by removal and replacement of testosterone, opiate receptor antagonism following systemic administration of naloxone and, in some cases at least, systemically administered dopaminergic receptor agonists and antagonists.

Dopamine, the ventral striatum and appetitive aspects of sexual behavior

As the wide-ranging behavioral effects of systemically administered dopaminergic drugs might suggest, manipulating dopamine in the ventral striatum profoundly affects appetitive aspects of sexual behavior, yet without affecting copulatory performance. Thus, infusing the dopamine releaser D-amphetamine into the nucleus accumbens region dose-dependently increased instrumental responses for an estrous female presented under a second-order schedule of reinforcement, reduced latencies to mount and intromit, but did not later copulatory parameters such as numbers of mounts and intromissions to ejaculation, "hit rate" or ejaculation latency (Everitt, 1990a,b). Conversely, infusing the dopamine D2 receptor antagonist raclopride, or lesioning the dopaminergic innervation of the ventral striatum by infusing the neurotoxin 6-hydroxydopamine, significantly lengthened mount and intromission latencies, also without altering parameters of copulatory performance (Everitt, 1990a,b). The impact of these manipulations on incentive motivational processes in a sexual context is emphasized by "devaluing" the incentive value of the receptive females with which treated males are interacting. This is achieved by injecting hormone-primed females systemically with a neuroleptic such as alpha-flupenthixol. This treatment selectively abolished proceptive, soliciting responses but actually enhances the display of immobile lordosis postures (Everitt, Fuxe, & Hökfelt, 1974; Everitt, Fuxe, Hökfelt, & Jonsson, 1975). In this condition, mounts and intromissions will only occur if the male himself initiates them and, even in normal males, mount and intromission latencies are prolonged by treating the female in this way (Everitt, 1990a,b). Males infused with raclopride in the ventral striatum were markedly affected by coincident treatment of females with alpha-flupenthixol such that latencies to mount and intromit were doubled or trebled - indeed, in some males, mounting and intromitting were actually prevented, even though this treatment caused relatively modest increases in these latencies when females were actively soliciting.

These and other data (see Everitt, 1990a for a more detailed review) demonstrate the importance of the mesolimbic dopaminergic innervation of the ventral striatum from the mesencephalic ventral tegmental area in the display of appetitive responses to incentive, in this case sexual, stimuli. Recent experiments by Phillips and Pfaus and their colleagues (Phillips, Blaha, & Fibiger, 1989; Phillips, Pfaus, Blaha, & Fibiger, 1990) have added support to this view by studying the conditions under which dopamine release in the ventral striatum is altered. They have demonstrated that, in the ventral but not the dorsal striatum, the dopaminergic oxidation signal measured using *in vivo* chronoamperometry is much increased when males are in the presence of, but are unable to interact with, a female in heat. The dopaminergic signal actually decreased during sexual interaction with a female, although it remained above baseline levels seen when the male was alone. At ejaculation, the dopamine signal decreased still further, but increased again *in advance* of the next mount or intromission. Thus, studies using drugs infused into the ventral striatum, dopaminergic lesions of the structure and also correlative *in vivo* measures of dopamine release there, all indicate a special relationship between the mesolimbic dopaminergic system innervating the ventral parts of the striatal complex and appetitive or preparatory behaviors. The results summarized here refer to studies of sexual behavior, but it would be incorrect to assume that it is only in this context that dopaminergic mechanisms are important. A wealth of data implicates the same system in responses to incentive stimuli across a wide range of motivated states as well as in the neural substrates of reward-related processes (Carr, Fibiger, & Phillips, 1989, Phillips & Fibiger, 1985; Phillips et al., 1989; Robbins, Cador, Taylor, & Everitt, 1989; Wise, 1989a).

The limbic forebrain and sexual behavior

In their original paper, Heimer and Larsson (1966/67) discussed the likely neural structures with which the mPOA/AHA interacts, such that lesions of the area result in the pattern of disrupted sexual behavior in males which they described. This discussion quite rightly emphasized the role of the mPOA/AHA in terms of its "strategic position in the limbic system" (p. 260) and especially mentioned its interactions with olfactory structures via the corticomedial amygdala and stria terminalis. Lennart Heimer, who collaborated with Knut Larsson in the 1960's, has also been responsible for elaborating, broadening, and even fundamentally changing our view of the "limbic system". In particular, Heimer and his colleagues (Alheid & Heimer, 1988; Heimer & Wilson, 1975) have recently demonstrated that allocortical and sub-cortical components of the limbic system, including the basolateral amygdala, project richly onto sub-cortical components of the motor system, notably the ventral striatum which receives a rich dopaminergic innervation arising from the mesencephalic ventral tegmental area. Thus, a route through which the results of limbic information processing, long thought to be involved with emotional behavior, might come to affect action has been described—a "limbic-motor interface" in the words of Gordon Mogenson (e.g., 1987). Does such a system have any significance for the display of sexual behavior?

The many reports of enhanced and aberrant sexual behavior following temporal lobe lesions, originally thought to be due to amygdaloid damage, have been shown more recently to be related to coincident damage to temporal neocortical structures (see Everitt, Cador, &

Robbins, 1989, and Sachs & Meisel, 1988, for references). Lesions restricted to the basolateral regions of the amygdala are without effect on copulation in male rats (see Everitt et al., 1989), although lesions which damage corticomedial regions impair sexual behavior, much as do lesions of the olfactory system (see Sachs & Meisel, 1988 for review). However, there are marked consequences for sexual behavior of damaging the amygdala, but to demonstrate them an appropriate paradigm must be employed. Thus, under the second-order schedule of sexual reinforcement, bilateral, excitotoxic, axon-sparing lesions of the basolateral parts of the amygdala in male rats permanently depressed instrumental responding for the conditioned reinforcer which maintains operant behavior during the session. However, when an estrous female ultimately entered the operant chamber, copulation was unimpaired (Everitt et al., 1989). Thus, the effects of mPOA/AHA and basolateral amygdala lesions are doubly dissociated using this behavioral procedure: mPOA/AHA lesions impair copulatory competence but not instrumental sexual responses while amygdala lesions have the opposite pattern of effects (see Everitt, 1990a).

In these same experiments, evidence of an important functional interaction between the basolateral amygdala and dopamine-dependent events in the ventral striatum was obtained. Thus, in lesioned animals, infusion of the dopamine releasing drug D-amphetamine into the ventral striatum enhanced responding under the second-order schedule, that is, it ameliorated to some extent the effects of the lesion. The same treatment also reduced subsequent mount and intromission latencies, but did not affect copulatory parameters. The broader implications of these observations are discussed in detail elsewhere (Everitt, 1990a; Everitt et al., 1989, Cador, Robbins, & Everitt, 1989, Robbins et al., 1989), but it may be said here that they indicate interactions between the basolateral amygdala and ventral striatum in reward-related processes—in these experiments they are sexual in nature, but the role of this system is, of course, not restricted to one such motivational state. This interaction is profoundly affected by dopamine transmission in the ventral striatum and this system, as we have seen above, is intimately involved with incentive motivational processes.

Summary

The picture emerging from the above discussion is that neural mechanisms operating at every level of the limbic system are involved in separable, though integrated, aspects of the pattern of sexual behavior. Associative mechanisms involved with stimulus-reward associations seem especially to be found in the basolateral amygdala and they affect instrumental response output via interactions with the ventral striatum. Here, dopaminergic transmission can both alter the level of such instrumental response output, but it also seems that this system is responsive to the presence of incentive motivational stimuli in the environment, such as an estrous female, and may also modulate a male's appetitive sexual acts as well as his sensitivity to alterations in the incentive value of a female. The purpose of this array of instrumental and other appetitive acts is to bring a male into direct, physical contact with a female, when he mounts, intromits, and ejaculates with her. It is in this context that hormone-dependent hypothalamic processes are of special and unique importance. What is unclear is whether these different neural mechanisms, although clearly separable if observed using appropriate techniques, actually operate separately and in parallel. It may also be that the complete masculine sexual behavior pattern reflects the complex interaction and integration of the activity of these systems—and there is both neuroanatomical and behavioral evidence that this might be the case (e.g., Swanson & Mogenson, 1981; Brackett & Edwards, 1984). An especially interesting aspect of the problem concerns situations where, for example, testosterone is a major determinant of particular aspects of sexual behavior, such as sexual reward as assessed in the CPP paradigm, or instrumental responding for an estrous female, yet its impact seems not to be mediated by the mPOA/AHA, which until now has dominated interest so far as neuroendocrine determinants of sexual behavior are concerned. This may prove to be both an interesting and complex area of experimentation in the future.

ACKNOWLEDGMENTS

The research summarized here was supported by project and program grants from the Medical Research Council. It is a pleasure to acknowledge the collaboration of Martine Cador, Andy Hughes, Joe Herbert, and Trevor Robbins on many of the experiments.

REFERENCES

Ågmo, A., & Berenfeld, R. (1990). Reinforcing properties of ejaculation in the male rat: The role of opioids and dopamine. *Behav. Neurosci., 104,* 177-182.

Alheid, G.F., & Heimer, L. (1988). New perspectives in basal forebrain organisation of special relevance for neuropsychiatric disorders: the striatopallidal, amygdaloid and corticopetal components of the substantia innominata. *Neuroscience, 27,* 1-39.

Beach, F.A. (1956). Characteristics of masculine "sex drive". In M.R. Jones (Ed.), *Nebraska Symposium on Motivation, 4 (pp. 1-31).* Lincoln: University of Nebraska Press.

Beach, F.A. (1967). Cerebral and hormonal control of reflexive mechanisms involved in copulatory behavior. *Physiol. Rev., 47,* 289-316.

Bermant, G., & Davidson, J.M. (1974). *Biological Bases of Sexual Behavior.* New York: Harper & Row.

Bitran, D., & Hull, E.M. (1987). Pharmacological analysis of male rat sexual behavior. *Neurosci. Biobehav. Rev.11,* 365-389.

Bitran, D., Hull, E.M., Holmes, G.M., & Lokkingland, K.J. (1988). Regulation of male rat copulatory behavior by preoptic incertohypothalamic dopamine neurons. *Brain Res. Bull. 20,* 323-331.

Brackett, N.L., & Edwards, D.A. (1984). Medial preoptic connections with the midbrain tegmentum are essential for male sexual behavior. *Physiol. Behav. 32,* 79-84.

Cador, M., Robbins, T.W., & Everitt, B.J. (1989). Involvement of the amygdala in stimulus-reward associations: interactions with ventral striatum. *Neuroscience, 30,* 77-86.

Carr, G.D., Fibiger, H.C., & Phillips, A.G. (1989). Conditioned place preference as a measure of drug reward. In J.M. Liebman and S.J. Cooper (Eds.), *The Neuropharmacological Basis of Reward* (pp. 264-319). Oxford University Press.

Davidson, J.M. (1966a). Activation of male rat's sexual behavior by intracerebral implantation of androgen. *Endocrinology 79,* 783-794.

Davidson, J.M. (1966b). Characteristics of sex behavior in male rats following castration. *Anim. Behav., 14,* 266-272.

Edwards, D.A., & Einhorn, L.C. (1986). Preoptic and midbrain control of sexual motivation. *Physiol. Behav. 37,* 329-335.

Everitt, B.J. (1990a). Sexual motivation: a neural and behavioral analysis of the mechanisms underlying appetitive and copulatory responses of male rats. *Neurosci. Biobehav. Rev.,* in press.

Everitt, B.J. (1990b). Manipulations of dopamine in the ventral striatum selectively affect appetitive components of sexual behavior in male rats. *Psychopharmacology,* submitted.

Everitt, B.J., & Stacey, P. (1987). Studies of instrumental behavior with sexual reinforcement in male rats (Rattus norvegicus): II. Effects of preoptic area lesions, castration and testosterone. *J. Comp. Psychol., 101,* 407-419.

Everitt, B.J., Fray, P.J., Kostarczyk, E., Taylor, S., & Stacey, P. (1987). Studies of instrumental behavior with sexual reinforcement in male rats (Rattus norvegicus): I. Control by brief visual stimuli paired with a receptive female. *J. Comp. Psychol., 101,* 395-406.

Everitt, B.J., Cador, M., & Robbins, T.W. (1989). Interactions between the amygdala and ventral striatum in stimulus-reward associations: studies using a second-order schedule of sexual reinforcement. *Neuroscience 30,* 63-75.

Everitt, B.J., Fuxe, K., & Hökfelt, T. (1974). Inhibitory role of dopamine and 5-hydroxytryptamine in the sexual behavior of female rats. *Eur. J. Pharmacol., 29,* 187-191.

Everitt, B.J., Fuxe, K., Hökfelt, T., & Jonsson, G. (1975). Role of monoamines in the control by hormones of sexual receptivity in the female rat. *J. Comp. Physiol. Psychol. 89,* 556-572.

Hansen, S., & Drake af Hagelsrum, L.J.K. (1984). Emergence of displacement activities in the male rat following the thwarting of sexual behavior. *Behav. Neurosci., 98,* 868-883.

Hansen, S., Köhler, C., Goldstein, M., & Steinbusch, H.W.M. (1982). Effects of ibotenic acid-induced neuronal degeneration in the medial preoptic area and lateral hypothalamic area on sexual behavior in the male rat. *Brain Res. 239,* 213-232.

Hart, B.L. (1974). Medial preoptic-anterior hypothalamic area and socio-sexual behavior of male dogs: a comparative neuropsychological analysis. *J. Comp. Physiol. Psychol., 86,* 328-349.

Hart, B.L. (1986). Medial preoptic-anterior hypothalamic area and socio-sexual behavior of male goats. *Physiol. Behav. 36,* 301-305.

Hart, B.L., Haugen, C.M., & Peterson, D.M. (1973). Effects of medial preoptic-anterior hypothalamic lesions on mating behavior of male cats. *Brain Res., 54,* 177-191.

Hart, B.L., & Voith, V.L. (1978). Changes in urine spraying, feeding and sleep behavior following medial preoptic-anterior hypothalamic lesions in cats. *Brain Res., 145,* 406-409.

Heimer, L., & Larsson, K. (1966/67). Impairment of mating behavior in male rats following lesions in the preoptic-anterior hypothalamic continuum. *Brain Res., 3,* 248-263.

Heimer, L., & Wilson, R.D. (1975). The subcortical projections of the allocortex: similarities in the neural associations of the hippocampus, the pyriform cortex and the neocortex. In M. Santini (Ed.), *Golgi Centennial Symposium* (pp. 177-192). New York: Raven Press.

Hughes, A.M., Everitt, B.J., & Herbert, J. (1988). The effects of simultaneous or separate infusions of some pro-opiomelanocortin-derived peptides (β-endorphin, melanocyte stimulating hormone and corticotrophin-like intermediate polypeptide) and their acetylated derivatives upon sexual and ingestive behavior of male rats. *Neuroscience 27,* 689-698.

Hughes, A.M., Everitt, B.J., & Herbert, J. (1987). Selective effects of β-endorphin infused into the hypothalamus, preoptic area and bed nucleus of the stria terminalis on the sexual and ingestive behavior of male rats. *Neuroscience 23,* 1063-1073.

Hughes, A.M., Herbert, J.H., & Everitt, B.J. (1990). Comparative effects of preoptic area infusions of opioid peptides, lesions and castration on sexual behavior in male rats: studies of instrumental behavior, conditioned place preference and partner preference. *Psychopharmacology,* submitted.

Hull, E.M., Bitran, D., Pehek, E.A., Warner, R.K., Band, L.C., & Holmes, G.M. (1986). Dopaminergic control of male sex behavior in rats: effects of an intracerebrally infused agonist. *Brain Res., 370,* 73-81.

Larsson, K. (1956). Conditioning and sexual behavior in the male albino rat. *Acta psychologica Gothoburgensia I.* (pp. 1-269). Stockholm: Almqvist & Wiksell.

Mehrara, B.J., & Baum, M.J. (1990). Naloxone disrupts the expression but not the acquisition by male rats of a conditioned place preference response for an oestrous female. *Psychopharmacology,* in press.

Miller, R.L., & Baum, M.J. (1987). Naloxone inhibits mating and conditioned place preference for an estrous female in male rats soon after castration. *Pharmacol. Biochem. Behav. 26,* 781-789.

Mogenson, G.J. (1987). Limbic-motor integration. In A.N. Epstein and A.R. Morrison (Eds.), *Progress in Psychobiology and Physiological Psychology, Vol. 12* (pp. 117-170). Orlando: Academic Press.

Pehek, E.A., Warner, R.K., Bazzett, T.J., Bitran, D., Band, L.C., Eaton, R.C., & Hull, E.M. (1988). Microinjection of cis-flupenthixol, a dopamine antagonist, into the medial preoptic area impairs sexual behavior of male rats. *Brain Res., 443,* 70-76.

Pehek, E.A., Thompson, J.T., Eaton, R.C., Bazzett, T.J., & Hull, E.M. (1990). Apomorphine and haloperidol, but not domperidone, affect penile reflexes in rats. *Pharmacol. Biochem. & Behav.,* in press.

Pfaus, J.G., & Phillips, A.G. (1989). Differential effects of dopamine receptor antagonists on sexual behavior of male rats. *Psychopharmacology, 98,* 363-368.

Phillips, A.G., Blaha, C.D., & Fibiger, H.C. (1989). Neurochemical correlates of brain-stimulation reward measured by ex vivo and in vivo analysis. *Neurosci. Biobehav. Rev. 13,* 99-104.

Phillips, A.G., Pfaus, J.G., Blaha, C.D., & Fibiger, H.C. (1990). Dopamine and motivated behavior: Insights provided by in vivo analyses. Manuscript in press.

Phillips, A.G., & Fibiger, H.C. (1985). Anatomical and neurochemical substrates of drug reward determined by conditioned place preference technique. In M.A. Bozarth (Ed.), *Methods of Assessing the Reinforcing Properties of Abused Drugs* (pp. 275-290). New York: Springer Verlag.

Phillips, A.G., Pfaus, J.G., Blaha, C.D., & Fibiger, H.C. (1989). Regional differences in dopamine efflux during specific phases of sexual behavior in the male rat: An in vivo electrochemical study. *Behav. Pharmacol., 1, (Suppl. 1),* 275.

Robbins, T.W., Cador, M., Taylor, J.R., & Everitt, B.J. (1989). Limbic-striatal interactions in reward-related processes. *Neurosci. Biobehav. Rev., 13,* 155-162.

Sachs, B.D. (1978). Conceptual and neural mechanisms of masculine copulatory behavior. In T.E. McGill, D.A. Dewsbury, and B.D. Sachs (Eds.), *Sex and Behavior* (pp. 267-295). New York: Plenum Press.

Sachs, B.D., & Barfield, R.J. (1976). Functional analysis of of masculine copulatory behavior in the rat. In J.S. Rosenblatt, R.A. Hinde, E. Shaw, and C. Beer (Eds.), *Advances in the Study of Behavior, Vol 7* (pp. 91-154). Orlando, FL: Academic Press.

Sachs, B.D., & Meisel, R.L. (1988). The physiology of male sexual behavior. In E. Knobil and Neill (Eds.), *The Physiology of Reproduction* (pp. 1393-1485). New York: Raven Press.

Slimp, J.C., Hart, B.L., & Goy, R.W. (1978). Heterosexual, autosexual and social behavior of adult male rhesus monkeys with medial preoptic-anterior hypothalamic lesions. *Brain Res., 142,* 105-122.

Stavy, M., & Herbert, J. (1989). Differential effects of β-endorphin infused into the hypothalamic preoptic area at various phases of the male rat's sexual behavior. *Neuroscience, 30,* 433-442.

Swanson, L.W., & Mogenson, G.J. (1981). Neural mechanisms for the functional coupling of autonomic, endocrine and somatomotor responses in adaptive behavior. *Brain Res. Rev., 3,* 1-34.

Ungerstedt, U. (1971). Adipsia and aphagia after 6-hydroxydopamine induced degeneration of the nigrostriatal dopamine system. *Acta Physiol. Scand., 82, Suppl 367,* 95-122.

Wise, R.A. (1989a). The brain & reward. In J.M. Liebman and S.J. Cooper (Eds.), *Neuropharmacological Basis of Reward* (pp 377-424). Oxford: Clarendon Press.

Wise, R.A. (1989b). Opiate reward: sites and substrates. *Neurosci. Biobehav. Rev., 13,* 129-134.

11 Vaginocervical Afference as a Trigger for Analgesic, Behavioral, Autonomic, and Neuroendocrine Processes

Barry R. Komisaruk
Institute of Animal Behavior
Rutgers, The State University of New Jersey
Newark, New Jersey

INTRODUCTION

Vaginocervical stimulation (VS) inhibits responses to noxious stimulation

In the rat, vaginocervical stimulation (VS) simultaneously facilitates the lordosis response (Komisaruk & Diakow, 1973; Komisaruk, Larsson, & Cooper, 1972) and produces immobilization (see Naggar & Komisaruk, 1977).* The immobilization antagonizes the righting reflex (Fig. 1). In collaboration with Knut Larsson, we challenged the VS-produced immobilization with foot pinch, and observed that VS completely blocked the leg withdrawal reflex (Komisaruk & Larsson, 1971). VS also inhibited cranial nerve nociceptive reflexes.

---

*The observations were made in the course of analyzing the effects of VS on neural activity related to neuroendocrine and behavioral processes (Komisaruk & Olds, 1968; Komisaruk, McDonald, Whitmoyer, & Sawyer, 1967; Ramirez, Komisaruk, Whitmoyer, & Sawyer, 1967). VS stimulates several reproductive neuroendocrine reflexes. Ovulation occurs in response to VS not only in "reflex" ovulating species (Rowlands, 1966), but also in the rat, a "spontaneous" ovulator, under appropriate conditions (e.g., Brown-Grant, Davidson, & Greig, 1973; Harrington, Eggert, Wilbur, & Linkenheimer, 1966; Johns, Feder, & Komisaruk, 1980). In the rat, this response is mediated by VS-induced by VS-induced release of hypothalamic LHRH (Takahashi, Ford, Yoshinaga, & Greep, 1975) and anterior pituitary luteinizing hormone (Moss & Cooper, 1973). Other neuroendocrine responses to VS include release of anterior pituitary prolactin, which induces pseudopregnancy (Terkel, 1986) and subsequent stimulation of ovarian progesterone secretion, which prepares the uterus for implantation of the fertilized ova in pregnancy (Adler, Resko, & Goy, 1970). Furthermore, VS releases posterior pituitary oxytocin (Moos & Richard, 1975), which stimulates uterine contractions that accelerate the delivery of the fetuses, i.e., the "fetus-ejection" or "Ferguson" reflex (Higuchi, Uchide, Honda, & Negoro, 1987).

In chronically-prepared spinal-transected rats, VS blocked withdrawal reflexes to noxious stimulation below the level of the transection (Fig. 2). This indicates that an intrinsic spinal mechanism mediates VS-produced inhibition of responses to noxious stimulation (Komisaruk & Larsson, 1971). The magnitude of the inhibitory effect is proportional to the force of VS (Crowley, Jacobs, Volpe, Rodriguez-Sierra, & Komisaruk, 1976).

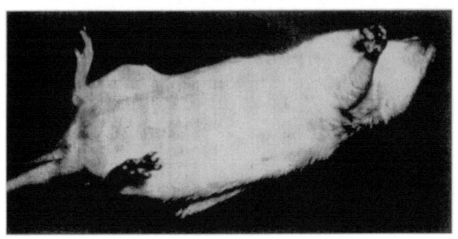

FIG. 11.1. VS (note probe) suppresses the righting response in awake, unanesthetized, unrestrained, intact rats.

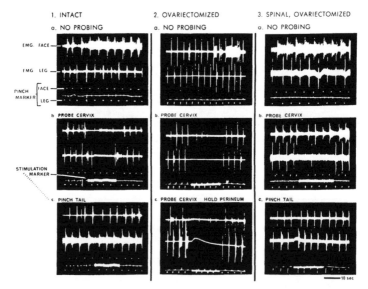

FIG. 11.2. VS inhibits EMG responses to noxious stimulation via an intraspinal mechanism. Leg withdrawal and facial twitch responses to pinch (applied at markers: (1.a.) were strongly attenuated by concurrent VS (1.b.) but not by the control stimulus, pinching the tail (1.c.). VS was similarly effective in ovariectomized, hormonally-untreated rats (2.). After chronic complete mid-thoracic spinal transection, VS blocked the response below, but not above, the level of the transection, and tail pinch again had no effect (3.).

124

VS inhibits sensory responses to noxious stimulation

Our initial hypothesis was that VS blocked withdrawal responses by inhibiting the motor capacity of the rats to show the response. However, blockage of the motor responses to noxious stimulation does not rule out blockage of pain input. We tested for pain blockage in several ways, first by eliciting vocalization using a noxious and a non-noxious stimulus, i.e., tail shock and sudden lifting off the table surface, respectively. VS inhibited the vocalization to tail shock, but not the vocalization to lifting (Komisaruk & Wallman, 1977). This suggested that although VS did not inhibit the motor capacity to vocalize, VS did suppress the noxious sensory component that stimulated vocalization.

As an alternative means of excluding a possible confounding effect of VS produced motor inhibition, we devised an operant procedure for rats to press a lever to obtain VS during inescapable skin shock. The rats performed the operant response that provided the VS prior to any possible immobilizing effect of the VS. During inescapable skin shock, the rats continued to do so for significantly more trials than rats that did not receive VS upon performance of the operant response. This indicates that the VS ameliorated the effects of the skin shock (Ross, Komisaruk, & O'Donnell, 1979).

As another means of excluding a possible confounding effect of VS-produced motor inhibition, we recorded neuronal responses in the somatosensory thalamus to noxious and innocuous stimulation and found that while VS had no effect on thalamic neuronal responses to innocuous (tactile) sensory stimulation, in contrast, it markedly suppressed the sensory neuronal response to noxious (pinch) stimulation. Based on these findings, we proposed that VS produces not anesthesia, but more specifically, analgesia, i.e., a differential suppression of responses to noxious, but not innocuous, sensory stimulation (Komisaruk & Wallman, 1973; 1977).

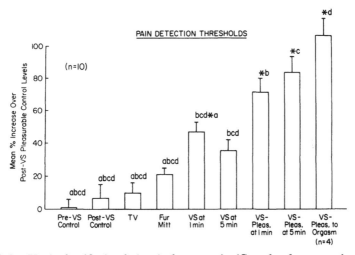

FIG. 11.3. Vaginal self-stimulation in humans significantly elevates pain detection threshold (finger compression test). The analgesia produced by continuous pressure self-applied to the anterior vaginal wall (VS at 1 min) adapted within several minutes (VS at 5 min). However, when VS was self-applied rhythmically so that it was pleasurable by self-report, the magnitude of the analgesia was greater, and increased with continued stimulation (VS Pleas at 1 min and 5 min). Four of the women went to orgasm, during which the increase in pain thresholds was still higher (VS Pleas to orgasm). No significant effects on pain threshold were reported in response to pleasurable, somatic surface non-genital self-stimulation (fur mitt) or a distracting stimulus (a diverting videotaped film segment: TV). Note: * before any letter indicates that the corresponding threshold value was significantly greater than the threshold values for which the same letter is not preceded by an asterisk.

However, conclusive evidence that VS suppresses pain *per se* required a verbal report. Consequently, in humans, we measured pain thresholds and tactile thresholds in response to vaginal self-stimulation. Vaginal self-stimulation elevated the pain detection threshold by 50% to more than 100% (Fig. 3)., while having no effect on the tactile threshold (Komisaruk & Whipple, 1984; 1986; Whipple & Komisaruk, 1985; 1988). Hence, we conclude that vaginal stimulation does indeed produce analgesia.

Our research has taken two paths: one, analyzing the possible functional significance of VS-produced analgesia and the other, ascertaining the mechanism of the analgesia.

Functional significance of VS-produced analgesia

We hypothesized that analgesia is produced naturally by the distention of the female reproductive tract that is produced by the mechanical stimulation resulting from growth and passage of the fetus during pregnancy and parturition, respectively. Gintzler (1980) had shown that in the rat, the pain threshold increases to reach a peak on the day of parturition. We hypothesized that this may be due, at least in part, to the uterine distention produced by the growing fetuses. Consequently, we denervated the uterus via bilateral hypogastric neurectomy and measured pain thresholds. We replicated the increase in pain threshold during pregnancy in the sham-operated controls, whereas the hypogastric-neurectomized rats showed a significantly lower increase in pain threshold (Gintzler, Peters, & Komisaruk, 1983).

We then measured the pain threshold in rats during parturition (Toniolo, Whipple, & Komisaruk, 1987). The rats gave birth on a small platform that rested on an open grillwork floor suspended over the laboratory table. This enabled the tail to be positioned vertically down through the grill without handling the parturient female. A beaker containing 55° C water was brought up to the tail so that it was immersed in the water, and the latency for the rat to withdraw the tail from the water was recorded. The latency for tail withdrawal was significantly greater during emergence of individual fetuses from the vaginal orifice than one minute after they emerged.

The finding led us to perform a parallel experiment in women during labor (Whipple, Komisaruk, & Josimovich, in press). Pain thresholds were measured on the fingers using a calibrated pressure-exerting instrument, as in our previous studies. Women who had undergone childbirth training showed a significantly higher pain threshold during labor than in the week prior to labor or one or 24 hours postpartum. By contrast, tactile thresholds showed no significant changes during these test periods. It was necessary to restrict our measurements to only stage 1 of labor, since in later stages the measurement procedure would have interfered with the activities of the hospital staff. Nevertheless, the significant increase in pain threshold during labor supports our hypothesis that birth canal distention generates analgesia during parturition (Komisaruk & Whipple, 1988). While parturient pain is one of the most severe forms of pain that is experienced (Melzack, 1984), it is likely that it would be more severe were it not for the activation of this endogenous analgesic process.

Mechanism of VS-produced analgesia

Evidence against a role of the autonomic nervous system

We ruled out an indirect role of blood pressure in the analgesia by the following analyses. Hypertension produces an analgesic effect in humans (Zamir & Segal, 1979), and baroceptor stimulation reduces responses to noxious stimuli in rats (Dworkin, Filewich, Miller, Craigmyle, & Pickering, 1979). We found that in rats, VS produces a strong activation of the sympathetic division of the autonomic system, leading within seconds to an increase in blood pressure and heart rate (Catelli, Sved, & Komisaruk, 1987) (Fig. 4) and pupillary dilatation (Szechtman, Adler, & Komisaruk, 1985).

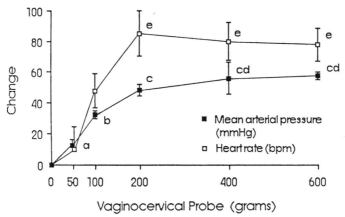

FIG. 11.4. VS increases heart rate and blood pressure in awake, unrestrained rats. The absolute magnitude of these autonomic responses increased as the force of VS was increased. For each parameter, values with different or no letters differed significantly from each other; those with the same letter did not.

Consequently, we hypothesized that the blood pressure mediates the analgesia, and tested this with the use of the ganglionic blocking agent, chlorisondamine. This drug blocked the VS-produced elevation in blood pressure and heart rate, yet did not attenuate the analgesia, as measured by tail flick latency. Moreover, we mimicked the magnitude of blood pressure elevation that is normally produced by VS, by administration of the alpha-adrenergic agonist, phenylephrine, but no analgesia resulted (Catelli et al., 1987).

Extending this analysis to humans, we observed that VS produced analgesia without concomitant elevation in sympathetic tone as measured by pulse rate. Conversely, induction of a comparable increase in the magnitude of sympathetic activation via exercise (deep knee bends), resulted in no significant increase in pain threshold (Martinez-Gomez, Whipple, Oliva-Zarate, Pacheco, & Komisaruk, 1988). Based on these studies, we conclude that the analgesia produced by VS is independent of the degree of activation of the sympathetic division of the autonomic nervous system.

We have analyzed the mechanism by which VS produces analgesia at the level of the spinal cord by pharmacological, neuroanatomical and neurochemical methodologies.

Pharmacological evidence

Our pharmacological analysis of spinal cord mechanisms mediating VS-produced analgesia (SPA) utilized the method of injecting pharmacological agents directly to the subarachnoid intrathecal space of the lumbosacral spinal cord via a chronically implanted catheter. We found that direct perispinal administration of the alpha adrenergic receptor antagonist, phentolamine, or the serotonin receptor antagonist, methysergide, each produced a significant attenuation of VSPA (Steinman, Komisaruk, Yaksh, & Tyce, 1983).

This suggested that a descending spinopetal system, with cells of origin in the lower brain stem, plays a role in VSPA (Fig. 5). These studies were confirmed by two independent experiments. First, we superfused the spinal cord with artificial cerebrospinal fluid and measured norepinephrine and serotonin levels in the superfusate before, during, and after VS. We found a significant increase in the concentration of both monoamines during VS (Steinman et al., 1983). In addition, when we measured norepinephrine turnover in the spinal cord, we found that VS increased its turnover rate, indicating an increase in activity of descending norepinephrine-containing pathways (Crowley, Rodriguez-Sierra, & Komisaruk,

1977). Second, we transected bilaterally the dorsolateral fasciculus, the tract through which these descending monoamine pathways project to the spinal cord, and found a significant attenuation of VSPA (Watkins, Faris, Komisaruk, & Mayer, 1984). This descending system has been shown by others to mediate other forms of stimulation-produced analgesia (Basbaum & Fields, 1978). However, at most about a 50% reduction in the effect of VS was obtained following either bilateral DLF transection or even complete spinal transection (at T2) (Watkins et al., 1984).

Putative Mechanisms of Vaginal Stimulation - Produced Analgesia

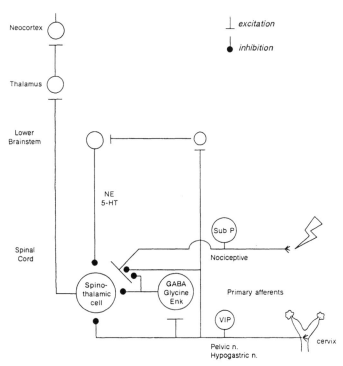

FIG. 11.5. Schematic diagram of a current concept of the neural basis of VS-produced analgesia. VS may inhibit nociception via one or more of the following mechanisms: 1) Release of primary afferent peptide(s) (e.g., vasoactive intestinal peptide [VIP]) in the spinal cord. This may stimulate inhibitory interneurons (e.g., enkephalinergic [Enk], GABAergic, and/or glycinergic) that in turn inhibit nociceptive responses of ascending (e.g., spinothalamic) pain pathways. 2) The primary afferent transmitter(s) may exert an inhibitory presynaptic and/or postsynaptic action on the pain afferent pathways. 3) VS-afferent activity ascends to the lower brainstem where it activates descending monoaminergic pathways that release pain-inhibitory norepinephrine [NE] and serotonin [5-HT] into the spinal cord.

Consequently, we searched for other neurotransmitter/neuromodulator systems that could mediate VSPA. Naloxone, a pure opiate receptor antagonist, produced a maximum of about a 30% reduction in VSPA measured by the tail flick latency to radiant heat when administered either systemically (Hill & Ayliffe, 1981) or intrathecally (Steinman, Roberts, & Komisaruk, 1982). However, on a different nociceptive test (vocalization threshold to tail shock), neither naloxone nor induction of tolerance to morphine antagonized VSPA, suggesting that VSPA produces analgesia in part via a non-opiate mechanism (Crowley, Rodriguez-

Sierra, & Komisaruk, 1977a). Additional evidence of a non-opiate component mediating VSPA is that after development of tolerance to a non-opiate form of analgesia (cold water swim), the analgesic effect of VS was significantly attenuated (Bodnar & Komisaruk, 1984).

The inhibitory amino acids, glycine and GABA, probably play a role in VSPA on the basis that their receptor antagonists, strychnine, and bicuculline (or picrotoxin), respectively, significantly attenuated VSPA when administered intrathecally (Beyer, Roberts, & Komisaruk, 1985; Roberts, Beyer, & Komisaruk, 1985; 1986).

Furthermore, the release of glycine into superfusates of the spinal cord was significantly increased by VS, but not by foot pinch (GABA levels were increased by VS in some cases, but were undetectable in most cases) (Masters, Beyer, Jordan, Steinman, & Komisaruk, 1988).

Since the above neurotransmitter/neuromodulators are all contained in second- or higher-order neurons, we then hypothesized that the transmitter(s) contained in the primary sensory neurons from the vagina and cervix should be capable of mimicking the effect of VS, by producing analgesia when applied directly to the spinal cord.

Neuroanatomical evidence

Support for the concept of a spinal site of blockage of noxious sensory stimulation, where the afferent activity would be blocked before it reached the thalamus, was provided by a C-14-2-Deoxyglucose ("2-DG") analysis (Johnson, Pott, Siegel, Adler, & Komisaruk, 1988). 2-DG is a form of glucose that is taken up by neurons in proportion to their metabolic activity, but they can not excrete it; consequently, if the activity of a neuron is increased by sensory stimulation, it accelerates its rate of accumulation of C14-2-DG. When subjected to autoradiography and quantitative densitometric analysis, increased neuronal activity is translated into an increased density of darkened grains on the photographic emulsion. Using this method, we analyzed the sites in the spinal cord where noxious stimulation activated neurons and where VS inhibited this activation. In laminae 1 and 2 of L5, unilateral foot pinch increased ipsilateral, compared to contralateral, neuronal activity. VS applied concurrently with unilateral foot pinch blocked this effect (Fig. 6). (Johnson et al., 1988). This indicates that VS inhibits noxious sensory input in the dorsal horn of the spinal cord.

Vaginal sensory input enters the spinal cord via the pelvic nerve (Peters, Kristal, & Komisaruk, 1987) at the L6-S1 level (Pacheco, Martinez-Gomez, Whipple, Beyer, & Komisaruk, 1989). We found that bilateral transection of the pelvic nerve plays a role in conveying the analgesia-producing message to the lumbo-sacral division of the spinal cord (Cunningham, Steinman, & Komisaruk, 1989).

Role of primary afferents

We further characterized the afferents that convey the analgesia-producing effect through the use of neonatal capsaicin administration. This procedure has been shown to irreversibly destroy small unmyelinated c fibers (Buck & Burks, 1986; Nagy, 1982). Since this is the predominant fiber type in the pelvic nerve, we hypothesized that neonatal capsaicin treatment would disrupt VSPA. The hypothesis was confirmed by our findings that neonatal capsaicin treatment abolished VSPA (Rodriguez-Sierra, Skofitsch, Komisaruk, & Jacobowitz, 1988). This finding suggests that analgesia-triggering primary afferent neurotransmitter/ neuromodulators are contained in c fibers. Furthermore, neonatal capsaicin treatment also reduced the release of glycine, which contributes significantly to VSPA (Beyer et al., 1985), into spinal cord superfusates (Masters, Beyer, Jordan, Steinman, & Komisaruk, 1989).

In an extension of the studies on capsaicin to humans, we found that women who have chronically consumed a diet high in hot chili peppers (capsaicin is the pungent constituent) daily since childhood, showed a significantly lower magnitude of vaginal self-stimulation produced analgesia than women who have chronically consumed a diet low in hot chili peppers (Whipple, Martinez-Gomez, Oliva-Zarate, Pacheco, & Komisaruk, 1989). We are currently ascertaining whether there is a difference in pain threshold elevation in these two groups of women during childbirth. The study raises the caveat that chronic ingestion of dietary sources of capsaicin may exert neurotoxic effects.

A variety of neuropeptides has been demonstrated to be present in primary afferent neurons (e.g., Kawatani, Nagel, & deGroat, 1986). However, it is virtually unknown whether they mediate different sensory modalities, and if so, which. Perhaps the best studied is substance P, a putative transmitter in the pain system (e.g., Piercey, Moon, Blinn, & Dobry-Schreur, 1986; Go & Yaksh, 1987). Substance P concentration in spinal cord superfusates was measured by radioimmunoassay. While hindpaw electrical shock released substance P into superfusate and VS failed to do so, VS significantly reduced hindpaw shock-induced release of substance P (Steinman, Banas, Hoffman, & Komisaruk, 1989). This suggests that VS presynaptically inhibits the release of substance P at the primary afferent synapse, a process that could produce analgesia.

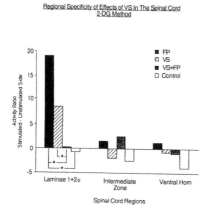

FIG. 11.6. Evidence that VS inhibits pain afferent activity in the (sensory) dorsal horn at the lower lumbar level of the spinal cord. Based on the C14-2-Deoxyglucose autoradiographic method, unilateral foot pinch (FP) significantly increased neuronal activity in the ipsilateral dorsal horn (laminae 1 and 2 [outer]) relative to the corresponding contralateral region, indicating an ipsilateral response to noxious stimulation. By contrast, VS applied concurrently with foot pinch significantly reduced the magnitude of the difference between the same ipsi- and contralateral regions. Other regions of the spinal cord were not significantly affected by these stimulus conditions.

Possible role of the primary afferent neuropeptide, Vasoactive Intestinal Peptide (VIP), in VS-produced analgesia

VIP, a peptide consisting of 28 amino acids, has been shown to exist in higher concentration in the sacral level of the human spinal cord than in any other level of the cord; by contrast, other neuropeptides showed essentially uniform distribution of all levels of the spinal cord (Anand, Gibson, McGregor, Blank, Ghatei, Bacarese-Hamilton, et al., 1983). The projection of the pelvic nerve into the sacral level of the spinal cord is similar in the rat and the cat (Kawatani et al., 1986; Pacheco et al., 1989). Based upon horseradish peroxidase staining of pelvic nerve afferent terminals in the spinal cord of the cat (Morgan, Nadelhaft, & deGroat, 1981), two groups have claimed independently that these terminals bear a "striking similarity" to the distribution of VIP (Anand et al., 1983; Basbaum & Glazer, 1983). Similarities have also been reported for the rat (Nadelhaft, 1983), rhesus monkey (Roppolo, Nadelhaft, & deGroat, 1983) and human (Anand et al., 1983; Kawatani, Lowe, Moossy, Martinez, Nadelhaft, Eskay, & deGroat, 1983). The terminal fields include the substantia gelatinosa and lamina V, sites at which pain modulation occurs (Basbaum & Fields, 1978).

Since the cell bodies of the dorsal root ganglion of the pelvic nerve synthesize VIP (Kawatani et al., 1986), we hypothesized that VS would stimulate the release of VIP from the primary afferent terminals of the pelvic nerve and thereby release VIP into superfusates of the sacral spinal cord. Consequently, using radioimmunoassay, we measured levels of VIP in superfusates of the spinal cord after VS and after foot pinch. VS significantly elevated the level of VIP in the superfusate, but foot pinch had no significant effect, indicating that the release of VIP in the spinal cord was specific to the VS (Komisaruk, Gintzler, Banas, & Blank, 1989).

This suggested that VIP may be capable of producing analgesia if administered directly to the spinal cord. In order to test this hypothesis, we administered VIP directly to the lumbosacral region of the spinal cord, intrathecally, via a chronically implanted catheter. We observed a dose-response effect, in which a significant increase in the latency of rats to flick the tail away from a radiant heat source occurred in response to 0.5 microgram or more of VIP (Komisaruk, Banas, Heller, Whipple, Barbato, & Jordan, 1988).

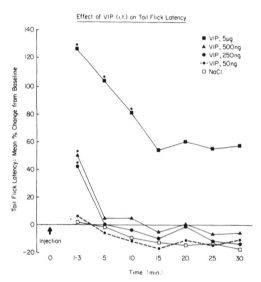

FIG. 11.7. Vasoactive intestinal peptide (VIP) administered directly to the spinal cord produces analgesia, as indicated by an increase in the tail flick latency to radiant heat.

On this test, pretreatment with naloxone (10 mg/kg, i.p.) did not affect the analgesia produced by VIP, indicating that this analgesic effect of VIP is independent of the opiate system. On a different measure, i.e., vocalization threshold to electrical tail shock, which uses a different nociceptive modality, the VIP-induced increase in threshold was significantly attenuated by naloxone. These findings indicate that VIP produces analgesia via an action on both opiate and non-opiate-modulated pain pathways.

Noting that the duration of the VIP-produced analgesia was only 10-20 minutes, we reasoned that this relatively short duration of action may due to the action of degradative enzymes on VIP. Therefore, if we first identified the site(s) of enzymatic cleavage, we could then inhibit the enzymatic action, and thereby prolong the duration of action of VIP. We had previously found that the analgesic effect of VS could be prolonged by the administration of the protease inhibitor, Leupeptin (Heller, Komisaruk, Gintzler, & Stracher, 1986). Consequently, we incubated VIP with spinal cord homogenates and then identified the enzymatic breakdown products using HPLC and subsequent amino acid analysis. Four major peptide fragments were generated by the incubation procedure, each produced by specific cleavage on the carboxy side of the tyrosine residues (Tyr 10 and Tyr 22) (Barbato, Jordan, &

Komisaruk, 1988). These four fragments were then synthesized by F. Jordan and P. Cash using solid phase methodology, and tested for analgesic effectiveness.

At dosages equimolar to 25 μg VIP 1-28, fragment 1-10 proved to be relatively ineffective in elevating pain thresholds. By contrast, fragment 11-28 produced a significant (70%) elevation in vocalization threshold by the first minute after injection. The vocalization threshold remained significantly elevated over the saline control group levels at each test for the duration of the 60 minute test period.

VIP 11-28 (dose equivalent to 25 μg VIP 1-28) was also significantly more effective in elevating vocalization thresholds than 25 μg morphine sulfate; the morphine in turn produced a significant elevation in vocalization threshold over the saline group. At 30 minutes, the elevation in vocalization threshold produced by the VIP 11-28 (170%) was significantly greater than that produced by the morphine sulfate 25 μg dose (60%), while both substances produced a significant elevation in vocalization threshold over the saline control group at this time. Thus, VIP 11-28 was significantly more effective in elevating vocalization thresholds than a seven-fold greater molar concentration of morphine.

The significant threshold elevating effect of the fragment persisted throughout the 60 minute test, at least 20 minutes longer than the analgesia produced by the parent compound (VIP 1-28).

On the tail flick latency test, VIP 11-28 (dose equivalent to 25 μg VIP 1-28) produced an elevation in TFL of 50% to 100% over the preinjection baseline levels. This was significantly greater than the change in the saline group (0-5%) at the same test periods.

In order to test the potential therapeutic value of VIP 11-28, i.e., by using a more readily applicable route than spinal administration, we administered the VIP 11-28 fragment via a systemic, intravenous, route (Komisaruk, Banas, & Jordan, unpublished observations). On the vocalization threshold test, the maximum threshold occurred at 1 minute after injection of VIP 11-28, approaching, but not attaining, significance. By contrast, the TFL Latency was significantly elevated over the saline control group at all test periods between 1 and 60 minutes postinjection (except at 15 minutes). At 1 minute after injection of VIP 11-28, the mean TFL elevation reached a maximum of 80% greater than the preinjection baseline level, after which it decreased gradually.

Thus, two breakdown products of VIP, both of which are produced by endogenous enzymatic action in the spinal cord but not the brain (Barbato et al., 1988) are differentially active in producing analgesia. One of the products, the 11-28 fragment of VIP, is more effective than VIP itself in producing analgesia, and more effective in producing analgesia than a 7-fold greater molar concentration of morphine.

We have not yet ascertained whether the VIP 11-28 fragment is normally present and functional in the intact spinal cord. If it is, this would suggest that endogenous spinal cord enzymes normally process the parent compound, VIP 1-28, into its analgesically more active 11-28 constituent. If this is in fact the case, then it would seem appropriate to consider VIP to be a pro-peptide relative to its analgesically more active 11-28 constituent.

CONCLUSION

There are three major classes of effect of vaginal stimulation: behavioral (immobilization, receptivity-lordosis, analgesia) (for review, see Komisaruk, 1978, 1982; Komisaruk & Whipple, 1988; Rodriguez-Sierra et al., 1975), neuroendocrine (luteinizing hormone, prolactin, and oxytocin release) (for review, see Komisaruk et al., 1981; Komisaruk & Steinman, 1986), and autonomic (acceleration of heart rate, increase in blood pressure, and pupillary dilatation) (Catelli et al., 1987; Szechtman et al., 1985). The present findings that analgesia is triggered by VIP, a neuropeptide in the pelvic nerve primary afferents, raised the question of whether the other effects of VS can be triggered similarly. Does one primary afferent neurotransmitter/neuromodulator elicit only certain of the effects, and/or does graded co-release of these neuropeptides produce specificity of response? This raises a more general question regarding the function of the numerous neuropeptides that in recent years have been identified in primary afferent nerves, but whose division of labor is virtually unknown. Addressing these questions is certain to be a fertile field for research in sensory processing and sensory stimulation-elicited behavioral, neuroendocrine, and autonomic responses in coming years. Furthermore, knowledge of the identity of the neurotransmitter(s)/neuromodulator(s) that trigger spinal cord-mediated analgesia in response to genital stimulation, could lead to development of novel therapeutic agents for control of pain of parturition and pain of pathological origin.

Dedicated to Knut Larsson for his presence, insights, influence and encouragement on this odyssey, that has led to explorations of which I never dreamed 20 years ago when we embarked on it.

ACKNOWLEDGMENTS

I gratefully acknowledge the critical reading of the manuscript by Dr. Beverly Whipple, the technical assistance of Ms.Cindy Banas with many of the experiments reported herein, and Ms. Esther Ronn with the preparation of this manuscript. Supported by: NIH Grant NLS 2 5 R01 NS 22948; NIH Minority Biomedical Research Support Grant GRS 5 SO6 RR08223; the Charles and Johanna Busch Foundation of Rutgers, The State University of New Jersey; and the Rutgers-CINVESTAV/University of Tlaxcala Mexican Exchange Program. Contribution number 511 from the Institute of Animal Behavior.

REFERENCES

Adler, N.T., Resko, J.A., & Goy, R.W. (1970). The effect of copulatory behavior on hormonal change in the female rat prior to implantation. *Physiol. Behav., 5,* 1003-1007.

Anand, P., Gibson, S.J., McGregor, G.P., Blank, M.A., Ghatei, M.A., Bacarese-Hamilton, A.J., Polak, J.M., & Bloom, S.R. (1983). A VIP-containing system concentrated in the lumbosacral region of the human spinal cord. *Nature, 305,* 143.

Barbato, G.F., Jordan, F., & Komisaruk, B.R. (1988). The *in vitro* proteolytic processing of vasoactive intestinal polypeptide by rat spinal cord homogenate. In S. Said and V. Mutt (Eds.), *Vasoactive Intestinal Peptide and Related Peptides, Vol. 527* (pp. 582-585). New York Academy of Science.

Basbaum, A.I., & Fields, H.L. (1978). Endogenous pain control mechanisms: review and hypothesis. *Ann. Neurol., 4,* 451-462.

Basbaum, A.I., & Glazer, E.J. (1983). Immunoreactive vasoactive intestinal polypeptide is concentrated in the sacral spinal cord: a possible marker for pelvic visceral afferent fibers. *Somatosensory Res., 1,* 69-82.

Beyer, C., Roberts, L.A., & Komisaruk, B.R. (1985) Hyperalgesia induced by altered glycinergic activity at the spinal cord. *Life Sci., 37,* 295-301.

Bodnar, R.J., & Komisaruk, B.R. (1984). Reduction in cervical probing analgesia by repeated prior exposure to cold-water swims. *Physiol. Behav., 32,* 653-655.

Brown-Grant, K., Davidson, J.M., & Greig, F. (1973). Induced ovulation in albino rats exposed to constant light. *J. Endocrinol., 57,* 7-22.

Buck, S.H., & Burks, T.F. (1986). The neuropharmacology of capsaicin: review of some recent observations. *Pharmacol. Rev., 38,* 179-226.

Catelli, J.J., Sved, A.F., & Komisaruk, B.R. (1987). Vaginocervical probing elevates blood pressure and induces analgesia by separate mechanisms. *Physiol. Behav., 41,* 609-612.

Crowley, W.R., Rodriguez-Sierra, J.F., & Komisaruk, B.R. (1977). Monoaminergic mediation of the antinociceptive effect of vaginal stimulation in rats. *Brain Res., 137,* 67-84.

Crowley, W.R., Rodriguez-Sierra, J.F., & Komisaruk, B.R. (1977a). Analgesia induced by vaginal stimulation in rats is apparently independent of a morphine-sensitive process. *Psychopharmacology, 54,* 223-225.

Crowley, W.R., Jacobs, R., Volpe, J., Rodriguez-Sierra, J.F., & Komisaruk, B.R. (1976). Analgesic effect of vaginal stimulation in rats: Modulation by graded stimulus intensity and hormones. *Physiol. Behav., 16,* 483-448.

Cunningham, S.T., Steinman, J.L., & Komisaruk, B.R. (1989). Hypogastric and pelvic neurectomy contribute differentially to vaginocervical stimulation-produced analgesia in rats. *Soc. Neurosci. Abst., 15,* 848.

Dworkin, B., Filewich, R.J., Miller, N.E., Craigmyle, N., & Pickering, T.G. (1979). Baroreceptor activation reduces reactivity to noxious stimulation: implications for hypertension. *Science, 205,* 1299-1301.

Gintzler, A.R. (1980) Endorphin-mediated increases in pain threshold during pregnancy. *Science, 210,* 193-195.

Gintzler, A.R., Peters, L.C., & Komisaruk, B.R. (1983). Attenuation of pregnancy-induced analgesia by hypogastric neurectomy in rats. *Brain Res., 227,* 186-188.

Go, V.L., & Yaksh, T.L. (1987). Release of substance P from the cat spinal cord. *J. Physiol. (London), 391,* 141-167.

Harrington, F.E., Eggert, R.R., Wilbur, R.D., & Linkenheimer, W.H. (1966). Effect of coitus on chlorpromazine inhibition of ovulation in the rat. *Endocrinology, 79,* 1130-1134.

Heller, S., Komisaruk, B.R., Gintzler, A., & Stracher, A. (1986). Leupeptin, a protease inhibitor, prolongs vaginal stimulation-produced analgesia in rats. *Ann. N.Y. Acad. Sci., 467,* 419-422.

Higuchi, T., Uchide, K., Honda, K., & Negoro, H. (1987). Pelvic neurectomy abolishes the fetus-expulsion reflex and induces dystocia in the rat. *Exp. Neurol., 96,* 443-455.

Hill, R.G., & Ayliffe, S.J. (1981). The antinociceptive effect of vaginal stimulation in the rat is reduced by naloxone. *Pharmacol. Biochem. Behav., 14,* 631-632.

Johns, M.A., Feder, H.H., & Komisaruk, B.R. (1980). Reflex ovulation in light-induced persistent estrus (LLPE) rats: role of sensory stimuli and the adrenals. *Horm. Behav., 14,* 7-19.

Johnson, B.M., Pott, C., Siegel, A., Adler, N.T., & Komisaruk, B.R. (1988). Vaginocervical stimulation suppresses noxious sensory input at the spinal cord; 2-DG autoradiographic evidence. *Soc. Neurosci. Abst., 14,* 709.

Kawatani, M., Lowe, I., Moossy, J., Martinez, J., Nadelhaft, I., Eskay, R., & deGroat, W.C. (1983). Vasoactive intestinal polypeptide (VIP) is localized to the lumbosacral segments of the human spinal cord. *Soc. Neurosci. Abst., 9,* 294.

Kawatani, M., Nagel, J., & deGroat, W.C. (1986). Identification of neuropeptides in pelvic and pudendal nerve afferent pathways to the sacral spinal cord of the cat. *J. Comp. Neurol., 249,* 117-132.

Komisaruk, B.R. (1978). The nature of the neural substrate of female sexual behaviour in mammals and its hormonal sensitivity: review and speculations. In J.B. Hutchison (Ed.), *Biological Determinants of Sexual Behaviour* (pp. 349-393). New York: John Wiley & Sons.

Komisaruk, B.R. (1982). The role of brain stem-spinal systems in genital stimulation-induced inhibition of sensory and motor responses to noxious stimulation. In B. Sjölund and A. Björklund (Eds.), *Brain Stem Control of Spinal Mechanisms* (pp. 493-508). New York: Elsevier Biomedical Press.

Komisaruk, B.R., Banas, C., Heller, S.B., Whipple, B.H., Barbato, G.F., & Jordan, F. (1988). Analgesia produced by vasoactive intestinal peptide administered directly to the spinal cord in rats. In S. Said and V. Mutt (Eds.), *Vasoactive Intestinal Peptide and Related Peptides, Vol. 527* (pp. 650-654). New York Academy of Sciences.

Komisaruk, B.R., & Diakow, C. (1973). Lordosis reflex intensity in rats in relation to the estrous cycle, ovariectomy, estrogen administration, and mating behavior. *Endocrinology, 93,* 32-41.

Komisaruk, B.R., Gintzler, A.R., Banas, C., & Blank, M.S. (1989). Vaginocervical stimulation releases vasoactive intestinal peptide-like immunoreactivity (VIP) into spinal cord superfusates in rats. *Soc. Neurosci. Abst., 15,* 216.

Komisaruk, B.R., & Larsson, K. (1971). Suppression of a spinal and a cranial nerve reflex by vaginal or rectal probing in rats. *Brain Res., 35,* 231-235.

Komisaruk, B.R., Larsson, K., & Cooper, R.L. (1972). Intense lordosis in the absence of ovarian hormones after septal ablation in rats. *Soc. Neurosci. Abst., 2,* 230.

Komisaruk, B.R., McDonald, P.G., Whitmoyer, D., & Sawyer, C.H. (1967). Effects of progesterone and sensory stimulation on EEG and neuronal activity in the rat. *Exp. Neurol., 19,* 494-507.

Komisaruk, B.R., & Olds, J. (1968). Neuronal correlates of behavior in freely moving rats. *Science, 161,* 810-813.

Komisaruk, B.R., & Steinman, J.L. (1986). Genital stimulation as a trigger for neuroendocrine and behavioral control of reproduction. In B.R. Komisaruk, H. Siegel, M-F. Cheng, and H.H. Feder (Eds.), *Reproduction: A Behavioral and Neuroendocrine Perspective* (pp. 64-75). New York Academy of Sciences.

Komisaruk, B.R., Terasawa, E., & Rodriguez-Sierra, J.F. (1981). How the brain mediates ovarian responses to environmental stimuli: neuroanatomy and neurophysiology In N.T. Adler (Ed.), *Neuroendocrinology and Reproduction* (pp. 349-376). New York: Plenum Publ. Co.

Komisaruk, B.R., & Wallman, J. (1973). Blockage of pain responses in thalamic neurons by mechanical stimulation of the vagina in rats. *Proc. Soc. Neurosci.,* 3rd Ann. Mtg., p. 315.

Komisaruk, B.R., & Wallman, J. (1977). Antinociceptive effects of vaginal stimulation in rats: neurophysiological and behavioral studies. *Brain Res., 137,* 85-107.

Komisaruk, B.R., & Whipple, B. (1984). Evidence that vaginal self-stimulation in women suppresses experimentally-induced finger pain. *Soc. Neurosci. Abst., 10,* 675.

Komisaruk, B.R., & Whipple, B. (1986). Vaginal stimulation-produced analgesia in rats and women. *Ann. N.Y. Acad. Sci., 467,* 30-39.

Komisaruk, B.R., & Whipple, B. (1988). The role of vaginal stimulation-produced analgesia in reproductive processes. In A.R. Genazzani, G. Nappi, F. Facchinetti, and E. Martignoni (Eds.), *Pain and Reproduction* (pp. 125-140). The Parthenon Publishing Group, Casterton Hall, U.K.

Martinez-Gomez, M., Whipple, B., Oliva-Zarate, L., Pacheco, P., & Komisaruk, B.R. (1988). Analgesia produced by vaginal self-stimulation in women is independent of heart rate acceleration. *Physiol. Behav., 43,* 849-850.

Masters, D.B., Beyer, C., Jordan, F., Steinman, J.L., & Komisaruk, B.R. (1988). Evidence that vaginocervical stimulation releases amino acids into superfusates of the spinal cord. *Soc. Neurosci. Abst., 14,* 349.

Masters, D.B., Beyer, C., Jordan, F., Steinman, J.L., & Komisaruk, B.R. (1989). Vaginocervical stimulation-produced analgesia and correlated spinal cord amino acid release are disrupted be neonatal capsaicin treatment. *Soc. Neurosci. Abst., 15,* 848.

Masters, D.B., Jordan, F., & Komisaruk, B.R. (1989). Regional in vivo superfusion of the spinal cord and KC1-induced amino acid release. *Pharmacol. Biochem. Behav., 34,* in press.

Melzack, R. (1984). The myth of painless childbirth. *Pain, 19,* 321-337.

Moos, F., & Richard, P. (1975). Level of oxytocin release induced by vaginal dilatation (Ferguson reflex) and vagal stimulation (vago-pituitary reflex) in lactating rats. *J. Physiol. (Paris), 70,* 307-314.

Morgan, C., Nadelfaft, I., & deGroat, W.C. (1981). The distribution of visceral primary afferents from the pelvic nerve to Lissauer's tract and the spinal gray matter and its relationship tothe sacral parasympathetic nucleus. *J. Comp. Neurol., 201,* 415-440.

Moss, R.L., & Cooper, K.J. (1973). Temporal relationship of spontaneous and coitus-induced release of luteinizing hormone in the normal cyclic rat. *Endocrinology, 92,* 1748-1753.

Nadelhaft, I. (1983). The distribution of vasoactive intestinal polypeptide (VIP) in the lumbosacral spinal cord of the rat. *Soc. Neurosci. Abst., 9,* 293.

Naggar, A.N., & Komisaruk, B.R. (1977). Facilitation of tonic immobility by stimulation of the vaginal cervix in the rat. *Physiol. Behav., 19,* 441-444.

Nagy, J.I. (1982). Capsaicin: a chemical probe for sensory neuron mechanisms. In L. Iverson, S.D. Iversen, and S.H. Snyder (Eds.), *Handbook of Psychopharmacology* (pp. 185-235). New York: Plenum Press.

Pacheco, P., Martinez-Gomez, M., Whipple, B., Beyer, C., & Komisaruk, B.R. (1989). Somato-motor components of the pelvic and pudendal nerves of the female rat. *Brain Res., 490,* 85-94.

Peters, L.C., Kristal, M.B., & Komisaruk, B.R. (1987). Sensory innervation of the external and internal genitalia of the female rat. *Brain Res., 408,* 199-204.

Piercey, M.F., Moon, M.W., Blinn, J.R., & Dobry-Schreur, P.J. (1986). Analgesic activities of spinal cord substance P antagonists implicate substance P as a neurotransmitter of pain sensation. *Brain Res., 385,* 74-85.

Ramirez, V.D., Komisaruk, B.R., Whitmoyer, D.I., & Sawyer, C.H. (1967). Effects of hormones and vaginal stimulation on the EEG and hypothalamic units in the rat. *Amer. J. Physiol., 212,* 1376-1384.

Roberts, L.A., Beyer, C., Komisaruk, B.R. (1985). Strychnine antagonizes vaginal stimulation-produced analgesia at the spinal cord. *Life Sci., 36,* 2017-2023.

Roberts, L.A., Beyer, C., Komisaruk, B.R. (1986). Nociceptive response to altered GABAergic activity in the spinal cord. *Life Sci., 39,* 1667-1674.

Rodriguez-Sierra, J.F., Crowley, W.R., & Komisaruk, B.R. (1975). Vaginal stimulation induces sexual receptivity to males, and prolonged lordosis responsiveness in rats. *J. Comp. Physiol. Psychol., 89,* 79-85.

Rodriguez-Sierra, J.F., Skofitsch, G., Komisaruk, B.R., & Jacobowitz, D.M. (1988). Abolition of vagino-cervical stimulation-induced analgesia by capsaicin administered to neonatal, but not adult rats. *Physiol. Behav., 44,* 267-272.

Roppolo, J.R., Nadelhaft, I., & deGroat, W.C. (1983). The preferential distribution of vasoactive intestinal polypeptide (VIP) in the sacral spinal cord of the rhesus monkey. *Soc. Neurosci. Abst., 9,* 293.

Ross, E., Komisaruk, B.R., & O'Donnell, D. (1979). Probing the vaginal cervix is analgesic in rats: evidence using an operant paradigm. *J. Comp. Physiol. Psychol., 93,* 330-336.

Rowlands, I.W. (1966). *Comparative Biology of Reproduction in Mammals. Proc. Zool. Soc. Lond. Symposia,* No. 15. New York: Academic Press.

Steinman, J.L., Banas, C., Hoffman, S.W., & Komisaruk, B.R. (1989). Vaginocervical stimulation (VS) reduces substance P release into spinal cord superfusates in rats. *Soc. Neurosci. Abst., 15,* 216.

Steinman, J.L., Komisaruk, B.R., Yaksh, T.L., & Tyce, G.M. (1983). Spinal cord monoamines modulate the antinociceptive effects of vaginal stimulation in rats. *Pain, 16,* 155-166.

Steinman, J.L., Roberts, L.A., & Komisaruk, B.R. (1982). Evidence that endogenous opiates contribute to the mediation of vaginal stimulation-produced anti-nociception in rats. *Soc. Neurosci. Abst., 8,* 265.

Szechtman, H., Adler, N.T., & Komisaruk, B.R. (1985). Mating induces pupillary dilatation in female rats: role of pelvic nerve. *Physiol. Behav., 35,* 295-301.

Takahashi, M., Ford, J.J., Yoshinaga, K., & Greep, R.O. (1975). Effects of cervical stimulation and anti-LH releasing hormone serum on LH releasing hormone content in the hypothalamus. *Endocrinology, 96,* 453-457.

Terkel, J. (1986). Neuroendocrinology of coitally and noncoitally induced pseudopregnancy. In B.R. Komisaruk, H.I. Siegel, M-F. Cheng, and H.H. Feder (Eds.), *Reproduction: A Behavioral and Neuroendocrine Perspective* (pp. 76-94). New York: The New York Academy of Sciences.

Toniolo, M.V., Whipple, B.H., & Komisaruk, B.R. (1987). Spontaneous maternal analgesia during birth in rats. *Proc. NIH Centennial MBRS-MARC Symp.,* p. 100.

Watkins, L.R., Faris, P.L., Komisaruk, B.R., & Mayer, D.J. (1984). Dorsolateral funiculus and intraspinal pathways mediate vaginal stimulation-induced suppression of nociceptive responding in rats. *Brain Res., 294,* 59-65.

Whipple, B., & Komisaruk, B.R. (1985). Elevation of pain threshold by vaginal stimulation in women. *Pain, 21,* 357-367.

Whipple, B., & Komisaruk, B.R. (1988). Analgesia produced in women by genital self-stimulation. *J. Sex Research, 24,* 130-140.

Whipple, B., Komisaruk, B.R., & Josimovich, J.B. Sensory thresholds during the antepartum, intrapartum, and postpartum periods, in press.

Whipple, B., Martinez-Gomez, M., Oliva-Zarate, L., Pacheco, P., & Komisaruk, B.R. (1989). Inverse relationship between intensity of vaginal self-stimulation-produced analgesia and level of chronic intake of a dietary source of capsaicin. *Physiol. Behav., 46,* 247-252.

Zamir, N., & Segal, M. (1979). Hypertension-induced analgesia: changes in pain sensitivity in experimental hypertensive rats. *Brain Res., 184,* 299-310.

12 Sexual Differentiation of a Neuropeptide Circuit Regulating Female Reproductive Behavior

Paul E. Micevych
University of California, Los Angeles

Gonadal steroid hormones influence reproductive events by modulating cellular and molecular events associated with intercellular communication in specific neuronal circuits in the central nervous system (CNS). Since gonadal steroids act through receptor molecules to affect transcription of specific genes, a signature of steroid action is the accumulation of estrogen, progesterone, testosterone, or dihydrotestosterone in cells thought to be steroid sensitive. Autoradiographic techniques have identified heavy concentrations of steroid accumulating cells in the posterior part of the medial amygdaloid nucleus (MeApd), encapsulated part of the bed nucleus of the stria terminalis (BSTe), medial preoptic area (MPO), and ventromedial nucleus of the hypothalamus (VMH). All of these regions are sexually dimorphic and have been implicated in the CNS regulation of sexual behavior. These interconnected nuclei will be referred to as the limbic-hypothalamic circuit.

Steroid effects on cholecystokinin

Immunologic techniques have demonstrated the neuroactive peptide cholecystokinin (CCK) in this hypothalamic-limbic system (Bloch, Gorski, Micevych, & Akesson, 1989; Frankfurt, Siegel, Sim, & Wuttke, 1986; Micevych, Park, Akesson, & Elde, 1987; Micevych, Akesson, & Elde, 1988; Simerly & Swanson, 1987). Estrogen regulates the levels of CCK-immunoreactivity (CCKi) in this circuit. Tissue levels vary during the estrous cycle (Frankfurt et al., 1986; Micevych et al., 1987; Micevych, Matt, & Go, 1988). Castration dramatically decreases the number of CCKi cells in every cell group in the circuit except the VMH where there are no CCKi cells. Testosterone restores the number of CCKi cells to that which is observed in the intact male (Oro, Simerly, & Swanson, 1988; Micevych et al., 1988) as does long-term treatment with estrogen (Micevych & Bloch, 1989). These results imply that estrogen and not dihydrotestosterone, the androgenic metabolite of testosterone, regulates the expression of CCK. Steroid accumulation in CCKi cells has been difficult to demonstrate (Akesson & Micevych, 1988a) which may indicate that estrogen may indirectly regulate CCK expression either by acting through other neurotransmitters or by altering the CCK receptor population. Indeed, we have demonstrated that several neuroactive peptide-containing cells accumulate estrogen, including methionine enkephalin (Akesson & Micevych, submitted) and substance P (Akesson & Micevych, 1988b) which may interact with the CCK circuit (Micevych, Yaksh, & Go, 1984) and galanin (Bloch, Dornan, Babcock, Gorski, & Micevych, in press). Estrogen has also been shown to modulate [125]ICCK binding sites in the VMH (Akesson, Mantyh, Mantyh, Matt, & Micevych, 1987). Specifically, these binding sites were observed to vary during the estrous cycle, implying a physiologic regulation of CCK receptors which may relate to the effects of sCCK-8 on lordosis.

Cholecystokinin regulation of reproduction

The sexually dimorphic, steroid-regulated distribution of CCK has been implicated in the regulation of estrogen-inducible reproductive behavior. Microinjections of sCCK-8 into the VMH inhibit estrogen-induced lordosis behavior in ovariectomized rats (Babcock, Bloch, & Micevych, 1988), while microinjections of sCCK-8 into the MPO of estrogen primed, ovariectomized females facilitate lordosis behavior (Dornan, Bloch, Priest, & Micevych, 1989). Both of these CCK effects on lordosis behavior can be produced by peripheral injections of CCK and are highly dependent on the females' level of receptivity as induced by estrogen. Sulphated cholecystokinin octapeptide facilitates lordosis when receptivity is low and inhibits it when receptivity is high (Bloch, Babcock, Gorski, & Micevych, 1987) through actions on the MPO and VMH, respectively.

Although sCCK-8 does not affect male copulatory behavior, sCCK-8 significantly increases lordosis behavior in castrated, estrogen-primed males (Bloch, Babcock, Gorski, & Micevych, 1988). The sCCK-8-facilitation of lordosis is mediated by the MPO in male (Bloch et al., 1989) as it is in female rats. In castrated, estrogen-primed, male rats in which sCCK-8 facilitates estrogen-induced lordosis behavior, the distribution of CCKi cell bodies in the MPO resembles the female pattern (Micevych & Bloch, 1989). These results indicate that estrogen-induced CCK expression in the castrated male is less sensitive than is the activation of lordosis. In the intact male, effects of testosterone on the CCK circuit appear to be dichotomous. Testosterone can maintain or restore the distribution of CCKi cell bodies in the hypothalamic-limbic system through the action of its aromatized metabolite estrogen. It may also interfere with, or block the action of, CCK on lordosis by acting through its 5-alpha-reduced metabolite, dihydrotestosterone. Our results in the castrated, estrogen-primed, male rat (Bloch et al., 1988; Bloch et al., 1989) imply that neural circuits necessary for lordosis behavior exist but may be inhibited by an androgen-mediated circuit in the intact male rat.

Development

Gonadal steroids act on the CNS during a perinatal critical period to sexually differentiate the structure and function of the hypothalamo-limbic system. Exposure to testosterone during the first week after birth decreases (Barraclough & Gorski, 1962; Feder & Whalen, 1964), while castration of males increases (Gerall, Hendricks, Johnson, & Bounds, 1967; Grady, Phoenix, & Young, 1965), the display of lordosis behavior in adulthood. Thus, exposing rats to androgen during the perinatal period defeminizes female sexual behavior and gonadotropin release. Because CCK is an important modulator of female reproductive behavior, we were interested in determining the ontogeny of the sexually differentiated limbic-hypothalamic CCK circuit.

Results of studies on the perinatal development of two components of the limbic-hypothalamic CCK circuit, the central part of the medial preoptic nucleus (MPNc) and the VMH, are presented. These areas were selected because each is of critical importance for the display of reproductive behaviors (for review, see Pfaff & Schwartz-Giblin, 1988) and because each has characteristics that can be studied to help elucidate the sexual differentiation of the CCK circuit. The MPNc is highly sexually dimorphic and the distribution of CCKi cell bodies parallels this dimorphism in adulthood. A great deal is known about the development of this region which will aid our interpretation of the ontogeny of the expression of prepro-CCK (pCCK) mRNA. The VMH does not have CCKi cell bodies; however, there is a substantial CCKi input to, and an abundance of steroid sensitive CCK-binding sites within the nucleus. In addition, the behavioral response to exogenous sCCK-8 appears to be sexually differentiated at the level of the VMH.

Ontogeny of pCCK mRNA expression

The expression of pCCK mRNA can be detected in the brain as early as embryonic day 14, but the vast majority of message and immunoreactive peptide appears postnatally (Cho, Shiotani, Shiosaka, Inagaki, Kubota, Kiyama, et al., 1983; Duchemin, Quach, Iadrola, Deschenes, Schwartz, & Wyatt, 1987). Since the sexually dimorphic parts of the MPO are organized by gonadal steroids during the first few days after birth (Döhler, Coquelin, Davis, Hines, Shryne, & Gorski, 1982; Döhler, Coquelin, Davis, Hines, Shryne, Sickmöller, et al., 1986; Jacobson, Csernus, Shryne, & Gorski, 1981), the ontogeny of pCCK mRNA was studied during this postnatal period.

Methods

Aldehyde fixed brains from rats on postnatal day 1 (PND 1 = day of birth), PND 5, 10, 15, 30, and 65 (adult) were sectioned in a cryostate at 12 mm and processed for *in situ* hybridization as described by Angerer, Stoler, and Angerer (1987). Briefly, T7 and SP6 RNA polymerases were used to synthesize asymmetric RNA probes complementary to rat pCCK mRNA from a full length 527 base-pair insert (supplied by Dr. J. E. Dixon, Purdue University). The cRNA was labeled by transcribing with ^{35}S-UTP (NEN,1320Ci/mmole). For *in situ* hybridization, sections were incubated with 4ng of the probe. To visualize labeled cells, the sections were dipped into photographic emulsion (NTB-2; Kodak), and photo-developed.

Results and Discussion

In both males and females on PND 1, there was a widespread distribution of pCCK mRNA. Cell bodies expressing pCCK mRNA were visualized in the neocortex, lateral amygdaloid complex, dentate gyrus and CA 1 - CA 3, endopiriform and piriform cortex, ventral and lateral thalamic nuclei, supraoptic, paraventricular and dorsomedial nuclei of the hypothalamus and the ventral mesencephalon. No labeling was detected in the medial preoptic area, periventricular preoptic nucleus, bed nucleus of the stria terminalis or the medial amygdaloid nucleus. The nuclei of the limbic-hypothalamic circuit all develop in parallel, thus the pattern of development of the MPNc is representative of the entire circuit (Fig. 1). By PND 5, scattered cells expressing pCCK mRNA were present in the nuclei of the limbic-hypothalamic circuit, but, as is illustrated in Fig. 2 for the MPNc, there was no discernible sex difference. At PND 10, there was already a significant ($p < 0.05$; Student's t-test) difference between the number of cells expressing pCCK mRNA in males compared to females. This sex difference continued to develop as the number of pCCK mRNA expressing cells increased during development. The sex difference was most pronounced in adulthood (Fig. 2).

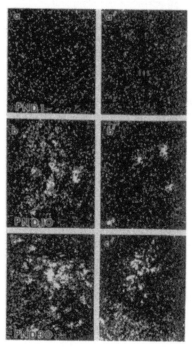

Fig. 12.1. Darkfield photomicrograph of the medial preoptic nucleus, central part showing the ontogeny of pCCK mRNA at PND 1, 10, and 30 in the male (a, b, c) and female (a´, b´, c´). III indicates the third ventricle in a´. Original magnification 82X.

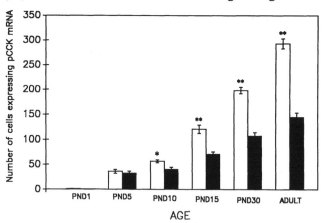

Fig. 12.2. Histogram illustrating the increase in the number of pCCK mRNA expressing cells in the central part of the medial preoptic nucleus (MPNc). At postnatal day (PND) 10, there is a significant difference ($p < 0.05$) between the number of hybridized cells in males (open bars) compared to females (solid bars). This difference increases at PND 15 and remains into adulthood ($p < 0.001$).

The ontogeny of the CCK system parallels the postnatal development of the MPNc, especially during the first 10 days postnatally. The number of cells and the volume are very similar in males and females (Dodson, Shryne, & Gorski, 1988). The male nucleus continues to grow in volume while the growth of the female nucleus is retarded. This divergence is observed at PND 5 first, and reaches statistical significance by PND 7. In the female MPNc, the number of cells decreases from PND 4 to PND 10 and remains constant until PND 30. The number of cells expressing pCCK mRNA continues to increase to adulthood in male and female rats. This result and the observation that estrogen increased the number of CCKi cells in the ovariectomized female to male levels (Oro et al., 1988) imply that cells which have the potential to express CCK may not be susceptible to programmed death in the absence of testosterone. It is, therefore, not the number of CCK cells that determines the male-ness or female-ness of the limbic-hypothalamic circuit, but the perinatal exposure to gonadal steroids that may establish the sensitivity of a CNS network regulating sexual behavior, which includes the limbic-hypothalamic CCK circuit, as is described below. This sensitivity along with tonic or cyclic gonadal hormone levels determines the reproductive network-response to steroids in adulthood. Indeed, estrogen stimulates the expression of CCKi in both males and females, although males have a constant, high level of endogenous steroids, as compared to females, and a high threshold to estrogen response (Micevych & Bloch, 1989). Additionally, androgens probably induce the formation of a specific male copulatory system and continue to stimulate these circuits during adult life. The existence of such an androgen-stimulated circuit is suggested by the distribution of androgen binding in the MPN.

In the adult, there is a specific distribution of estrogen and androgen accumulating cells within the medial preoptic nucleus (MPN; Micevych & Akesson, 1987). The most sexually dimorphic parts of the MPN, the central and anteroventral cell groups, preferentially accumulate testosterone and dihydrotestosterone (Fig. 3). These results suggest that estrogen may stimulate circuits common to both sexes, while androgen may activate a specific male circuit in the MPN that stimulates circuits controlling male copulatory behavior.

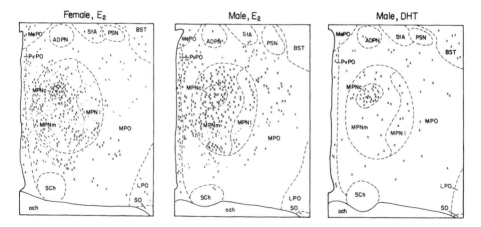

Fig. 12.3. Camera lucida drawing of steroid autoradiograms at the level of the central part of the medial preoptic nucleus in male and female rats illustrating the distribution of estradiol- (E_2) and dihydrotestosterone- (DHT) accumulating cells. Abbreviations: ADPN = anterodorsal preoptic nucleus, BST = bed nucleus of the stria terminalis, LPO = lateral preoptic area, MePN = median preoptic nucleus, MPN = medial preoptic nucleus, MPNc = central part of MPN, MPNl = lateral part of MPN, MPNm = medial part of the MPN, MPO = medial preoptic area, och = optic chiasm, PSN = parastrial nucleus, PvPO = periventricular preoptic nucleus, Sch = suprachiasmatic nucleus, SO = supraoptic nucleus, StA = strial part of preoptic area.

Sexual differentiation of adult response to cholecystokinin

The VMH is an essential part of the CNS circuitry that regulates lordosis behavior. Although estrogen can elicit lordosis behavior from both females and and castrated males after implants into the VMH (Barfield & Chen, 1977; Davis & Barfield, 1979), the postnatal androgen milieu reduces the sensitivity of neurons in the VMH to estrogen and attenuates the CCK-induced inhibition of lordosis. Thus, in adulthood, sCCK-8 inhibits the frequency of lordosis in ovariectomized, estrogen-treated female rats but not in castrated male rats primed with estrogen. The response to sCCK-8 in adulthood can be used to study the sexual differentiation of a significant component of the CNS circuitry regulating the display of lordosis behavior.

Methods

Male pups were castrated or sham operated on PND 1, and all females were injected with 100 mg or 0 mg testosterone propionate on PND 5. All animals were allowed to survive until adulthood, then implanted with injection cannulae aimed at the VMH and castrated as necessary. Behavior was elicited by a subcutaneous injection of 2 mg estradiol benzoate (EB), in subsequent experiments with 5 mg EB and finally with 2 mg EB and 500 mg progesterone (Ulibarri & Micevych, submitted). Ten minutes before the behavioral tests, sCCK-8 (0 ng, 5 ng, 50 ng, 100 ng) was microinjected into the VMH in a 0.3 ml volume of aCSF (Babcock et al., 1988).

Results and Discussion

Neonatal castration of male rats prevented the defeminization of female sexual behavior both in terms of these animals' ability to respond to sCCK-8 (Fig. 4) and the potentiation of their lordosis behavior with EB and progesterone (LQ = 88.9 ± 11.1) as compared to control males (LQ 0 12.73 ± 8.9) and control females (LQ 0 97.1 ± 2.9). Testosterone treatment of female rats on PND 5 produced a slightly defeminized animal that exhibited lordosis and a sCCK-8-induced inhibition of lordosis behavior only after the higher dose of EB (Fig. 4) but does not respond to progesterone (LQ 0 45.0 ± 11.9). Sexual differentiation of the response to sCCK-8 resembles the sexual differentiation of female sexual behavior, implying that the development of the CCK circuit is a fundamental component of the network that regulates female reproductive behavior by inhibiting its expression. Although males can be induced to express lordosis behavior, our results suggested that the male VMH does not respond to sCCK-8 (Bloch et al., 1989). There is a critical period after which the male is relatively insensitive to estrogen induction of lordosis (Whalen, Luttge, & Gorzalka, 1971). One line of evidence is that although estrogen will stimulate the expression of CCK in a male, 5 mg EB for three days does not increase the number of CCKi cells in the limbic-hypothalamic circuit (Micevych & Bloch, 1989).

The site of the estrogen-induction of the CCK response in the VMH is probably through modulation of the CCK receptor population in the VMH. Both males and females have CCK binding sites in the VMH. In females, these binding sites are sensitive to estrogen (Akesson, Simerly, & Micevych, 1987; 1988). The time-course of the development of CCK binding sites follows the initial appearance of scattered CCKi fibers in the VMH at PND 1 (Yamano, Inagaki, Tateishi, Hamaoka, & Tohyama, 1984) with the first detectable ^{125}I-sCCK-8 binding at PND 5 (Fig. 5, Micevych, Ulibarri, & Popper, 1989). Interestingly, the CCKi innervation of the VMH does not reach the adult pattern until PND 30 while the intensity and distribution of ^{125}I-sCCK-8 binding sites appear fully developed at PND 10 (Pelaprat, Dusart, & Peschanski, 1988). In females, the level of binding is regulated by estrogen (Akesson et al., 1987). Estradiol dramatically decreases the CCK binding 24 hrs after administration. Significantly, in females, behavioral testing for lordosis is done 48 hrs after estradiol priming and there is a CCK-induced inhibition. When females are tested for lordosis 24 hrs after estradiol administration, sCCK-8 does not attenuate lordosis (Ulibarri &

Micevych, submitted). Our lordosis tests in males were done 24 hrs after the last injection of estradiol (Bloch et al., in press). We hypothesize that the CCK binding sites are down-regulated and because of this, inhibition does not result. In the intact male, the tonic levels of testosterone maintain the CCK system in a maximally active state, resulting in a low incidence of lordosis behavior in adult males. We speculate that castrated, estrogen-treated male rats will respond to exogenous sCCK-8 in the VMH like females do if the estrogen dose is sufficiently high and if they are tested 48 hours after the last estrogen injection. In the female, the estrogen surge on the morning of proestrus down-regulates CCK receptors in the VMH. This down-regulation of lordosis-inhibiting CCK receptors, along with estrogen-activation of the lordosis-facilitatory elements of the circuit, result in the display of lordosis. As the CCK receptor population of the VMH returns to pre-estrogen surge levels, CCK release which has been stimulated by estrogen (Micevych et al., 1988) binds to its receptors and terminates the behavior. Receptors remain high until the next proestrous surge of estrogen.

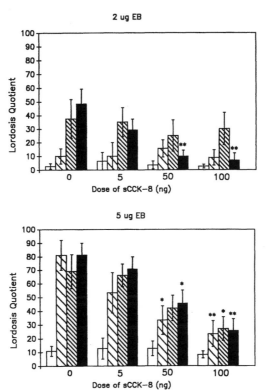

Fig. 12.4. A. Receptivity scores after priming with 2 mg EB 48 hrs before testing and sCCK-8 10 min before testing. Control male (open bars), male gonadectomized on PND 1 (large stripes), TP-treated females (small stripes), control females (solid bars). Double asterisks indicate p < 0.001 difference from mean LQ after infusions of 10 ng sCCK-8 using Dunnett's method for comparison to a control mean. Doses of sCCK-8 administered in 0.3 ml artificial CSF. Only control females show inhibition of behavior after sCCK infusions. B. Receptivity scores after priming with 5 mg EB 48 hrs before testing and sCCK-8 10 min before testing. Control male (open bars), male gonadectomized on PND 1 (large stripes), TP-treated females (small stripes), control females (solid bars). Single asterisks indicate p < 0.01 difference from mean LQ after infusions of 0 ng sCCK-8 using Dunnett's method of comparison. Neonatally castrated males and androgenized females are equally sensitive to sCCK-8 infusions, while TP-treated females only respond to the highest dose of sCCK-8.

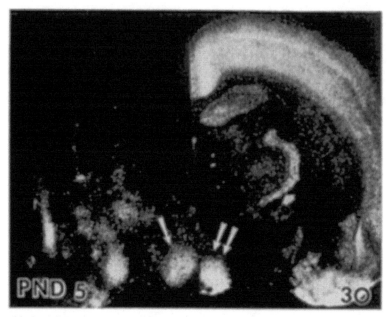

Fig. 12.5. Autoradiograph of ^{125}I-sCCK-8 binding sites in the rat brain at the level of the ventromedial nucleus of the hypothalamus (VMH). This photo-montage illustrates the binding in the VMH of a postnatal day (PND) 5 rat on the left (single arrow) and the ^{125}I-sCCK-8 in PND 30 on the right (double arrows). The development of binding in neocortex, piriform cortex and reticular thalamic nucleus (arrowhead on the right) lag behind binding in the VMH.

CONCLUSION

The limbic-hypothalamic CCK is postnatally organized to modulate lordosis behavior. Its primary function in the VMH appears to be the termination of this behavior in females, and the static inhibition of lordosis in males. The action of estrogen in both females and males (aromatized from testosterone) on the CCK circuit is activation. In males, postnatal androgens drive the level of expression pCCK mRNA. In the female, the levels of estrogens is low and the ontogeny of expression lags behind the male (Fig. 2). In adulthood, females appear to be able to increase the expression of pCCK mRNA to adult male levels. There is no sex difference in the postnatal development of CCK receptors, but they are modulated by the adult gonadal steroid environment. The VMH receptors are down-modulated in the female by the proestrous estrogen surge to allow lordosis behavior to occur. In the male, they remain at a high level, thus inhibiting lordosis. The male or female response of this important circuit is determined by the postnatal gonadal steroid environment which sets the sensitivity to estrogen and organizes the network regulating sexual behavior. Although the final outcome of CCK action on lordosis in females and males appears to be sexually differentiated, it is the network in which the CCK circuit acts, and not the CCK circuit itself, that is differentiated.

ACKNOWLEDGEMENTS

The author would like to thank all the colleagues involved with the studies reviewed above. In particular, the author thanks Dr. C. Ulibarri, and Paul Popper who had helpful suggestions, L. Abelson for her excellent technical assistance, and C. Priest for her help in the preparation of this manuscript. This research was supported by grant NS-21220.

REFERENCES

Akesson, T.R., Mantyh, P., Mantyh, C., Matt, D., & Micevych, P.E. (1987). Estrous cyclicity of ^{125}I-sCCK-8 octapeptide binding in the hypothalamic ventromedial nucleus: Evidence for down modulation by estrogen. *Neuroendocrinology, 45,* 254-262.

Akesson, T.R., & Micevych, P.E. (1988a). Cholecystokinin immunoreactive neurons do not concentrate estrogen. *J. Neurobiol., 19,* 3-16.

Akesson, T.R., & Micevych, P.E. (1988b). Substance P neurons of the ventromedial and arcuate hypothalamic nuclei concentrate estradiol. *J. Neurosci. Res., 19,* 412-419.

Akesson, T.R., & Micevych, P.E. Enkephalin-immunoreactive neurons of the hypothalamic ventromedial nucleus concentrate estrogen in male and female rats. Submitted

Akesson, T.R., Simerly, R.B., & Micevych, P.E. (1987). Estradiol concentration by hypothalamic and limbic neurons which project to the medial preoptic nucleus. *Soc. Neurosci. Abst., 13,* 1164.

Akesson, T.R., Simerly, R.B., & Micevych, P.E. (1988). Estrogen-concentrating hypothalamic and limbic neurons project to the medial preoptic nucleus. *Brain Res., 451,* 381-385.

Angerer, L.M., Stoler, M.H., & Angerer, R.C. (1987). In situ hybridization with RNA probes: An annotated recipe. In K.L. Valentine, J.H. Eberwine, & J.D. Barchez (Eds.), *In situ Hybridization: Applications to Neurobiology* (pp. 42-70). New York: Oxford Univ. Press.

Babcock, A.M., Bloch, G.J., & Micevych, P.E. (1988). Injections of CCK into the ventromedial hypothalamic nucleus inhibit lordosis behavior in rat. *Physiol. Behav., 43,* 195-199.

Barfield, R.J., & Chen, J.J. (1977). Activation of estrous behavior in ovariectomized rats by intracerebral implants of estradiol benzoate. *Endocrinology, 101,* 1716-1725.

Barraclough, C.A., & Gorski, R.A. (1962) Studies on mating behavior in the androgen-sterilized rat and their relations to the hypothalamic regulation of sexual behavior in the female rat. *J. Endocrinol., 25,* 175-182.

Bloch, G.J., Babcock, A.M., Gorski, R.A., & Micevych, P.E. (1987). Cholecystokinin facilitates and inhibits lordosis behavior. *Physiol. Behav., 39,* 217-224.

Bloch, G.J., Babcock, A.M., Gorski, R.A., & Micevych, P.E. (1988). Cholecystokinin facilitates lordosis behavior but does not alter male copulatory behavior in rat. *Physiol. Behav., 43,* 351-357.

Bloch, G.J., Dornan, W.A., Babcock, A.M., Gorski, R.A., & Micevych, P.E. (In press). Site specific microinjections of CCK into MPO facilitates lordosis in the male rat. *Physiol. Behav.*

Bloch, G.J., Gorski, R.A., Micevych, P.E., & Akesson, T.R. (1989). Estrogen-concentrating cells within the medial preoptic area: Sex differences and colocalization with galanin-immunoreactive (GAL-I) cells. *Soc. Neurosci. Abst., 15,* 577.

Cho, H.J., Shiotani, Y., Shiosaka, S., Inagaki, S., Kubota, Y., Kiyama, H., Umegaki, K., Tateishi, K., Hashimura, E., Hamaoka, T., & Tohyama, M. (1983). Ontogeny of cholecystokinin-8-containing neuron system of the rat; An immunohistochemical analysis. I. Forebrain and upper brainstem. *J. Comp. Neurol., 218,* 25-41.

Davis, P.G., & Barfield, R.J. (1979). Activation of feminine sexual behavior in castrated male rats by intrahypothalamic implants of estradiol benzoate. *Neuroendocrinology, 28,* 228-233.

Dodson, R.E., Shryne, J.E., & Gorski, R.A. (1988). Hormonal modification of the number of total and late-arising neurons in the central part of the medial preoptic nucleus of the rat. *J. Comp. Neurol., 275,* 623-629.

Döhler, K.D., Coquelin, A., Davis, F., Hines, M., Shryne, J.E., & Gorski, R.A. (1982). Differentiation of the sexually dimorphic nucleus of the preoptic area of the rat brain is determined by the perinatal hormone environment. *Neurosci. Lett., 33,* 295-298.

Döhler, K.D., Coquelin, A., Davis, E., Hines, M., Shryne, J.E., Sickmöller, P.M., Jarzab, B., & Gorski, R.A. (1986). Pre- & post-natal influences of an estrogen antagonist and an androgen antagonist on differentiation of the sexually dimorphic nucleus of the preoptic area in male and female rats. *Neuroendocrinology, 42,* 443-448.

Dornan, W.A., Bloch, G.J., Priest. C.A., & Micevych, P.E. (1989). Microinjection of cholecystokinin into the medial preoptic nucleus facilitates lordosis behavior in the female rat. *Physiol. Behav., 45,* 969-976.

Duchemin, A.M., Quach, T.T., Iadarola, M.J., Deschenes, R.J., Schwartz, J.P., & Wyatt, R.J. (1987). Expression of the cholecystokinin gene in rat brain during development. *Dev. Neurosci., 9,* 61-67.

Feder, H.H., & Whalen, R.E. (1964). Feminine behavior in neonatally castrated and estrogen-treated male rats. *Science, 142,* 306-307.

Frankfurt, M., Siegel, R.A., Sim, I., & Wuttke, W. (1986). Estrous cycle variations in cholecystokinin and substance P concentrations in discrete areas of the rat brain. *Neuroendocrinology, 42,* 226-231.

Gerall, A.A., Hendricks, S.E., Johnson, L.L., & Bounds, T.W. (1967). Effects of early castration in male rats on adult sexual behavior. *J. Comp. & Physiol. Psychol., 64, (2),* 206-212.

Grady, K.L.G., Phoenix, C.H., & Young, W.C. (1965). Role of the developing rat testis in differentiation of the neuronal tissues mediating mating behavior. *J. Comp. Physiol. Psychol., 59, (2),* 176-182.

Jacobson, C.D., Csernus, V.J., Shryne, J.E., & Gorski, R.A. (1981). The influence of gonadectomy, androgen exposure or gonadal grafts in the neonatal rat on the volume of the sexually dimorphic nucleus of the preoptic area. *J. Neurosci., 1,* 1142-1147.

Micevych, P.E., & Akesson, T.R. (1987). Differential distribution of estrogen and androgen concentrating cells in the rat medial preoptic area. *Soc. Neurosci. Abstr., 13,* 1164.

Micevych, P.E., Akesson, T.A., & Elde, R.P. (1988). Distribution of cholecystokinin-immunoreactive cell bodies in the male and female rat: II. Bed nucleus and amygdaloid complex. *J. Comp. Neurol., 269,* 381-391.

Micevych, P.E., Park, S.S., Akesson, T.R., & Elde, R.P. (1987). The distribution of cholecystokinin immunoreactive cell bodies in the male and female rat: I. Hypothalamus. *J. Comp. Neurol., 255,* 124-136.

Micevych, P.E., & Bloch, G.J. (1989). Estrogen regulation of a reproductively relevant cholecystokinin circuit in the hypothalamus and limbic system of the rat. In J. Hughes, D. Dockray, & G. Woodruff (Eds.), *The Neuropeptide Cholecystokinin* (pp. 68-73). Chichester: Ellis Horwood Limited.

Micevych, P.E., Matt, D.W., & Go, V.L.W. (1988). Brain distribution of CCK, sP and bombesin in male and female rats during the estrous cycle. *Exp. Neurol., 100,* 416-425.

Micevych, P.E., Ulibarri, C., & Popper, P. (1989). Sexual differentiation of adult lordotic response to cholecystokinin in rats. *Soc. Neurosci. Abst., 15,* 577.

Micevych, P.E., Yaksh, T.L., & Go, V.L.W. (1984). Dose dependent inhibition K+ stimulated CCK and sP from cat hypothalamus, in vitro. *Brain Res., 290,* 87-94.

Oro, A.E., Simerly, R.B., & Swanson, L.W. (1988). Estrous cycle variations in levels of cholecystokinin immunoreactivity within cells of three interconnected sexually dimorphic forebrain nuclei: Evidence for a regulatory role for estrogen *Neuroendocrinology, 47,* 225-235.

Pelaprat, D., Dusart, I., & Peschanski, M. (1988). Postnatal development of cholecystokinin (CCK) binding sites in the rat forebrain and midbrain: An autoradiographic study. *Dev. Brain Res., 44,* 119-132.

Pfaff, D.W., & Schwartz-Giblin, S. (1988). Cellular mechanisms of female reproductive behaviors. In E. Knobil, J.D. Neill, L.L. Ewing, G.S. Greenwald, C.L. Markert, & D.W. Pfaff (Eds.), *The Physiology of Reproduction, Vol. II* (pp. 1487-1568). New York: Raven Press.

Ulibarri, C., & Micevych, P.E. (Submitted). Cholecystokinin inhibits lordosis in gonadectomized female and male rats after estrogen priming. *Brain Res.*

Varro, A., Bulock, A.J., William, R.G., & Dockray, G.J. (1983). Regional differences in the development of cholecystokinin-like immunoreactivity in rat brain. *Dev. Brain Res., 9,* 347-352.

Whalen, R.E., Luttge, W.G., & Gorzalka, B.B. (1971). Neonatal androgenization and the development of estrogen responsivity in male and female rats. *Horm. Behav., 2,* 83-90.

Yamano, M., Inagaki, S., Tateishi, N., Hamaoka, T., & Tohyama, M. (1984). Ontogeny of neuropeptides in the nucleus ventromedialis hypothalami of the rat: An immunohistochemical analysis. *Dev. Brain Res., 16,* 253-262.

13 Functional Implications of Progesterone Metabolism: Effects on Psychosexual Development, Brain Sexual Differentiation, and Perception

Carlos Beyer and Gabriela González-Mariscal
CINVESTAV-Universidad Autónoma de Tlaxcala
Tlaxcala

METABOLISM OF PROGESTERONE

Progesterone (P) is metabolized in the Central Nervous System (CNS) to a variety of pregnanes (Karavolas, Bertics, Hodges, & Rudie, 1984) that possess different properties than those of the parent hormone. All metabolic changes occurring to P, i.e., 20α- or 20β-reduction, etc., result in a loss of progestational potency due to a decrease in the affinity of these metabolites for an intracellular receptor (progestin receptor, PR; Kontula, Jänne, Vihko, de Jager, de Visser, & Zeelen, 1975; Smith, Smith, Toft, Nergaard, Burrows, & O'Malley, 1974). However, new properties emerge in these P metabolites. Thus, ring A-reduced progestins acquire to a greater extent than delta-4-3-keto pregnanes the property of interacting with the membrane and, therefore, of modifying neuronal excitability (see below).

Structure-activity relationships

From the analysis of a large series of progestins, Duax, Cody, Griffin, Rohrer, & Weeks (1978) have shown that the delta-4-3-keto structure is essential for high affinity binding to the PR. Furthermore, this affinity can be enhanced by several chemical modifications, e.g., elimination of the C19 methyl (Raynaud, Ojasoo, Pottier, & Salmon, 1982). Progestin interaction with the PR triggers in target cells a genomic mechanism that results in protein synthesis (Kato, 1985; Parsons & Pfaff, 1985). Progestational actions in the brain may not involve growth or secretory responses as in peripheral tissues but, rather, the synthesis of a discrete number of specific proteins involved in neuronal excitability (channels, enzymes related to neurotransmitter synthesis or action, etc.). Reduction of C5, at either 5α- or 5β-position, results in a loss of the delta-4-3-keto structure and, consequently, in a marked reduction in the binding affinity of these 5α- or 5β-pregnanediones for PR (Raynaud et al., 1982; Smith et al., 1974). On the other hand, this structural change allows pregnanes to stabilize membrane (Seeman, 1972) and decrease neuronal excitability (Gyermek, 1975; Kubli, Cervantes, & Beyer, 1976; P'an & Laubach, 1964; Selye, 1941). Further reduction of 5α- or 5β- pregnanediones at C3 by a 3α-hydroxysteroid oxidoreductase yields 3-hydroxypregnanes (3α, 5α- or 3α, 5β-pregnanolone) capable of interacting with the GABA-A receptor at nanomolar concentrations (Harrison, Majewska, Meyers, & Barker, 1989; Turner & Simmonds, 1989).

This effect enhances the inhibitory action of GABA on CNS neurons (Smith, Waterhouse, & Woodward, 1987). By contrast, 3β-progestins lack or have only a weak effect on the GABA-A receptor (Harrison et al., 1989; López-Colomé, McCarthy, & Beyer, 1990; Majewska, Mienville, Vicini, 1988; Turner & Simmonds, 1989). Nonetheless, 3β-reduced progestins alter membrane excitability by as yet unidentified mechanisms. Thus, 3β-pregnanolones are the most potent progestins for inhibiting uterine motility *in vitro* (Kubli et al., 1979) and for releasing LHRH from hypothalamic explants (Park & Ramírez, 1987).

Behavioral responses to progestins

Progestins stimulate or inhibit a number of stereotyped behavioral patterns related to reproductive processes, i.e., estrous behavior (Morali & Beyer, 1979), maternal behavior (Rosenblatt, Mayer, & Siegel, 1985), aggression (de Jonge, Eerland, & van de Poll, 1986; Fraile, McEwen, & Pfaff, 1988), territorial marking (Hudson, Gonzalez-Mariscal, & Beyer, 1989). Most of these behaviors are integrated in relatively well.circumscribed brain areas (hypothalamus, preoptic area) possessing high concentrations of estrogen-dependent PRs (Kato, 1985). On the other hand, progestins modify emotional reactivity and vigilance by acting on a widespread system of neurons at all levels of the CNS (Gyermek, 1975; Kubli et al., 1976; P'an & Laubach, 1964; Selye, 1941). Two progestin responsive neural systems can be proposed: 1) a limbic-diencephalic discrete system in which they produce trophic progestational effects by interacting with PRs and 2) a diffuse system in which they act as modulators in various types of synapses - GABAergic (see above), cholinergic (Klangkalya & Chan, 1988), glutamatergic (Halpain & McEwen, 1988), purinergic (Phillis, 1986), opiatergic (Stu, Lodnon, & Jaffe, 1988) - through membrane interactions. From the above-mentioned characteristics of these two systems, we could propose that P as such (a delta-4-3-keto pregnane) would act on the former system while its ring A-reduced metabolites would act on the latter. In the present paper, we will review some experimental data pertinent to this proposal.

1. Behaviors linked to the action of delta-4-3-keto progestins

1.1 Lordosis behavior

The best-studied behavioral process activated by P is the lordosis behavior of estrogen-primed rodents (Etgen & Barfield, 1986; Morali & Beyer, 1979). It is generally believed that the delta-4-3-keto structure plays an essential role in the facilitation of this behavior. Pharmacological results support this interpretation, at least in estrogen-primed rats. A good, though not perfect, correlation between the binding affinity for PR and the lordogenic potency of pregnanes (Glaser, Etgen, & Barfield, 1985; Vathy, Etgen, & Barfield, 1987) has been reported. Moreover, in a study where several—both natural and synthetic—pregnanes with well-characterized PR affinities were systemically administered, we found that l-norgestrel, possessing the highest binding affinity for PR, showed the highest lordogenic potency (Fig. 1).

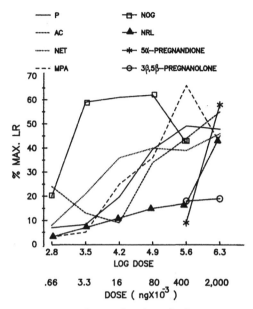

FIG. 13.1. Lordogenic potency of natural and synthetic pregnanes administered s.c. to ovariectomized, estrogen primed rats. % MAXLR=per cent maximal lordosis response. P=progesterone; AC=chlormadione acetate; NET=norethysterone; MPA=medroxyproges-terone acetate; NOG=norgestrel; NRL=norethynodrel.

Conversely, norethynodrel and the natural ring A-reduced progestins (5β-pregnanedione and 3β, 5β-pregnanolone), all having a low affinity for PR (see above), showed low lordogenic potencies. Additional support for the participation of the PR in the facilitation of lordosis by P comes from studies in which RU486, an antiprogestin competing for the binding of P to PR, diminished or suppressed the lordogenic action of P when administered either systemically or into the brain (Brown, Moore, Blaustein, 1987; Etgen & Barfield, 1986). Moreover, protein synthesis inhibitors infused into the hypothalamus block the effect of P on lordosis (for review, see Morali & Beyer, 1979), a finding indirectly supporting the participation of a genomic mechanism in the behavioral action of progestins.

1.2 Proceptive behavior

Figure 2 shows a comparison of the potencies of various pregnanes, systemically injected, for inducing proceptivity. These data show that the presence of a delta-4-3-keto structure, and, therefore, the binding to PR, is important for the display of this behavior. Again, norgestrel was the most potent pregnane for inducing proceptivity, followed by medrox-yprogesterone. Interestingly, these progestins induced a dualistic response, in which larger dosages were less effective for inducing proceptivity. Progestins lacking the delta-4-3-keto structure (norethynodrel and the natural ring A-reduced progestins) were less potent than those having this structure. Yet the fact that 5α-pregnanedione, which cannot be converted back to P, induced proceptivity at the highest dose used, indicates that the delta-4-3-keto structure is not essential for this effect.

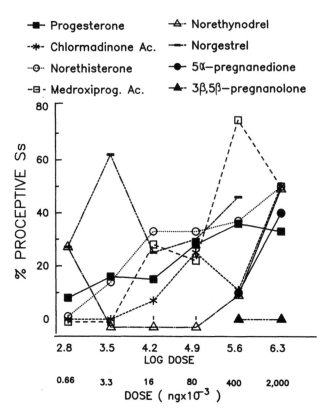

Progesterone
Chlormadinone Ac.
Norethisterone
Medroxiprog. Ac.
Norethynodrel
Norgestrel
5α–pregnanedione
3β,5β–pregnanolone

FIG. 13.2. Induction of proceptivity by the s.c. administration of natural and synthetic pregnanes to ovariectomized, estrogen primed rats. See subjects.

1.3 Sequential inhibition

P, following its initial facilitatory action, induces a refractory period during which additional injections of P fail to stimulate lordosis (Sequential Inhibition, SI; Marrone, Rodríguez-Sierra & Feder, 1977; Morali & Beyer, 1979; Nadler, 1970). Pharmacological studies using a variety of pregnanes suggest that the delta-4-3-keto structure is important, if not essential, for the induction of SI. Thus, as seen in Table 1, the order of potency for inhibiting lordosis was similar to the one observed for its facilitation and was, again, related to the affinity of progestins for PR. Neither norethynodrel nor the ring A-reduced natural progestins induced significant SI, in accordance with other workers (Czaja, Goldfoot, & Karavolas, 1974; Pleim & de Bold, 1984) while norgestrel was the most potent pregnane for inducing this effect. Similarly, Blaustein and Wade (1978) found R5020 more potent for inducing SI than P. In spite of the above-mentioned data, the mechanisms related to this action of P and of other delta-4-3-keto pregnanes are uncertain. Thus, the administration of RU486, which competes for the binding of P to PR, failed to interfere with the SI induced by P (Vathy, Etgen, & Barfield, 1989). Moreover, protein synthesis inhibitors failed to interfere with P induced SI (Shivers, Harlan, Parker, & Moss, 1980). Furthermore, the inhibitory action of P may not be circumscribed to a decrease in PRs (Parsons, McGinnis, & McEwen, 1981; Schwartz, Blaustein, & Wade, 1979;), since the lordogenic responses to LHRH and to prostaglandin E2, which are PR-independent, are also inhibited following the initial facilitatory action of progestins (Melo, Gonzalez-Mariscal, & Beyer, 1989).

Table 13. 1. Sequential inhibition induced by natural and synthetic progestins with different affinities for PR.

	percent inhibition doses (ng x 10^{-3})					
	0.66	3.3	16	80	400	2000
progesterone	18	5	13	41	74	78
chlormadinone acetate	15	8	22	43	95	94
norethysterone	43	45	13˙	38	70	41
norgestrel	0	37	25	45	100	96
norethynodrel	8	48	51	76	88	-
5a-pregnanedione	34	0	15	21	0	34
3ʙ. 5b-pregnanolone	-	-	-	-	35	22

2. Behaviors linked to the action of ring A-reduced progestins

2.1 Lordosis behavior

It has been proposed that an important component of lordosis facilitation is a process of disinhibition, in which the activity of neurons inhibitory to lordosis would be suppressed (Pfaff & Schwartz-Giblin, 1988). Since ring A-reduced pregnanes drastically depress neuronal firing (see above), it could be proposed that their action on lordosis was exerted through this disinhibition process. Indeed, the administration of various ring A-reduced pregnanes with low binding affinities for PR can, under some conditions, facilitate intense lordosis in estrogen-primed rodents (Gorzalka & Whalen, 1977; Kubli-Garfias & Whalen, 1977; Meyerson, 1972; Whalen & Gorzalka, 1972). The lordogenic potency and efficacy of these pregnanes varies widely according to the species, strain, and, particularly, the solvent and route of administration. It has been suggested that the low lordogenic potency of ring A-reduced progestins injected systemically is due to bioavailability problems (Czaja et al., 1974) from a higher clearance rate and/or a higher binding to serum proteins (Westphal, 1971).

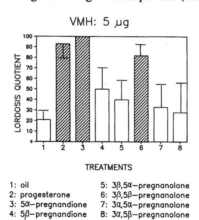

FIG. 13.3. Lordosis behavior induced by the administration of 5 mg (bilaterally, in 0.5 ml oil) of natural progestins into the ventromedial hypothalamus (VMH) of ovariectomized, estrogen primed rats. Hatched bars=significantly greater than oil group (p < 0.05).

155

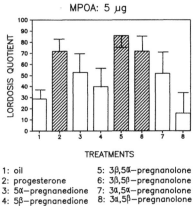

FIG. 13.4. Lordosis behavior induced by the administration of 5 mg (bilaterally, in 0.5 ml oil) of natural progestins into the medial preoptic (MPOA) of ovariectomized, estrogen primed rats. Hatched bars=significantly greater than oil group (p < 0.05).

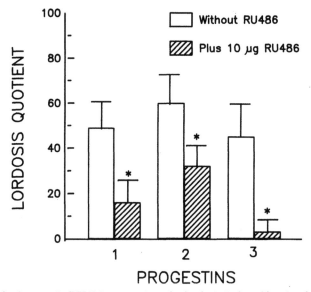

FIG. 13.5. Antiprogestin RU486 antagonizes the lordosis induced by simultaneous injections of natural progestins (1 mg bilaterally, in oil) into the ventromedial hypothalamus (VMH) of ovariectomized, estrogen primed rats.

156

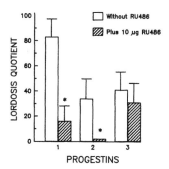

MPOA

1: Progesterone
2: 5α−Pregnandione
3: 3β,5β−pregnanolone

FIG. 13.6. Antiprogestin RU486 antagonizes the lordosis induced by simultaneous injections of progesterone or 5α (1 mg bilaterally, in oil) into the medial preoptic area (MPOA) of ovariectomized, estrogen primed rats. RU486 did not antagonize the lordosis induced by 3β, 5β-pregnanolone.

In order to circumvent this problem, the lordogenic potency and efficacy of P and its metabolites were assessed by implanting them directly into two brain areas considered to control lordosis through opposite mechanisms: the ventromedial hypothalamus (VMH), related to lordosis facilitation, and the medial preoptic area (MPOA), related to its inhibition. As shown in Figs 3 and 4, some ring A-reduced progestins were as effective as P when administered intracerebrally at either VMH or MPOA (Beyer, Gonzalez-Mariscal, & Eguibarm, & Gomora, 1988a). Since in this study the dose used (5 mg, bilaterally) could have been so large to allow progestins with a low affinity for PR to stimulate lordosis, two of the most effective progestins besides P (i.e., 5α-pregnanedione and 3β, 5β-pregnanolone) were tested at much lower dosages at the VMH. These three progestins facilitated lordosis even when a minute dosage (40 ng) was injected into the VMH (Beyer et al., in preparation). To assess the possible participation of PRs in the lordogenic action of the various progestins, we tested the capacity of the antiprogestin RU486 to antagonize the lordogenic action of P, 5α-pregnanedione and 3β, 5β-pregnanolone when infused along with them into either VMH or MPOA (González-Mariscal, Gonzalez-Flores, & Beyer, 1989). As shown in Figure 5, RU486 significantly depressed the lordogenic responses of the three progestins in the VMH and those of P and 5α-pregnanedione in the MPOA. By contrast, the lordogenic action of 3β, 5β-pregnanolone at the MPOA was not significantly affected by RU486 (González-Mariscal, González-Flores & Beyer, unpublished results).

The facilitation of lordosis by 3β-OH pregnanolones at the MPOA may involve a depression of the firing of MPOA neurons, an idea that would be consistent with the proposition that this area regulates lordosis in a different way than does the VMH, i.e., by exerting a tonic inhibition (Pfaff & Schwartz-Giblin, 1988). This inhibition of MPOA neurons by progestins could be achieved by an enhancement of GABAergic activity at this area, an idea supported by the existence of high GABA concentrations (Tappaz, Brownstein, & Palkovits, 1986; Wallis & Luttge, 1980) and high density of GABAergic terminals (Flügge, Oertel, & Wuttke, 1986) in this region. However, an analysis of the effective pregnanes to facilitate lordosis when infused at either VMH or MPOA indicates that no correlation exists between their lordogenic potency and their reported capacity for enhancing GABAergic activity. Thus 3α, 5β- and 3α, 5α-pregnanolone, two effective modulators of GABA-A receptors, were without effect on lordosis, while 3β, 5β-pregnanolone stimulated intense lordosis behavior in spite of having weak or no effects on GABAergic function (see above). Interestingly, 3β, 5β-pregnanolone is extremely potent in facilitating LHRH release in the female rat (Park &

Ramírez, 1987). Therefore, it is possible that the lordogenic action of this pregnane is mediated through release of this decapeptide, since an important concentration of LHRH is found in the MPOA (Hoffman, 1985; Palkovits, 1984).

2.2 Excitability to somatosensory stimulation

There is evidence that P or its metabolites modulate brain excitability under normal and pathological conditions. Thus, this hormone increases the electroshock seizure threshold in several species (Woodbury & Vernadakis, 1967) and decreases the frequency of interictal spikes from penicillin foci (Landgren, Aasly, Bäckström, Dubrovsky, & Danielsson, 1987). Conversely, a higher incidence of epileptic seizures has been reported in some women during the declining phase of P plasma concentrations in the menstrual cycle (catamenial epilepsy; Bäckström, 1976; Laidlaw, 1965).

From these and other studies, it has been suggested that, under normal conditions, P modulates the excitability of neural circuits associated to mood, perception, and emotion (see Datta, 1986). A common characteristic of these progestin effects is depression of neuronal firing (Kubli et al., 1976), a fact strongly suggesting the participation of ring A-reduced pregnanes.

2.2.1 Brain sexual differentiation

The protective effect of P or progestins on the potentially damaging effects of excessive neuronal activation is particularly clear in the developing brain if the infantile rat. Thus, Holmes and Weber (1984) have found that extremely low dosages of P, that would have no effect on the adult brain, prevent occurrence of kindling induced by amygdala stimulation. This protective action of P or its metabolites in the developing brain may be a factor in brain sexual differentiation. P is known, since many years ago, to exert a protective action against the defeminizing effect of neonatal androgen administration to female rats (Dorfman, 1967; Kincl & Maqueo, 1965). Arai and Gorski (1968) found that 5β-pregnanedione was more potent than P at counteracting testosterone-induced brain defeminization, a finding supporting the role of ring-A reduction in this protective action on the brain. More recently, we tested the effects of a variety of natural and synthetic pregnanes for their ability to protect against androgen-induced defeminization (González-Mariscal, Fernandez-Guasti, & Beyer, 1982). As shown in Figure 7, pregnanes, specifically those reduced in the 5β-position, were the most effective ones in protecting against defeminization. This action of the pregnanes was clearly correlated with their capacity to induce a state of behavioral depression in neonatal rats, evidenced by the disappearance of EEG and EMG activities. It is possible that pregnanes, either secreted by the ovaries or adrenals, or directly synthetized in the brain, may exert some normal function in brain differentiation. The fact that 3α, 5β-pregnanolone, a potent enhancer of GABAergic activity, was the most effective progestin for protecting against testosterone induced brain defeminization suggests the participation of this inhibitory amino acid in the process of regulating brain sexual differentiation.

2.2.2 Modulation of perception

Fluctuations in P plasma levels influence the responsivity of females to environmental stimuli. In ovariectomized cats, chronic administration of both P or 5α-pregnanedione facilitates the appearance of relaxation behavior in response to milk drinking and other pleasant stimulation (Cervantes, Ruelas, & Beyer, 1979). Relaxation behavior is characterized by the appearance of a pattern of EEG synchronization in the parieto-occipital cortex associated to a decrease in brain stem neuronal firing. Unfortunately, in this study, no comparison of the potencies between P and its 5α-reduced derivatives was made. Benzodiazepines and

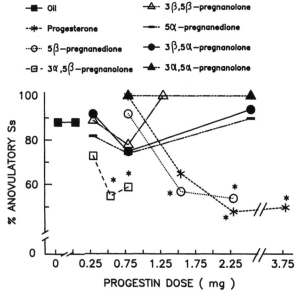

FIG. 13.7. Protection against testosterone-induced defeminization in neonatal rats by the concurrent administration of natural progestins.
* Significantly smaller than control oil group.

TAIL—SHOCK TEST

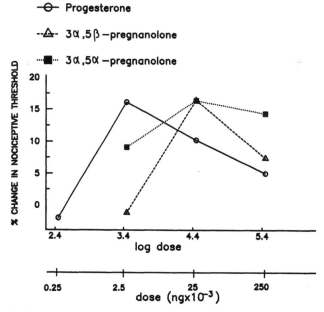

FIG. 13.8. Analgesia induced by the administration of natural progestins along with 1 μg muscimol into the perispinal space of ovariectomized rats.

barbiturates have also been found to facilitate relaxation behavior (Beyer et al., unpublished observations), findings pointing to an involvement of the GABAergic system in this response. Progestins can also modulate the reactivity of the animals to negative or aversive stimulation. McCarthy, Caba, Beyer, and Komisaruk (1989) reported that the analgesic effect of the perispinal injection of 1 μg muscimol, a GABA-A agonist, varied according to the stage of the estrous cycle in rats. Muscimol induced hyperalgesia during diestrus while provoking analgesia during late proestrus and estrus. Ovariectomized rats consistently showed hyperalgesia in response to perispinal muscimol administration. Systemic administration of P to castrated rats reverted this tendency, since muscimol induced in these animals clear analgesia (Beyer, Caba, McCarthy, Gonzalez-Mariscal, & Komisaruk, 1988). The presumption that variations in the response to muscimol are due to interactions of P with the GABA-A receptor points to 3α-OH ring A-reduced P metabolites as mediating this effect. Indeed, the perispinal administration of either 3α, 5β- or 3α, 5α-pregnanolone along with muscimol, consistently induces analgesia (Beyer et al., in preparation). However, to our surprise, as shown in Figure 8, P, P *per se* was more potent than its ring A-reduced derivatives. This last result suggests that delta-4-3-keto pregnanes like P may enhance GABAergic activity by themselves, through an alternative mechanism than that used by the 3α-OH derivatives. In fact, some *in vitro* studies support the idea that pregnanes other than the 3α-OH 20-keto pregnanes (for example, corticosterone, pregnanolone) can modify GABAergic function (Majewska, 1985; Majewska et al., 1988; Ong, Kerr, & Johnston, 1987).

Moreover, P as such, when pre-incubated along with brain membranes for somewhat longer periods than ring A-reduced pregnanes, can enhance muscimol binding as effectively as its ring A-reduced derivatives (López Colomé et al., 1990).

SUMMARY AND CONCLUSIONS

As can be seen from Table 2, the results agree with the idea that P as such is involved in the modulation of estrous behavior (lordosis, proceptivity, and SI). Chemical transformations that enhance the affinity of the pregnane molecule for PR, in general, increase their lordogenic potency. Since estrous behavior is a prolonged event in most species, it appears energetically more efficient to alter the excitability of the neural substrate for lordosis through trophic, persistent changes induced by "progestational" pregnanes than to maintain a continuous synaptic bombardment, through the action of "neuromodulatory pregnanes". This proposition agrees well with the lordogenic action of delta-4-3-keto pregnanes, which can induce "trophic" changes through their binding to PR. However, it remains puzzling that ring A-reduced pregnanes, which cannot use this genomic mechanism, can nonetheless induce estrous behavior for periods comparable in duration to those induced by delta-4-3-keto progestins. It is possible that the initial excitability changes produced by the "neuromodulatory" pregnanes can secondarily lead to "trophic" events similar to those induced by the "progestational" pregnanes. This idea coincides with the proposition (Beyer & González-Mariscal, 1986; Whalen & Lauber, 1986) that second messengers, notably cyclic nucleotides, may mediate the facilitation of lordosis by P.

C3-reduction of ring A-reduced pregnanediones yields a series of pregnanes with different pharmacological properties regarding lordosis. Thus, 3α-OH-pregnanolones lacked any lordogenic effects in spite of being potent modulators of GABAergic action. This finding suggests that this aminoacid neurotransmitter may not be directly involved in pregnane facilitation of lordosis. On the other hand, 3β-OH-pregnanolones were potent activators of lordosis, particularly when infused into the MPOA. This lordogenic effect, rather than ensuing from a disinhibition process—i.e., from the inhibition of MPOA neurons—may be due to LHRH release since the 3β-conformation is associated with a great potency for eliciting the release of this decapeptide (Park & Ramírez, 1987). Interestingly, no ring A-reduced progestin was consistently found to induce SI.

Table 13. 2. Cellular, neuroendocrine and behavioral effects of progestins with various ring A conformations.

Ring A structure	Estrous behavior*	Sequential inihibition	GABA analgesia	Protection vs. defemini-zation	LHRH release	Affinity for PR	GABA-A receptor modulation	Membrane stabilization
delta-4-3-keto	S MPOA xxx / VMH xxx	xxx	xxx	x	xx	xxx	0-xx&	x
5α-pregnanedione	S VMH xxx x / MPOA x	0	-	0	x	x	x	xx
5β-pregnanedione	S MPOA 0 / VMH 0	0	-	xx	-	0	x	xx
3α-OH-pregnanolones	0	0	5α xx / 5β x	5α 0? / 5β xxx	5α - / 5β 0	0	5α, 5β xxx	xxx
3β-OH pregnanolones	5α S MPOA 0 xxx / 5β S MPOA/VMH 0 xxx	0	-	0	5α xx / 5β xxx	0	0	x

* includes receptivity and proceptivity
& depending on pre-incubation conditions
S=systemic administration

As also shown in Table 2, ring A-reduced pregnanes, particularly 3α, 5β-pregnanolone, were the most potent ones for protecting against testosterone-induced defeminization in neonatal rats. Most likely, this action is associated to an inhibition of neuronal firing. The possibility that, under normal conditions, pregnanes may participate in brain differentiation by antagonizing the effects of excitatory neurotransmitters like glutamate has been recently highlighted by us (Beyer & Feder, 1987). The weak protective effect of P was probably related to its conversion into ring A-reduced pregnanes. As could be anticipated on the basis of recent biochemical studies, 3α-OH pregnanolones synergized with subthreshold doses of muscimol to induce analgesia. However, it was rather surprising that P was much more potent than these pregnanes to enhance muscimol action. This enhancing effect of P might be related to an indirect increase in the number of GABA binding sites, as recently reported by López-Colomé et al. (1990).

REFERENCES

Arai, Y., & Gorski, R.A. (1986). Protection against the neural organizing effect of exogenous androgen in the neonatal female rat. *Endocrinology, 82,* 1005-1009.

Bäckström, T. (1976). Epileptic seizures in women related to plasma estrogen and progesterone during the menstrual cycle. *Acta Neurol. Scand., 54,* 321-347.

Beyer, C., & González-Mariscal, G. (1986). Elevation in hypothalamic cyclic AMP as a common factor in the facilitation of lordosis in rodents: a working hypothesis. *Ann. N.Y. Acad. Sci., 474,* 270-281.

Beyer, C., & Feder, H.H. (1987). Sex steroids and afferent input: Their roles in brain sexual differentiation. *Ann. Rev. Physiol., 49,* 349-364.

Beyer, C., Caba, M., McCarthy, M., González-Mariscal, G., & Komisaruk, B.R. (1988). Modulación de la acción analgésica de agonistas GABA-A por progesterona. *XXXI Congreso Nacional de Cienciac Fisiológicas. Querétaro Qro. Mexico. Abst. #C 156.*

Beyer, C., González-Mariscal, G., Eguíbarm J.R., & Gómora, P. (1988a). Lordosis facilitation in estrogen primed rats by intrabrain injection of pregnanes. *Pharmacol. Biochem. Behav., 31,* 919-926.

Blaustein, J.D., & Wade, G.N. (1978). Progestin binding by brain and pituitary cell nuclei and female rat sexual behavior. *Brain Res., 140,* 360-367.

Brown, T.J., Moore, M.J., & Blaustein, J.D. (1987). Maintenance of progesterone-facilitated sexual behavior in female rats requires continued hypothalamic protein synthesis and nuclear progestin receptor occupation. *Endocrinology, 121,* 298-304.

Cervantes, M., Ruelas, R., & Beyer, C. (1979). Progesterone facilitation of EEG synchronization in response to milk drinking in female cats. *Psychoneuroendocrinology, 4,* 245-251.

Czaja, N.A., Goldfoot, D.A., & Karavolas, H.J. (1974). Comparative facilitation and inhibition of lordosis in the guinea pig with progesterone, 5α-pregnane-3,20-dione, or 3α-hydroxy-5α-pregnan-20-one. *Horm. Behav., 5,* 261-274.

Datta, S. (1986). Sex hormone effects on excitable membranes. In P.J. Goldstein (Ed.), *Neurological Disorder of Pregnancy* (pp. 265-277). New York: Futura Publ. Co.

De Jonge, F., Eerland, E.M.J., & van de Poll, N.E. (1986). Sex-specific interactions between aggressive and sexual behavior in the rat: effects of testosterone and progesterone. *Horm. Behav., 20,* 432-444.

Dorfman, R.Y. (1967). The antiestrogenic and antiandrogenic activities of progesterone in the defense of a normal fetus. *Anat. Record, 157,* 547-558.

Duax, W.L., Cody, V., Griffin, J.F., Rohrer, D.C., & Weeks, C.M. (1978). Molecular conformation and protein binding affinity of progestins. *J. Toxicol. Environ. Health, 4,* 205-227.

Etgen, A.M., & Barfield, R.J. (1986). Antagonism of female sexual behavior with intracerebral implants of antiprogestin RU38486: correlation with binding to neural progestin receptors. *Endocrinology, 119,* 1610-1617.

Flügge, G., Oertel, W.H., & Wuttke, W. (1986). Evidence for estrogen-receptive GABA-ergic neurons in the preoptic/anterior hypothalamic area of the rat brain. *Neuroendocrinolgy, 43,* 1-5.

Fraile, I.G., McEwen, B.S., & Pfaff, D.W. (1988). Comparative effects of progesterone and alphaxalone on aggressive, reproductive and locomotor behaviors. *Pharmacol. Biochem. Behav., 30,* 729-735.

Glaser, J.H., Etgen, A.M., & Barfield, R.J. (1985). Intrahypothalamic effects of progestin agonists on estrous behavior and progestin receptor binding. *Physiol. Behav., 34,* 871-877.

González-Mariscal, G., Fernandez.Guasti, A., & Beyer, C. (1982). Anesthetic pregnanes counteract androgen-induced defeminization. *Neuroendocrinology, 34,* 357-362.

González-Mariscal, G., González-Flores, O., & Beyer, C. (1989). Intrahypothalamic injection of RU 486 antagonizes the lordosis induced by ring A-reduced progestins. *Physiol. Behav., 46,* 435-438.

Gorzalka, B.B., & Whalen, R.E. (1977). The effects of progestins, mineralocorticoids, glucocorticoids and steroid solubility on the induction of sexual receptivity in rats. *Horm. Behav., 8,* 94-99.

Gyermek, L., & Soyka, L.F. (1975). Steroid anesthetics. *Anesthesiology, 42,* 331-344.

Halpain, S., & McEwen, B.S. (1988). Corticosterone decreases ^{3}H-glutamate binding in the rat hippocampal formation. *Neuroendocrinology, 48,* 235-241.

Harrison, N.L., Majewska, M.D., Meyers, D.E.R., & Barker, J.L. (1989). Rapid actions of steroids on CNS neurons. In J.M. Lakoski, J.R., Pérez-Polo, and D.K. Rassin (Eds.), *Neural Control of Reproductive Function* (pp. 137-166). New York: Alan R. Liss, Inc.

Hoffman, G. (1985). Organization of LHRH cells: differential apposition of neurotensin, substance P and catecholamine axons. *Peptides, 6,* 439-461.

Holmes, G.L., & Weber, D.A. (1984). The effect of progesterone on kindling: a developmental study. *Dev. Brain Res., 16,* 45-53.

Hudson, R., González-Mariscal, G., & Beyer, C. (1989). Chin marking behavior, sexual receptivity and pheromone emission in steroid-treated ovariectomized rabbits. *Horm. Behav.,* in press.

Karavolas, H.J., Bertics, P.J., Hodges, D., & Rudie, N. (1984). Progesterone processing by neuroendocrine tissues. In F. Celotti, F. Naftolin, and L. Martini (Eds.), *Metabolism of Hormonal Steroids in the Neuroendocrine Structures* (pp. 149-170). New York: Raven Press.

Kato, J. (1985). Progesterone receptors in brain and hypophysis. In D. Ganten and D. Pfaff (Eds.), *Actions of Progesterone on the Brain* (pp. 31-81). Berlin-Heidelberg: Springer-Verlag.

Kincl, F.A., & Maqueo, M. (1965). Prevention by progesterone of steroid-induced sterility in neonatal male and female rats. *Endocrinology, 77,* 859-862.

Klangkalya, B., & Chan, A. (1988). Inhibition of hypothalamic and pituitary muscarinic receptor binding by progesterone. *Neuroendocrinology, 47,* 294-302.

Kubli, C., Cervantes, M., & Beyer, C. (1976). Changes in multiunit activity and EEG induced by the administration of natural progestins to flaxedil immobilized cats. *Brain Res., 114,* 71-81.

Kontula, K., Jänne, O., Vihko, R., de Jager, E., de Visser, J., & Zeelen, F. (1975). Progesterone-binding proteins: *in vitro* binding and biological activity of different steroidal ligands. *Acta Endocrinol., 78,* 574-592.

Kubli, C., Medrano-Conde, L., Beyer, C., & Bondani, A. (1979). In vitro inhibition of rat uterine contractility induced by 5α- and 5β progestins. *Steroids, 34,* 609-617.

Kubli-Garfias, C., & Whalen, R.E. (1977). Induction of lordosis behavior in female rats by intravenous administration of progestins. *Horm. Behav., 9,* 380-386.

Laidlaw, J. (1965). Catamenial epilepsy. *Lancet, 271,* 1235-1237.

Landgren, S., Aasly, J., Bäckström, T., Dubrovsky, B., & Danielsson, E. (1987). The effect of progesterone and its metabolites on the interictal epileptiform discharge in the cat's cerebral cortex. *Acta Physiol. Scand., 131,* 33-42.

163

López-Colomé, A.M., McCarthy, M., & Beyer, C. (1990). Enhancement of ^{3}H-muscimol binding to brain synaptic membranes by progesterone and related pregnanes. *Eur. J. Pharmacol.*, in press.

Majewska, M.D. (1985). Glucocorticoids are modulators of GABA-A receptors in brain. *Brain Res., 339*, 178-182.

Majewska, M.D., Mienville, J.M., & Vicini, S. (1988). Neurosteroid pregnanolone sulfate antagonizes electrophysiological responses to GABA in neurons. *Neurosci. Lett., 90*, 279-284.

Marrone, B.L., Rodríguez-Sierra, J.F., & Feder, H.H. (1977). Lordosis: inhibiting effects of progesterone in the female rat. *Horm. Behav., 8*, 391-402.

McCarthy, M.M., Caba, M., Beyer, C., & Komisaruk, B. (1989). Steroid-modulation of GABA-induced analgesia. *Society for Neuroscience 19th Annual Meeting. Phoenix, Arizona, U.S.A. Abst. # 304.*

Melo, A., González-Mariscal, G., & Beyer, C. (1989). Caraterización del fenómeno de Inhibición Secuencial desencadenado por progesterona en la rata hembra. *XXXII Congreso Nacional de Ciencias Fisiológicas. Oaxtepac, Mor. México. Abst. # M6.*

Meyerson, B. (1972). Latency between intravenous injection of progestins and the appearance of estrous behavior in estrogen treated ovariectomized rats. *Horm. Behav., 3*, 1-9.

Moralí, G., & Beyer, C. (1979). Neuroendocrine control of mammalian estrous behavior. In C. Beyer (Ed.), *Endocrine Control of Sexual Behavior* (pp. 33-75). New York: Raven Press.

Nadler, R.D. (1970). A biphasic influence of progesterone on sexual receptivity of spayed female rats. *Physiol. Behav., 5*, 95-97.

Ong, J., Kerr, D.I.B., & Johnston, G.A.R. (1987). Cortisol: a potent biphasic modulator at GABA$_A$-receptor complexes in the guinea pig isolated ileum. *Neurosci. Lett., 82*, 101-106.

Palkovits, M. (1984). Distribution of neuropeptides in the central nervous system: a review of biochemical mapping studies. *Prog. Neurobiol., 23*, 151-189.

P'an, S.Y., & Laubach, G.D. (1964). Steroid central depressants. In R.I. Dorfman (Ed.), *Methods in Hormone Research, Vol. III* (pp. 415-475). New York-London: Academic Press.

Park, O.K., & Ramírez, V.D. (1987). Pregnanolone, a metabolite of progesterone, stimulates LHRH release: *in vitro* and *in vivo* studies. *Brain Res., 437*, 245-252.

Parsons, B., McGinnis, M.Y., & McEwen, B.S. (1981). Sequential inhibition by progesterone: effects on sexual receptivity and associated changes in brain cytosol progestin binding in the female rat. *Brain Res., 221*, 149-160.

Parsons, B., & Pfaff, D. (1985). Progesterone receptors in CNS correlated with reproductive behavior. In D. Ganton and D. Pfaff (Eds.), *Actions of Progesterone on the Brain* (pp. 103-140). Berlin-Heidelberg: Springer-Verlag.

Pfaff, D.W.,, & Schwartz-Giblin, S. (1988). Cellular mechanisms of female reproductive behaviors. In E. Knobil and J. Neill (Eds.), *The Physiology of Reproduction, Vol. 2* (pp. 1487-1568). New York: Raven Press.

Phillis, J.W. (1986). Potentiation of the depression by adenosine of rat cerebral cortical neurons by progestational agents. *Br. J. Pharmacol., 89*, 693-702.

Pleim, E.T., & Debold, J.F. (1984). The relative effectiveness of progestins for facilitation and inhibition of sexual receptivity in hamsters. *Physiol. Behav., 32*, 743-747.

Raynaud, J.P., Ojasoo, T., Pottier, J., Salmon, J. (1982). Chemical substitution of steroid hormones: Effect on receptor binding and pharmacokinetics. In *Biochemical Actions of Hormones, Vol. IX* (pp. 305-342). New York: Academic Press.

Rosenblatt, L., Mayer, A.D., & Siegel,, H.I. (1985). Maternal behavior among the nonprimate animals. In N. Adler, D. Pfaff, and R.W. Goy (Eds.), *Handbook of Behavioral Neurobiology. Vol. 7, Reproduction* (pp. 229-298). New York-London: Plenum Press.

Schwartz, S.M., Blaustein, J.D., & Wade, G.N. (1979). Inhibition of estrous behavior by progesterone in rats: role of neural estrogen and progestin receptors. *Endocrinology, 105*, 1078-1082.

Seeman, P. (1972). Membrane actions of anesthetics and tranquilizers. *Pharmacol. Rev.,* *24,* 583-655.

Selye, P. (1941). Studies concerning the anesthetic action of steroid hormones. *J.* *Pharmacol. Exp. Ther., 73,* 127-141.

Shivers, B.D., Harlan, R.E., Parker, C.R., Jr., & Moss, R.L. (1980). Sequential inhibitory effect of progesterone on lordotic responsiveness in rats: time course, estrogenic nullification and actinomycin D-insensitivity. *Biol. Reprod., 23,* 963-973.

Smith, H.E., Smith, R.G., Toft, D.O., Neergaard, J.R., Burrows, E., & O'Malley, B.W. (1974). Binding of steroids to progesterone receptor chick proteins in chick oviduct and human uterus. *J. Biol. Chem., 249,* 5924-5932.

Smith, S.S., Waterhouse, B.D., & Woodward, D.J. (1987). Sex steroid effects on extrahypothalamic CNS. II. Progesterone, alone and in combination with estrogen, modulates cerebellar responses to aminoacid neurotransmitters. *Brain Res., 422,* 52-62.

Stu, T.P., London, E.D., & Jaffe, J.H. (1988). Steroid binding of σ receptors suggests a link between endocrine, nervous and immune systems. *Science, 240,* 219-221.

Tappaz, M.L., Brownstein, M.J., & Palkovits, M. (1976). Distribution of glutamate decarboxylase in discrete brain nuclei. *Brain Res., 108,* 371-379.

Turner, J.P., & Simmonds, M.A. (1989). Modulation of the GABA$_A$ receptor by steroids in slices of rat cuneate nucleus. *Br. J. Pharmacol., 96,* 409-417.

Vathy, I.U., Etgen, A.M., & Barfield, R.J. (1987). Actions of progestins on estrous behavior in female rats. *Physiol. Behav., 40,* 591-595.

Vathy, I.U., Etgen, A.M., & Barfield, R.J. (1989). Actions of RU38486 on progesterone facilitation and sequential inhibition of rat estrous behavior: correlation with neural progestin receptor levels. *Horm. Behav., 23,* 43-56.

Wallis, C.J., & Luttge, W.G. (1980). Influence of estrogen and progesterone on glutamic acid decarboxylase activity in discrete regions of rat brain. *J. Neurochem., 34,* 609-613.

Westphal, U. (1971). *Steroid-protein interactions.* New York: Springer-Verlag.

Whalen, R.E., & Gorzalka, B.B. (1972). The effects of progesterone and its metabolites on the induction of sexual receptivity in rats. *Horm. Behav., 3,* 221-226.

Whalen, R.E., & Lauber, A.H. (1986). Progesterone substitutes: cyclic GMP mediation. *Neurosci. Biobehav. Rev., 10,* 47-53.

Woodbury, D.M., & Vernadakis, A. (1967). Influence of hormones on brain activity. In L. Martini and W.F. Ganong (Eds.), *Neuroendocrinology, Vol. II* (pp. 335-375).

IV Clinical Psychobiology

Frank O. Ödberg
University of Gent

My initial reaction as chairman having to comment on these four papers was rather one of awe: "How shall I manage to find some unifying trend in so diverse topics?". While reading them, I realizes that all my colleagues were dealing with problems of interdisciplinary, either on the theoretical level, either on the methodological one.

Originally trained as an ethologist, I soon realized how indispensable interdisciplinarity was to put together the puzzle of behavioural phenomenons. Interdisciplinarity can be achieved in two ways: either one man integrates in his brain different sciences, either several people with distinct specializations start working together. Due to the tremendous scientific development, the first solution can be developed in only a restricted way nowadays. One man can integrate only a very limited series of sciences. The encyclopedical idea of the renaissance is an utopia now. It is more practical, reasonable and efficient to educate scientists in such a way that they never succumb to the pitfall of reductionism, i.e. to believe that they will explain everything from level of their speciality alone. In other words, educate them so that they keep an open mind for other fields by giving them just a very general survey of what relevant sciences can offer. Such a scientist will readily go and see colleagues and wonder how they would approach a given problem, suggest a solution and eventually start together a very concrete research project. We should remember that scientists exist to study and solve problems and not to parade through their lives as ambassadors of a particular science.

One problem is that, even when different specialists want to cooperate, they are often thwarted by academical barriers (e.g. "territorial behaviours" of faculties, classification codes of granting agencies, etc.). Some amongst us have even experienced aggressive reactions, or have suffered administrative discrimination ("What are you in fact?"). Sitting between two chairs is not a very stable position.

Even within the scope of psychobiology, cooperations should develop and quit according to the specific problems being studied. Of course, now and then, highly synthetic minds appear who are able to present a view on life and behaviour which is as comprehensive as possible, from the molecular, through the physiological, neuropsychological and social till the philosophical. However, in order to create a mosaic one needs the people carving accurately each little stone. May the contacts at this meeting with colleagues from different backgrounds result into improved experimental designs and theoretical models.

Mikael Heimann's work on neonatal imitation in human babies is a nice example of integration of ethological concepts and methods into developmental child psychology, together with hints for neurophysiological investigations.

One danger of interdisciplinarity is that somebody trained essentially in science A does not integrate well the concepts and methodology of science B, or does not consult the right knowledgeable colleague. This is, by the way, something which affects for the moment to some extent behavioural research in animal production (i.e. a part of "applied ethology"). Ethological concepts are often misused. One must therefore appreciate Heimann's critical sense when he stresses the importance of very precise ethological definitions of overt behaviour and even of very discrete bouts of behaviour (see "Methodological issues"). Apparently contradictory results could be due to the fact people have used different behavioural parameters or are even dealing with different behaviours. On the other hand, I wonder whether it is wise and useful to bring ethological antiquities down from the attic. The dust they are covered with is more blinding our understanding than bringing some light. The concept of the IRM has been the subject of much criticism and changes all along the development of ethology since the early fifties. Without any precise definition, its use does not mean very much when opposing it (Abravanel & Sigafoos; Jacobsen) to the Meltzoff & Moore's hypothesis of "active intermodal mapping".

Heimann demonstrates the complexity of behavioural ontogeny. Never excluding the possibility of phylogenetic predispositions, he carries out a lucid analysis, taking into account CNS maturational processes, individual differences, various functional hypotheses on a short or long term scale affecting motor, cognitive or social development. Once again, the inadequacy of the Lorenzian dissection of behaviour into innate and learned elements becomes apparent.

Ivan Divac's paper shows how important it is not to stare oneself blind on brain structures which may be appropriate for a given behaviour. I appreciated his way of making the notion of specific functional systems compatible with more general less specific mechanisms. Forgetting about the intertwining of both has probably complicated a.o. psychopharmacological approaches of pathological behaviour - a subject I am particularly interested in, as we will see. It happened to me whenever I thought to have found an explanation for the involvement of some neurotransmitter or brain area for the development of the stereotypies I am studying: talking it over with a colleague more knowledgeable than I am in that field mostly resulted in the more cautious enumeration of other possible alternatives. In any case, the study of the basal ganglia is a fascinating one as they are involved in the regulation of so many behavioural systems. By the way, I am wondering whether we should do some brainstorming together about the value of the term "interface" in general. We have been influenced by computer science and the term has been useful. I have used it many times myself. However, I wonder whether we are not running the risk of dealing sometimes with different things under the same flag.

Unfortunately laymen will not read this book. People often think that scientific research consists of daily exciting discoveries. They ignore how often we (for the lucky ones amongst us: our technicians) have to go through dull and repetitive procedures. They surely do not realize, as Divac's and Larsson's experience shows, to what extent the fate of precious material depends from devoted and healthy technicians, as well as from the efficiency of the post office!

One must sympathise with Philip Hwang endeavours to find trends through the results of many enquiries on ill-defined concepts or at least using very little operationalized ones. One can have a subjective notion of "shyness", "inhibition", "sociability", "dependency", which is sufficient to communicate in daily life. A scientific approach requires more precise definitions. As I have stated before, the most important thing is to give a precise operational definition before starting the observations. One could still disagree about the meaning of a given term, without affecting the scientific quality of the observation itself if everybody can find out exactly which behaviour patterns were measured. Only one of such definitions is reported (and even - rightly - criticised by Hwang), namely "dependency": "prolonged clinging or remaining proximal to the mother". One can still wonder how much is "prolonged" and "proximal".

In the early stages of a science one has inevitably the tendency to oversimplify phenomena. One discovers gradually that things are more complex than originally thought.

Some concepts are subsequently reluctantly used, not because they do not refer any more to a given reality, but because one should add new caveats. Hence I would like to warn briefly the reader about two concepts which have been mentioned: "domination" and "inborn". Warnings about the latter have been expressed earlier in this text. Without entering into details, the essential thing to remember is that dominance does not imply the existence of a stable hierarchy, neither a linear one, neither one which encompasses all contexts. Several questions arise when one reads that one child dominates the other one:

- has this been observed in different conflictual situations (access to toys, space, to the --- attention of an adult, verbal predominance, etc.)?
- has it been the subject of a follow-up study?
- have different dyadic relations been observed? Is the same child subordinate to A but dominant over B?

It seems more fruitful to analyse social relations in terms of roles carried out at different developmental stages than in terms of hierarchies. From this point of view, the Göteborg Child Care Project contains positive aspects: follow-up, different situations, quality of peer play.

This reminds us of a new approach of stress research, i.e. could an organism cope with stress in different ways? The Groningen school talks about "active and passive copers". According to its neurophysiological set-up, one individual seems to be more enclined to respond by more stereotyped behaviours or behavioural sequences which can be less modulated by environmental cues, while the other one remains more passive and "scans" the environment for additional data. When we apply such an approach to "shyness", one can wonder whether the dichotomy shy/non-shy hides in fact more subtile strategies. One child could initially need the safety of e.g. the mother to gather as much as possible information about the physical and social surroundings in order to control better the situation later on. One could hence ask the question who is really the dominant one. I am not enclined to label as "subordinate" or "inhibited" the child who does not initiates the contacts or the play, or does not respond to aggression by aggression, but who manoeuvres in such a way that he gets what he wants (and it could be e.g. the desire to play alone). His temperamental and cognitive set-up allows him to use other (and sometimes more efficient) tactics.

Maybe the Göteborg project will yield in the end rather series of delicate personality nuances than just an inhibited/uninhibited black-or-white classification.

Finally, I am wondering to what extent the (eventually erroneous) idea of having an "inhibited" child can affect the way parents behave towards him and maybe moulds that child (more) into that personality trait. Furthermore, ethological research has demonstrated in several primate species the importance of the mother (or experimental dummies) as a safety platform from which the baby explores the world. Work in The Netherlands has analysed more in detail the relevance of this factor at a given age for the development of specific phobias. Observation of the parent's behaviour could thus be important for the understanding of the causality of some forms of "shyness", i.e. how far are parents themselves exposed to various physical and social stimuli in the presence of their children at given ages?

Sven Carlsson's description of the evolution of the theoretical model supporting biofeedback training under influence of Gary Schwartz and the report of his own experience with mandibular dysfunction and dental fear represents on a small scale the opposition between the old behaviouristic school, cognitive psychology and ethology. Against the equipotentiality of stimuli principle of the first (any stimuli can be as easily associated with any other one), both latter sciences opposed the consideration that one has to investigate the meaning of a given stimulus for an animal species or for each individual. Hence it is absurd to test e.g. learning ability of different species in exactly the same experimental set-up; it must be relevant for that animal. Furthermore, there are so-called "constraints on learning". One species will form more easily a given association than another species. Predispositions to develop phobias for a certain range of stimuli could be present in the human species. This is something clinicians will have to take into account when trying to explore the relations between emotional problems, psychophysical activation and symptoms in order to understand better the treatment mechanisms. H.G. Wolff's methodology is a nice way of objectivating more the exploration of man's subjectivity and should be a source of inspiration for those involved with the study of "animal awareness" or "cognitive ethology".

Myself I have always tried to approach from different sides the problems I was studying. For the moment, my main interest goes to one of the so-called abnormal behaviours appearing in lasting conflict situations (this saves me from using the controversial term "stress"!): stereotypies. Stereotypies induced by dopamine agonists have since long been used as an animal model for the study of schizophrenia. However, as Axel Randrup once said, the problem with the amphetamine model is the amphetamine. Potential neuroleptics are currently screened through their capacity to antagonise drug-induced stereotypies. However, I am wondering whether we are closer to reality when using stereotypies induced by environmental manipulations. On the other hand, such behaviours are being used in the field of applied ethology as indicators for the assessment of the welfare of animals in captivity (modern husbandry systems, zoological gardens). Up to now, we have no certainty that welfare is - or has been - affected whenever stereotypies are performed, although I think in many cases it is very likely. It is hence worthwhile to understand more about the mechanisms and causality of these behaviours.

It should first be mentioned that one should describe clearly the stereotypy one is studying and the situation in which it appears, because we do not know yet whether we are dealing with the same phenomenon when observing e.g. baby primates isolated from their mother, heifers tied up in stalls or polar bears kept in a zoo.

As far as I am concerned, after an initial ethological approach I integrated a neurobiochemical one. I now wonder whether a cognitive-neuropsychological one would not be profitable.

Key questions are:
-    Which are the precise causal factors of each conflict-induced stereotopy? This can be multifactorial or monofactorial. Such investigations are important from the pragmatic point of view, in order to design environments which are not stereotypogenic. The trouble is that stereotypies can become emancipated from the original causal factor(s), their frequency level being thereafter influenced by merely modulating factors.
-    Which underlying neurophysiological mechanisms are responsible? How do they evolve during the development of the behaviour?
-    Why do some individuals develop stereotypies and other do not while living in the same environment?
-    Are such behaviours functional? I.e. test the hypothesis that they are arousal-reducing.

Besides working with pigs, I have developed a kind of cheaper laboratory model using bank voles (*Clethrionomys glareolus*). These rodents develop easily stereotypies when simply bred in barren laboratory cages. Most individuals jump up-and-down at a rate of 3.2 jumps/sec. There are large individual differences ranging from no jumps to more than 45.000/day. Group averages usually lie around 5.000-7.000 jumps/day.

More voles develop stereotypies because of the barreness of the cage than because of its size.

Recent work in my laboratory performed by J. Cooper shows that older stereotypies tend to have more chances to get emancipated, i.e. the animal continues to perform stereotypies even when transferred from a barren small cage to a big enriched one.

There is a positive correlation between general activity and stereotypy level. However, individuals stopping to present stereotypies when e.g. transferred to an enriched environment also decrease other activities (except at very high levels). This means that in these animals stereotypies are not usually performed at the cost of other behaviours, but that some individuals are more prone to react actively in conflict situations than others (all activities increase, stereotypies included).

Work with another collaborator, D. Kennes, demonstrated that dopamine (use of haloperidol and interferences with the biosynthesis) and opioids (naloxone) are involved in the expression of these stereotypies. A developmental study during the first 10 months of life showed that haloperidol keeps inhibiting these stereotypies, while the inhibitory effect of naloxone disappears after 4 months. Could it be that emotional factors (opioid-mediated limbic functions?) are responsible in the first stages of development, while stereotypies would become rather pure motor automatisms later on? R. Dantzer (Bordeaux) advocates that stereotypies could result from pure neurophysiological processes without even any emotional implication. In any case it shows how important it is to mention whenever possible since how long the stereotypy one is working with has been performed.

An amount of data exists suggesting that stereotypies are (part of) a de-arousing homoeostatic mechanism. It would take too long to discuss this matter here. I will only state that it is too early to affirm anything with certainty concerning the causal or correlational nature of the relation between the stereotypy performance and the decrease or the absence of different physiological stress parameters.

14 Neonatal Imitation: A Social and Biological Phenomenon

Mikael Heimann
University of Göteborg

INTRODUCTION

The human capacity to imitate is well known, very basic, and almost as common as our use of language for communicative purposes (see Speidel & Nelson, 1989). Imitation is used by children when they play, by the adult trying to learn tennis, as well as by the student observing an experienced teacher when learning about a complicated experiment. These are only a few examples of situations when the capacity to imitate can be recognized. In reality, imitation can be observed within almost any ongoing interaction between two or more persons.

Until recently, however, the capacity to imitate has been thought of as a slowly growing skill. Guillaume's (1926/1971) and Piaget's (1951/1962) theories concerning the development of imitation have been dominant. Both Guillaume and Piaget had concluded, after observing infants and young children, that the capacity to imitate developed slowly over time. Neither the essential cognitive nor perceptual abilities were assumed to be present in the very young child. Piaget's theory describes the development of imitation as a process following the same invariant sequences for all children. No special consideration is given to socio-emotional factors or to individual differences.

This view of imitation has been challenged during the last decades. Since the mid-70's, several reports have claimed that imitation can be observed as early as during the first weeks or month of life (e.g., Dunkeld, 1978; Field, Woodson, Greenberg, & Cohen, 1982; Heimann & Schaller, 1985; Heimann, Nelson, & Schaller, 1989; Kugiumutzakis, 1985; Maratos, 1973; Meltzoff, & Moore, 1977, 1983a, 1989; Wolff, 1987). If this is so, several questions arise: 1) How is this capacity mediated? 2) Do all children imitate? 3) What is the function of this early imitative capability?

The following sections provide a review of the evidence to date for the existence of a neonatal imitative capacity. Furthermore, the issue of individual differences is discussed at some length. The chapter also includes a discussion of methodological differences between studies reviewed as well as suggestions that have been made for the function of neonatal imitation.

Support for the existence of an early imitative capacity

The earliest findings concerning neonatal imitation were anecdotal case studies (Gardner & Gardner, 1970; Zazzo, 1957) reporting one single infant showing imitative reactions. It was not until some years later that more elaborate studies were conducted, Maratos' (1973, 1982) and Meltzoff and Moore's (1977) reports being the first two.

Maratos (1973, 1982) observed 12 infant girls during the first six months of life and found evidence of tongue protrusion and mouth opening during the second month of life. Regrettably, her method for collecting data makes it difficult to draw any definite conclusions. The main weakness is that her subjects were never videorecorded or filmed. Maratos' own observations and conclusions could, therefore, never be validated by independent observers unaware of the experimental procedure.

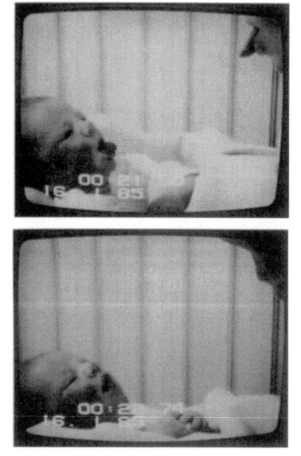

Fig. 14.1. A 2-day old female infant from the Heimann et al. (1989) study displaying imitiative responses to tongue protrusion and mouth opening.

174

Dunkeld (1978) reported that she observed imitation of tongue protrusion among her six subjects (3-11 weeks of age). Similar findings had been reported one year earlier by Meltzoff and Moore (1977). Their data (12 infants; 2 - 3 weeks of age) strongly suggest that a general capacity to imitate mouth opening and tongue protrusion exists in infancy. Weaker support was obtained for imitation of sequential finger movements. However, both Meltzoff and Moore's and Dunkeld's studies used more stringent experimental designs than Maratos' did, including both videorecording of the sessions and the use of "blind" observers for data coding.

In their second study, Meltzoff and Moore (1983a, b) report results on neonatal imitation after using an even more stringent procedure and a large sample of very young children. Forty neonates, ranging in age from 42 minutes to 72 hours, were observed in a dimly lighted room. The infant's behavior was recorded with an infrared-sensitive video camera focused so that it would give a close-up picture. The results of the 1983-study by Meltzoff and Moore are compelling: strong significant effects were found for imitation of both mouth opening and tongue protrusion. Only 4 of the 40 children displayed no imitative responses at all. Meltzoff and Moore concluded that the results from both their studies (1977; 1983a) show that the very young infant has the capacity to match input through one modality with output via another modality. This capacity is named "active intermodal mapping" (AIM) by Meltzoff and Moore.

In their third report on neonatal imitation, Meltzoff and Moore (1989) expand the range of behaviors that can be imitated in finding that neonates not older than 72 hours also imitate head turning.

In an attempt to replicate Meltzoff and Moore's first study, Jacobson (1979) found that children between 6 and 14 weeks of age did show matching behavior, but that this response was unspecific. No significant difference was obtained between the number of tongue protrusions after modeling of tongue protrusion and the frequency displayed when a pen was moved towards the infant's face. Jacobson concluded that the children's responses most likely were due to the activation of an innate releasing mechanism. Hence, according to Jacobson, no true imitation is possible at this early age; only imitative-like behaviors can be seen.

Evidence of neonatal imitation can also be found in a study by Field, Woodson, Greenberg and Cohen (1982) on the ability of infants (N=72; age: not over 36 hrs) to discriminate between sad, happy, and surprised expressions. In a subsequent study, Field, Woodson, Cohen, Greenberg, Garcia, and Collins (1983) replicated the findings for a sample, including term as well as preterm infants.

Additional support for the neonates' imitative capacity has been presented by eleven recent studies reporting positive findings from infants observed in the USA (Abravanel & Sigafoos, 1984; Wolff, 1987), Nepal (Reissland, 1988), Israel (Kaitz, Meschulach-Sarfaty, Auerbach, & Eidelman, 1988), France (Fontaine, 1984), Greece (Kugiumutzakis, 1985), Japan (Ikegami, 1984), Italy (Vinter, 1986), and Sweden (Heimann, 1989b; Heimann, & Schaller, 1985; Heimann, Nelson, & Schaller, 1989).

Fontaine (1984) observed infants cross-sectionally between 3 and 24 weeks of age. Imitation of tongue protrusion and mouth opening were observed in 2-month-old infants. He also found that the infants of this age imitated swelling of the cheeks as well as closing of the eyes.

Kugiumutzakis (1985) presents evidence of imitation of tongue protrusion and mouth opening in infants no more than 40 hours old. The study was conducted on Crete, Greece, and the sample included both full-term babies with complicated and uncomplicated vaginal delivery as well as children born with Caesarean section. A small separate sample (N = 12) of infants born prematurely (age in gestational weeks: 35-36) was also studied with the same results: clear imitative responses of mouth opening and tongue protrusion.

Although Kugiumutzakis' study shows impressive results, it has at least two important draw-backs: 1) Kugiumutzakis himself was the only person to code all of the tapes. A blind coder was used in only 33 percent of the observations. 2) Since he was the experimenter modeling the behaviors, the possibility exists that he unintentionally adjusted his modeling to the infants' behavior.

Heimann, Nelson, & Schaller (1989) observed infants longitudinally from birth to three months of age. Their results indicate that an overall imitative capability exists at 2-3 days of age and at 3 weeks of age for imitation of tongue protrusion. The result was clearly significant and almost 70 percent of the infants actually displayed imitative responses (see Table 1). The tendency to imitate tongue protrusions was especially strong at 3 weeks of age. However, no similar effect was noted for imitation of tongue protrusion at 3 months of age.

Table 14.1. Percentage of Children Displaying Imitative Responses (+) at Observation 1, 2, and 3.

	Observation		
	---	---	---
	1	2	3
Behavior:	(n=23)	(n=23)	(n=24)
Mouth opening	44.0	39.0	46.0
Tongue protrusion	61.0*	69.5**	29.0

* = $p<.05$; ** = $p<.01$ (from Heimann et al., 1989)

The marked decrease in imitative responses after modeling of tongue protrusion at three months of age indicates a downward trend in imitation over the first three months of life. This finding is in line with what has been reported by others (Abravanel & Sigafoos, 1984; Fontaine, 1984; Kugiumutzakis, 1985; Maratos, 1973).

In sum, a substantial number of studies have reported that imitation can be observed during the first two months of life. These studies have not always used exactly the same methodology nor always the same behavior, making neonatal imitation a robust phenomenon. Furthermore, the studies reviewed also represent findings from at least eight different countries and more than ten different research teams.

Methodological issues

Although the majority of findings presented has been in favor of an early imitative capability, a substantial number of studies have failed to replicate these results (Hayes and Watson, 1981; Kleiner & Fagan, 1984; Koepke, Hamm, Legerstee, & Russell, 1983a, b; Lewis & Sullivan, 1985; McKenzie & Over, 1983a,b; Neuberger, Merz, & Selg, 1983). Furthermore, the studies claiming that neonatal imitation is a robust phenomenon differ in several respects. Differences among the studies reviewed can be found in sample size, modeled behavior, way of presentation, length of the response period, definition and coding of responses, and setting.

Coding of responses

Some of these methodological factors were addressed by Heimann et al. (1989) by coding the responses into different categories. For tongue protrusion, three categories were used: weak, medium, or strong (see Table 2). The result for imitation of tongue protrusion showed a clearly significant group effect. This observed significance was, however, only observed as long as weak responses were included in the analysis. No overall effect was seen if medium or strong responses were used.

Table 14.2. Definition of imitative responses used by Heimann et al. (1989)

Type of response	Definition
Tongue protrusion (TP):	
1) Weak TP	The tongue was not protruded beyond the lips, although a clear forward movement of the tongue could be seen. The tongue had to pass the posterior part of the lip.
2) Medium TP	Coded whenever the tongue was protruded beyond the lips, but not longer than a distance equal to the width of the lip.
3) Strong TP	Noted if the tongue was protruded beyond the outer part of the lip with at least the length of the lip's width.
Mouth opening (MO):	
1) Weak MO	Defined as a clearly visible separation of the lips, although the end state was only a minor change (never exceeding the width of the lips).
2) Strong MO	Coded if the lips were clearly and strongly separated from each other. The opening had to be wider than the width of lips. (In order to be coded as a strong or weak MO, no simultaneous protrusion of the tongue was accepted. Yawning was not included as an adequate mouth opening response).

Large differences in rules used for judging the infants' responses can be observed among studies investigating neonatal imitation. Meltzoff and Moore's most recent experiments (1983a; 1989) used a criterion for tongue protrusion similar to weak responses in the Heimann et al. (1989) report. Few studies are, however, sufficiently explicit about how they have defined their responses. For example, Neuberger, Merz, and Selg (1983), and McKenzie and Over (1983a) state that the tongue has to "pass clearly beyond the lips", Maratos (1982) talks about "distinct tongue protrusions", whereas Kugiumutzakis (1985) writes that the "tongue must be seen to leave the mouth". It is difficult to judge from these definitions if the authors have accepted only strong, only strong and medium, or strong, medium, and weak responses.

Partial responses have been used by Vinter (1986), Lewis, and Sullivan (1985), Abravanel and Sigafoos (1984), and Wolff (1987). Both Vinter and Lewis and Sullivan report that their results did not differ due to whether "exact" or "partial" responses were coded. However, both Abravanel and Sigafoos and Wolff did, however, find that an overall imitative tendency only existed for partial responses. A comparison between the criteria used by Abravanel and Sigafoos and the criteria used by Heimann et al. (1989) reveals that both "weak" and "medium" responses, as defined by Heimann et al., are included in what Abravanel & Sigafoos have called "partial responses". Furthermore, Wolff (1987) studied 10 infants of various ages (range: 4-7 weeks) and found that imitation of tongue protrusion does occur but that the result depends on how the responses have been defined. In contrast, Lewis and Sullivan (1985) report no evidence of imitation of tongue protrusion regardless of whether partial or exact responses are used.

Response time

Differences among the studies have also been noticed for response time used and for length of presentation. Heimann et al. found the strongest indication of imitation when the complete response period of 60 seconds was used. There was no indication that the reponse pattern during the first 30 seconds differed from what was observed for the last 30 seconds of the response period. However, a clear difference was noted as to whether responses were analyzed during or after modeling. For example, the observed overall imitative reaction in response to tongue protrusion was only noted for the children's behavior after modeling.

The work by Heimann et al. demonstrates that whether imitation is observed at a group level depends on the criteria used for defining adequate responses. It is also shown that a strong difference in response pattern exists between responses emitted during or after modeling.

Stimulus features

Another important factor to consider when studying neonatal imitation was demonstrated by Vinter (1986). She studied imitation of tongue protrusion during the first week of life in 36 Italian children with a median age of 4 days. In Vinter's study, the stimulus was manipulated so that some children were presented with a moving tongue while another group of children were shown a static tongue. She found that imitation only occurred among those children presented with a moving tongue.

Stimulus factors necessary for neonatal imitation of tongue protrusion have, furthermore, been analyzed in a study conducted in Japan (Ikegami, 1984), reporting "that the complete face, whether real, schematic or mirror image, was efficient for 1-month-old infants..." (p. 127). Presenting only the mouth pattern itself did not, however, render any imitative response.

Type of behavior

Tongue protrusion has been used in 21 studies reviewed, mouth opening in 12 studies. These are by far the two most commonly used. They are also two examples of behaviors that are displayed frequently and spontaneously by infants.

All other kinds of behavior are either used by only a small number of researchers or are unique for the study in focus. The following is a list of additional types of behavior used among the studies reported in this review: facial expressions (5 studies), hand opening (4), sequential finger movement (3), eye blinking (3), lip protrusion (3), head turning (3), hand movement (2), chest tapping (1), chin tapping (1), arm waving (1), swelling of cheeks (1), index finger protrusion (1), nodding head (1), shaking head (1), and sounds "A", "M", and "ANG" (1).

It may be worth noting that of the 21 studies in which tongue protrusion was used, 16 report positive findings. Note that one study (Abravanel & Sigafoos, 1984) reports both positive and negative results depending on whether exact or partial responses are used. For mouth opening, six studies observed imitative responses among their subjects.

The studies also differ in using either an experimenter or the mother as the person presenting the gestures to the infant. Most studies (N = 21) have used an experimenter, two have used both an experimenter and the mother (Field, Goldstein, Vega-Lahr, & Porter, 1986; Lewis & Sullivan, 1985), and two have used only the mother (Dunkeld, 1978; Heimann, & Schaller, 1985).

Individual differences

Field (1982) stressed that, although her data favor the conclusion that neonatal imitation is a real phenomenon, she observed large individual differences among the infants' responses. Roughly one-third of her sample displayed a strong tendency to imitate, one third showed weak imitative responses, and one third displayed no imitation. In a subsequent study, Field, Goldstein, Vega-Lahr, and Porter (1986) observed imitation of happy, sad, and surprised expressions among 40 infants longitudinally from 2 to 6 months of age. In short, they found that imitation occurred over the whole time-period although they detected a decreasing trend over time. Furthermore, they did find that 40 percent of their sample were consistently good imitators.

In a study on neonatal imitation of tongue protrusion and mouth opening, Heimann and Schaller (1985) reported that approximately half of their group displayed imitative responses as judged both by independent coders and the mothers' verbal reports.

Recently, Heimann et al. (1989) reported findings on individual differences and neonatal imitation after having studied infants longitudinally over the first three months of life (observations were made at 2-3 days, 3 weeks, and 3 months of age). The results indicate a short-time stability in imitative behavior over the first three weeks of life (see Fig. 2). Several imitation index scores at 2-3 days of age and 3 weeks of age displayed a significant correlation: of 14 observed correlations, four reached statistical significance (range: .49 - .68).

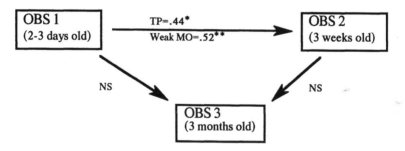

Fig. 14.2. Some examples of observed correlations between imitation of tongue protrusion (TP) and mouth opening (M=) at 2-3 days of age (Observation 1), 3 weeks of age (Observation 2), and 3 months of age (Observation 3). From Heimann, Nelson, & Schaller, 1989.

Using these same index scores, no similar indication of stability in imitative response patterns was observed between observation 1 (at 2-3 days of age) and 3 (3 months of age) or observation 2 (3-weeks-old) and 3. Hence, the observed stability seems to disappear as the children grow older. This is to be expected, since our knowledge of early brain development suggests that stability in response patterns over the first three months of life ought to be rare. Bronson (1982) has pointed out that major neural development takes place during this early phase of life. There is a marked increase in the development of connections between neocortex and subcortical areas as well as of connections between different cortical areas during these first months. These changes in neural organization suggest, according to

Bronson, that it would be difficult to find stability between the behavior of neonates measured around birth and later behavioral observations at 3 months of age.

Field (1982) put forward a hypothesis suggesting a possible relationship between early imitative responses and temperamental constructs (externalizers vs. internalizers). In a subsequent study, Field et al. (1986) reported that 40 percent of her sample were consistently good imitators, indicating a stability over time when imitation of facial expressions are observed. Stability in expressiveness per se has also been reported (Field, 1986; Fox, 1986), further supporting the existence of important early individual differences.

Following the suggestion by Field (1982), Heimann (1989a) assessed temperament at 3 months of age using the Baby Behavior Questionnaire (BBQ; see Bohlin, Hagekull, & Lindhagen, 1981; Hagekull, Lindhagen, & Bohlin, 1980). A possible relationship between the subscale Intensity/Activity and imitation was found: imitation at 3 weeks and 3 months of age displayed positive correlations with this subscale. This observation seems to fit Field's notion that more expressive infants are to be found among infants that imitate. Overall, however, very few correlations between the BBQ and imitation measures were obtained, suggesting the possibility of the results being due to chance probabilities. In conclusion, the existence of short-time stability in imitative behavior has been demonstrated partially by the results presented by Heimann, Nelson, & Schaller (1989). Exactly how this stability develops over time is, however, difficult to interpret. For a more conclusive analysis of early stability, further investigations are needed. Furthermore, Field's hypothesis concerning a possible relationship between early temperamental constructs and neonatal imitation needs to be explored further. The evidence to date is mostly speculative.

Neonatal imitation and early cognitive and social development

Cognitive development

Piaget's theory (1951/1962) is still the most complete and detailed theory of imitation available. The implication of Piaget's standpoint is that in order to be able to imitate, the child's schemata must be capable of differentiation when they are confronted with information or experiences from the external world. Furthermore, the behavior of the model must be analogous to the result the child has acquired. The first signs of true imitation are therefore, according to Piaget, not seen before stage four (approx. 8-12 months of age) of the sensori-motor period. Piaget's position is supported by the studies reporting failures to replicate, whereas those investigators having found support for neonatal imitation tend to argue that the standpoint of Piaget has to be reevaluated.

Meltzoff and Moore (1977; 1983a; 1989) strongly argue that the human infant is capable of matching input through one modality (e.g., vision) with output via another modality (e.g., a motor act that is invisible to the infant). They call this process "active intermodal mapping" (AIM) and propose that it "is mediated by a representational system that allows the infant to unite, within one common framework, his own body-transformations and those by others" (Meltzoff & Moore, 1983a; p. 295). In essence, Meltzoff and Moore suggest that the infant, already at birth, possesses a representational system which makes it possible to imitate types of behavior that the infant cannot see itself perform (e.g., tongue protrusion). In contrast, others have claimed that neonatal imitation is governed by an innate releasing mechanism (Abravanel & Sigafoos, 1984; Jacobson, 1979) or is simply an artifact (Hayes & Watson, 1981).

To date, a vast majority of the studies having investigated neonatal imitation have reported positive findings. Hayes and Watson's (1981) conclusion that neonatal imitation is an artifact is, therefore, not accepted. It is, however, more difficult to judge whether Jacobson's notion of an innate releasing mechanism is responsible for the observed results or if Meltzoff and Moore's (1983, 1985) suggestion of an "active intermodal mapping"-process is more correct. The observation by Meltzoff and Moore (1985, p. 153) that "some infants converge toward more accurate imitation matches over successive efforts" fits the hypothesis that neonatal imitation is governed by a more active "supramodal" process and not by an innate releasing mechanism.

Social development

Several authors have suggested that neonatal imitation must be understood within the early social development (Abravanel & Sigafoos, 1984; Ikegami, 1984; Kugiumutzakis, 1985; Maratos, 1982). For instance, neonatal imitation is labeled "social reflex" by Abravanel and Sigafoos (1984) and is explained as an innate releasing mechanism. This perspective is similar to what Jacobson (1979) suggested, and also similar to what Marcus (1984) proposed after studying imitation within mother-infant interactions. Both Kugiumutzakis (1985) and Maratos (1982) avoid talking about imitation as a reflex. Instead, they argue that neonatal imitation has to be understood within the framework of early social interaction. As example, Maratos (1982) writes (p. 98):

> *Furthermore, it is the author's belief that imitation is functional in eliciting and maintaining social interaction of the infant with the human beings in his environment. Mutual imitation is a privileged form of communication before smiling and vocal responses develop.*

In a relatively recent chapter, Meltzoff (1985) discussed the implications of neonatal imitation. He goes beyond what has earlier been described for the process of active intermodal mapping and suggests that the infant's early cognitive abilities make the infant "socially attuned". According to Meltzoff, the neonate's imitative competence "constitutes important roots for subsequent cognitive and social development" (p. 29).

Heimann (1989b) has presented the first evidence to date of a relationship between neonatal imitation and early social behavior. A correlational analysis between categories coded in mother-infant interaction (when the child was 3 months old) and the infants' imitative responses when 2-3 days, 3-weeks, or 3-months old revealed interesting findings (see Fig. 3). It was found that the category Avert (brief gaze aversion) displayed strong negative correlations with imitation at 3 months, 3 weeks, and 2-3 days of age. Of 27 observed correlations between Avert and imitation, 24 were in a negative direction of which eight were strong enough to reach statistical significance.

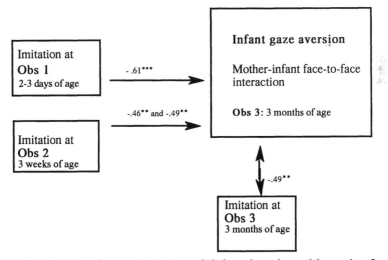

Fig. 14.3. Correlations between imitation at 2-3 days, 3 weeks, and 3 months of age and brief gaze aversion measured in a face-to-face interaction at 3 months of age (from Heimann, 1989).** = $p < .05$; *** = $p < .01$

This finding indicates that children displaying a high level of imitation tend to show fewer instances of brief gaze aversion when observed in spontaneous interaction with their mothers. Gaze aversion reflects eye-to-eye contact which has been suggested to be indicative of the quality of the emerging relationship between the mother and her child (Field, 1977; Greenspan, 1981).

Field (1977) demonstrated that an increase in eye-to-eye contact could be obtained by asking the mothers to imitate their children. In light of the findings by Heimann (1989b), one might speculate about a possible regulative function for imitation in the early mother-infant relationship. Regardless of who imitates whom, imitation can be viewed as a factor that facilitates each participating person's sensitivity to social cues within the interaction.

Psychobiological considerations

In the present author's view, the data on neonatal imitation support the hypothesis that neonates can act upon visual information in order to perform a motor response they cannot see themselves perform. This capacity has been observed during the first days of life, making it possible to suggest an innate capacity for amodal perception. Vinter (1986) further suggests that the young infant's supramodal competence is mediated through subcortical structures. It is well known that cortical areas are relatively immature at birth (Bronson, 1982) and experiments with animals have demonstrated that subcortical areas might control early behavior. In a series of experiments investigating the performance of rhesus monkeys on a delayed-response task, Goldman-Rakic (1985) has shown that the behavior of the monkeys is under subcortical control during infancy. Frontal cortical areas do not take over until the monkey is approximately two years old. Goldman-Rakic writes (p. 292):

> Thus, the appearance of an ability (e.g., imitation behavior, attention, visual pursuit), however adult-like in its phenotypic expression, cannot be taken as evidence that the neural system in question is fully differentiated.

Vinter (1986) further suggests that the subcortical area responsible for early imitative behavior might be found within the tectopulvinar visual system. More specifically, she proposes that the superior colliculus is a possible candidate for mediating the intermodal responses of the neonate. It is not possible, on the grounds of our knowledge to date, to draw any definite conclusions about which subcortical structure should be considered. Recent findings about multisensory connections within the superior colliculus (Meredith & Stein, 1985) do, however, give additional support to Vinter's hypothesis. Furthermore, the findings, reported by Heimann et al. (1989), that stability in imitation can be observed over the first three weeks, but not the first three months of life, could indicate an age shift due to changes in neural organization. It may be that subcortical areas are responsible for imitative reactions seen during the first weeks of life while cortical structures play a more active role when the child is 3 months old.

Meltzoff and Moore (1977, 1983a, 1989) propose that a process they call "active intermodal mapping" might explain the astonishing capacity by the human neonate to imitate several different behaviors. Moreover, they are never explicit as to whether this process is governed mainly by subcortical or cortical areas during the first months of life. It is the present author's view that our knowledge of early brain development makes it difficult to assume that cortical areas should dominate. We know that the visual cortex is immature at birth (Atkinson & Braddick, 1989; Bronson, 1982) and we know from the studies by Goldman-Rakic (1985) that the same behavior can be governed by different parts of the brain at different stages in development. Thus, it is not necessary to postulate a cortical mechanism in order to understand the biological substrate for the neonates capacity to imitate. The "active intermodal mapping"-process (AIM) is most probably governed by subcortical structures during the first months of life.

CONCLUSION

The review presented has demonstrated 1) an overall imitative capacity during the first weeks of life (e.g., Meltzoff & Moore, 1977, 1983, 1989; Field et al., 1982; Heimann, Nelson, & Schaller, 1989; Kugiumutzakis, 1985; Vinter, 1986), 2) some evidence for individual differences in neonatal imitation (Field, 1982; Field et al., 1986; Heimann & Schaller, 1985; Heimann et al., 1989), and 3) short-time stability in imitation during the first month of life (Heimann et al., 1989). Furthermore, evidence of a significant relationship between imitation during the first three months of life and observed mother-infant interaction has been presented (Heimann, 1989b).

It is always difficult to judge the long-term consequences of individual differences observed very early in life (Stratton, 1982). Early behavior and early individual differences might be important in themselves even if no long-term continuity can be found. Hence, there is no need to concentrate only on continuity. This is also stressed by Hall and Oppenheim (1987) in reviewing psychobiological aspects of early behavioral development. They list three categories that, by themselves or in combination with each other, can help explain the significance of behavior seen during the prenatal, perinatal, or postnatal period (Hall & Oppenheim, 1987, p. 116):

> A. *For any early behavior... the initial emergence of a behavior pattern may simply represent an epiphenomenon reflecting the 'anticipatory' onset of a function that will only become adaptive at some later age...*
> B. *Early behavior may... represent necessary antecedants to later behavior which, when suppressed... result in ... atypical development...*
> C. *Finally, early behavior may serve some immediate adaptive role for the ... infant (ontogenic adaption).*

Bjorklund (1987) suggests that neonatal imitation has to be understood within Hall and Oppenheim's category C, as an example of ontogenic adaption. According to Bjorklund, neonatal imitation helps to maintain the social interaction between the young infant and an adult. He further proposes that this capacity is biologically "wired-in" to the infants' visual system. In his view, neonatal imitation has a survival value in promoting the social contact with the caretaker. The observed relationship between early imitation and infant gaze behavior (Heimann, 1989b) lend support to the idea that neonatal imitation may play an important role in early social development.

Following the ideas proposed by Bjorklund, Hall and Oppenheim, and others, there is no need to expect a long-term relationship between neonatal imitation and imitative skills emerging later in life. It may be that imitation at different stages is governed by different processes and that imitation serves different purposes at different ages (Tharp & Burns, 1989; Nadel & Fontaine, 1989). It is, however, important to continue to study the area. Future studies have to adopt a long-term longitudinal approach including both imitation, socio-emotional, and interactive measures. A longitudinal approach is the only one that may provide us with answers to the most important questions remaining: 1) How (if at all) does neonatal imitation relate to later imitation seen around 9-18 months of age? 2) What is the function of the observed relationship between neonatal imitation and mother-infant interaction when data from both areas are related to long-term development? and 3) Do imitative behaviors play a lesser or different role in social interaction and skill development for low imitators as compared with medium imitators or high imitators?

ACKNOWLEDGEMENTS

Preparation of this report was supported by a grant from the Bank of Sweden Tercentenary Foundation to Mikael Heimann (#89/313).

REFERENCES

Abravanel, E., & Sigafoos, A.D. (1984). Exploring the presence of imitation during early infancy. *Child Develop., 55,* 381-392.

Atkinson, J., & Braddick, O. (1989). Development of basic visual functions. In A. Slater and G. Bremner (Eds.), *Infant Development* (pp. 7-41). London: Erlbaum.

Bjorklund, D.F. (1987). A note on neonatal imitation. *Develop. Rev., 7,* 86-92.

Bohlin, G, Hagekull, B., & Lindhagen, K. (1981). Dimensions of infant behavior. *Infant Behav. Develop., 4,* 83-96.

Bronson, G.W. (1982). Structure, status, and characteristics of the nervous system at birth. In P. Stratton (Ed.), *Psychobiology of the Human Newborn* (pp. 99-118). New York: Wiley.

Dunkeld, J. (1978). *The function of imitation in infancy.* Unpublished doctoral dissertation. University of Edinburgh, UK.

Field, T.M. (1977). Effects of early separation, interaction deficits and experimental manipulation on infant-mother face-to-face interaction. *Child Develop., 48,* 763-771.

Field, T.M. (1982). Individual differences in the expressivity of neonates and young infants. In R.S. Feldman (Ed.), *Development of nonverbal behavior in children.* New York: Springer-Verlag.

Field, T.M. (1986, April). *Individual differences in facial expressivity among singelton, monozygotic and dyzygotic twin newborn.* Paper presented at the Fifth International Conference on Infant Studies, Los Angeles, CA.

Field, T.M., Goldstein, S., Vega-Lahr, N., & Porter, K. (1986). Changes in imitative behavior during early infancy. *Infant Behav. Develop., 9,* 415-421.

Field, T.M., Woodson, R., Cohen, D., Greenberg, R., Garcia, R., & Collins, K. (1983). Discrimination and imitation of facial expressions by term and preterm neonates. *Infant Behav. Develop., 6,* 485-490.

Field, T.M., Woodson, R., Greenberg, R., & Cohen, D. (1982). Discrimination and imitation of facial expressions by neonates. *Science, 218,* 179-181.

Fontaine, R. (1984). Imitative skills between birth and six months. *Infant Behav. Develop., 7,* 323-333.

Fox, N. (1986, April). *Individual differences in facial expression.* Paper presented at the Fifth International Conference on Infant Studies, Los Angeles, CA.

Gardner, J., & Gardner, H. (1970). A note on selective imitation by a six week old infant. *Child Develop., 41,* 1209-1213.

Goldman-Rakic, P.S. (1985). Toward a neurobiology of cognitive development. In J. Meehler and R. Fox (Eds.), *Neonate Cognition: Beyond the Blooming Buzzing Confusion* (pp. 285-306). Hillsdale, N.J.: Erlbaum.

Greenspan, S.I. (1982). *Psychopathology and Adaption in Infancy and Early Childhood: Principles of Clinical Diagnosis and Preventive Intervention.* Clinical Infant Reports, No. 1. New York: International Universities Press.

Guillaume, D. (1971). *Imitation in Children* (E.P. Halperin, Trans.). Chicago: University of Chicago Press (Original work published 1926).

Hagekull, B., Lindhagen, K., & Bohlin, G. (1980). Behavioral dimensions in one-year-olds and dimensional stability in infancy. *Int. J. Behav. Develop., 3,* 351-364.

Hall, W.G., & Oppenheim, R.W. (1987). Developmental psychobiology: Prenatal, perinatal and early postnatal aspects of behavioral development. *Ann. Rev. Psychol., 38,* 91-128.

Hayes, L.A., & Watson, J.G. (1981). Neonatal imitation: Fact or artifact? *Develop. Psychol., 17 (5),* 655-660.

Heimann, M. (1989a). Imitation during the first months of life - what we know and what we don't know. *Nordisk Psykologi, 41,* 193-203 (in Swedish).

Heimann, M. (1989b). Neonatal imitation, gaze aversion, and mother-infant interaction. *Infant Behav. Develop., 12,* 495-505.

Heimann, M., Nelson, K.E., & Schaller, J. (1989). Neonatal imitation of tongue protrusion and mouth opening: Methodological aspects and evidence of early individual differences. *Scand. J. Psychol., 90,* 90-101.

Heimann, M., & Schaller, J. (1985). Imitative reactions among 14-21 days old infants. *Infant Ment. Health J., 6,* 31-39.

Ikegami, K. (1984). Experimental analysis of stimulus factors in tongue-protruding imitation in early infancy. *Japan. J. Ed. Psych., 32,* 117-127.

Jacobson, S.W. (1979). Matching behavior in the young infant. *Child Develop., 50,* 425-430.

Kaitz, M., Meschulach-Sarfaty, O., Auerbach, J., & Eidelman, A. (1988). A reexamination of newborns' ability to imitate facial expressions. *Develop. Psych., 24 (1),* 3-7.

Kleiner, K.A., & Fagan III, J.F. (1984, April). *Neonatal discrimination and imitation of facial expression: A failure to replicate.* Paper presented at the Fourth International Conference on Infant Studies, New York, NY.

Koepke, J.E., Hamm, M., Legerstee, M., & Russell, M. (1983a). Neonatal imitation: Two failures to replicate. *Infant Behav. Develop., 6,* 97-102.

Koepke, J.E., Hamm, M., Legerstee, M., & Russell, M. (1983b). Methodological issues in studies of imitation: A reply to Meltzoff & Moore. *Infant Behav. Develop., 6,* 113-116.

Kugiumutzakis, J. (1985). *The origin, development, and function of the early infant imitation.* Unpublished doctoral dissertation, University of Uppsala, Uppsala, Sweden.

Lewis, M., & Sullivan, M.W. (1985). Imitation in the first six months of life. *Merrill-Palmer Quart., 31,* 315-333.

Maratos, O. (1973). *The origin and development of imitation in the first six months of life.* Paper presented at the Annual Meeting of the British Psychological Society, Liverpool, UK.

Maratos, O. (1982). Trends in the development of imitation in early infancy. In T.G. Beaver (Ed.), *Regressions in Mental Development: Basic Phenomena and Theories* (pp. 81-101). Hillsdale, N.J.: Erlbaum.

Marcus, J. (1984, April). *Imitation during mother-infant interactions in the first six months of life.* Paper presented at the Fourth International Conference on Infant Studies, New York.

McKenzie, B., & Over, R. (1983a). Young infants fail to imitate manual and facial gestures. *Infant Behav. Develop., 6,* 85-95.

McKenzie, B., & Over, R. (1983b). Do neonatal infants imitate? A reply to Meltzoff & Moore. *Infant Behav. Develop., 6,* 109-111.

Meltzoff, A.N. (1985). The roots of social and cognitive development: Models of man's original nature. In T.M. Field and N. Fox (Eds.), *Social Perception in Infants* (pp. 11-30). Norwood, N.J.: Ablex.

Meltzoff, A.N., & Moore, M.K. (1977). Imitation of facial and manual gestures. *Science, 198,* 75-80.

Meltzoff, A.N., & Moore, M.K. (1983a). Newborn infants imitate adult facial gestures. *Child Develop., 54,* 702-709.

Meltzoff, A.N., & Moore, M.K. (1983b). The origin of imitation in infancy: Paradigm, phenomena, and theories. In L. Lipsitt and C. Rovee-Collier (Eds.), *Advances in infancy research, Vol II* (pp. 265-301). Norwood, NJ: Ablex.

Meltzoff, A.N., & Moore, M.K. (1985). Cognitive foundations and social functions of imitation and intermodal representation in infancy. In J. Meehler and R. Fox (Eds.), *Neonate Cognition: Beyond the Blooming Buzzing Confusion* (pp. 139-156). Hillsdale, N.J.: Erlbaum.

Meltzoff, A.N., & Moore, M.K. (1989). Imitation in newborn infants: Exploring the range of gestures imitated and the underlying mechanisms. *Develop. Psychol., 25 (6),* 954-962.

Meredith, M.A., & Stein, B.E. (1985). Descending efferents from the superior colliculus relay integrated multisensory information. *Science, 227,* 657-659.

Nadel, J., & Fontaine, M. (1989, September). *The communicative function of gestural and verbal synchronized imitation in normal prelinguistic toddlers and autistic children: A comparative study.* Poster presented at the Fourth Congress of the World Association of Infant Psychiatry and Allied Disciplines. Lugano, Switzerland.

Neuberger, H., Merz, J., & Selg, H. (1983). Imitation bei Neugeboreren - eine kontroverse Befundlage. *Zeit. Entwicklungspsychol. Pädagog. Psychol., XV,* 267-276.

Piaget, J. (1962). *Play, dreams and imitation in childhood* (S. Gattegno and F.M. Hodgson, Trans.). London: Routledge & Kegan (Original work published 1951).

Reissland, N. (1988). Neonatal imitation in the first hour of life: Observations in rural Nepal. *Develop. Psychol., 24 (4),* 464-469.

Speidel, G.E., & Nelson, K.E. (1989). A fresh look at imitation in language learning. In G.E. Speidel and K.E. Nelson (Eds.), *The Many Faces of Imitation in Language Learning* (pp. 1-21). New York: Springer-Verlag.

Stratton, P. (1982). Newborn individuality. In P. Stratton (Ed.), *Psychobiology of the Human Newborn.* New York: Wiley.

Tharp, R.G., & Burns, C.E.S. (1989). Phylogenetic processes in verbal language imitation. In G.E. Speidel and K.E. Nelson (Eds.), *The Many Faces of Imitation in Language Learning* (pp. 231-250). New York: Springer-Verlag.

Vinter, A. (1986). The role of movement in eliciting early imitations. *Child Develop., 57,* 66-71.

Wolff, P.H. (1987). *Behavioral States and Expression of Emotions in Early Infancy.* Chicago: University of Chicago Press.

Zazzo, R. (1957). Le problème de l'imitation chez noveau-né. *Enfance (2),* 135-142.

15 Normal Brain Functions and Signs of Dysfunction

Ivan Divac
University of Copenhagen

On one occasion, decades ago, at the beginning of our friendship, Knut Larsson and I discussed our stands in science. For me, nothing was more interesting than to understand normal functioning of the basal ganglia. This was an essential step, I believed, toward understanding the organization of the forebrain. During our discussion, I hoped to get Knut interested in this problem. Sincerely, kindly, and succinctly, he told me: "Ivan, I am only interested in sex". There we were, representatives of two approaches to the study of brain and behavior; he, looking for brain mechanisms underlying a complex, but well defined, behavior, and I, looking for the behavioral roles of a brain system. In the years which followed, Knut studied sexual consummatory behavior in male rats (see, Larsson, this volume), and I pursued, for a while, two lines: either mapping the prefrontal system (e.g., Divac & Diemer, 1980) and other 'vertical forebrain systems' (e.g., Collins & Divac, 1984), or analyzing the prefrontal impairment by means of varied behavioral tests (Divac, Gade, & Wikmark, 1975; Divac, Widmark, & Gade, 1975; Rosenkilde & Divac, 1976; Ursin & Divac, 1975).

With our different approaches, Knut and I attempted joint experiments on two occasions. We studied consequences on sexual performance of male rats, first by making small electrolytic lesions in different neostriatal regions and later after ablations of varying areas of the frontal cortex separately or together. Unfortunately, the results were not published because in neither study could the lesions be reconstructed. The technician in charge of the brains in the first experiment died without leaving useful codes to identify different brains. The second set of brains were lost somewhere in the triangle Gothenburg-Uppsala-Copenhagen. We, nevertheless, reported in an abstract (Larsson, Öberg, & Divac, 1980) that cortical ablations, regardless of their location and size, had no effect on male sexual performance in rats. That statement was made on the basis of histological results in other contemporary studies involving frontal cortical ablations made by the same surgeon (Wikmark, Divac, & Weiss, 1973; Divac, Gade, & Wikmark, 1975).

My first attempt to understand the organization of the forebrain showed that each of three groups of monkeys with lesions in one of three different neostriatal regions was impaired in one of three behavioral tasks. The impairments following any neostriatal region were qualitatively the same as those seen after the ablation of the anatomically related cortical area (Divac, Rosvold, & Szwarcbart, 1967). These results suggested a view very different from the concepts expounded by leading contemporary authorities in the field of the basal ganglia. For example, it seemed to me that according to Denny-Brown (1962), the entire brain was involved only in regulation of posture and elementary approach-avoidance motor patterns. There was scarcely any neural tissue left for complex functions. If Denny-Brown was right, we would have to believe in the existence of the soul.

Thus, from the beginning of my involvement in research I was confronted with two classes of observations, each indicating a different function of the basal ganglia, "cognitive" and "motor". Two explanations seemed possible: that artifacts were responsible for one or both of these classes, or that there was an explanation for their coexistence. My attempts to falsify the notion of involvement of the basal ganglia in cognitive functions failed (e.g., Divac & Diemer, 1980; Divac, Markowitsch, & Pritzel, 1978), but akinesia, rigidity and tremor, seemed also attributable to dysfunctions of the basal ganglia. Therefore, I attempted to develop a concept which could explain pathogenesis of both the disappearance of complex behaviors and the appearance of motor symptoms in organisms with basal ganglia dysfunctions (Divac, 1984; Divac, Öberg, & Rosenkilde, 1987). The rest of this chapter is mainly devoted to a summary of the ensuing concept.

We must start with a brief description of the anatomo-functional relations in the forebrain. Different data suggest that the basal ganglia complex can be understood as an interface between the cerebral cortex and both pyramidal and some extrapyramidal motor pathways (e.g., Divac, 1977). Furthermore, different cortical areas are related to different neostriatal regions, forming so-called vertical forebrain systems which also include at least the respective specific thalamic nuclei. Each vertical system has a unique function, (e.g., Divac, 1984). Secondly, unit recording studies in the cerebral cortex and neostriatum strongly suggest that an important aspect of the functioning of these structures lies in the patterning of their neuronal discharges: different neurons discharge only under specific circumstances (Rolls & Williams, 1987). Obviously, the pattern of discharges normally gets from one to the next set of neurons in the chain of basal ganglia and cerebro-spinal paths by means of transmitters.

This concept of organization of the forebrain has a considerable explanatory power. First, if patterned discharges are the basis of the complex functions of the thalamo-cortico-neostriatal 'vertical systems', then each agent, pathological or experimental, which interferes with this pattern will abolish normal function. Indeed, impairment of delayed alternation in animals was found when the neostriatal region related to the prefrontal cortex was destroyed, or faradically stimulated, or cooled, or injected with different drugs. Similar manipulations carried out in other parts of the neostriatum affected other complex behaviors (references in Öberg & Divac, 1979). The chapters by Everitt and Ahlenius in this volume give further examples of functional heterogeneity of the neostriatum.

According to the same concept, motor signs obtained with manipulations in the neostriatum may be generated in two ways. First, one part of the neostriatum is related to the motor cortical areas and has unique motor functions which have recently been revealed (Frysz & Stepien, 1983; Pisa, 1988a, 1988b; Whishaw, O´Connor, & Dunnett, 1986). Secondly, also the role of the basal ganglia as an interface between the cerebral cortex and cerebro-bulbospinal pathways seems responsible for some motor symptoms seen in dysfunctions of the basal ganglia (Divac et al., 1987). The agents which interfere with patterning of neostriatal discharges not only impair complex behaviors, but in addition enhance, diminish or modify the transmitter release in the targets of neostriatal efferents. In this way, electrical stimulation or inactivation of large volumes of the neostriatum, regardless of the locus of action, induces motor signs such as body curving or circling. While stimulation and inactivation have *similar* effects on complex behaviors because of the dependence of these behaviors on patterned neuronal discharges of respective 'vertical systems', these agents have *opposite* effects on the motor symptoms which can be described as artifacts induced by abnormal, indiscriminate quantitative changes of transmitter release in the targets of neostriatal output. These motor signs are scarcely informative about normal functions of the basal ganglia but retain their importance as signals of dysfunction.

In conclusion, separation of the message coded by patterned neuronal discharges, on the one hand, and the synaptic transmission which carries this message from structure to structure can explain the occurrence of both impairments of complex behaviors and so-called motor signs. In other words, clinically relevant observations of changes of muscle tone in basal ganglia dysfunctions leave the possibility open for involvement of the basal ganglia in much more complex functions, e.g., regulation of behavior.

This way of looking at the forebrain organization, although far from complete, gives me a feeling of having resolved a puzzle which confronted me at the start of my research. Thus, Knut (as evident from his chapter in this volume) and I, while pursuing our common goal along different lines of approach, have both been lucky to believe that our efforts were worthwhile.

REFERENCES

Collins, R.C., & Divac, I. (1984) Neostriatal participation in prosencephalic systems. Evidence from deoxyglucose autoradiograph. In R.G. Hassler and J.F. Christ (Eds.), *Advances in Neurology* (pp. 117-122). New York: Raven Press).

Denny-Brown, D. (1962). *The Basal Ganglia*. Oxford University Press.

Divac, I. (1977). Does the neostriatum operate as a functional entity? In A.R. Cools, A.H.M. Lohman, and J.H.L. van der Bercken (Eds.), *Psychobiology of the Striatum* (pp. 21-30). Amsterdam: Elsevier.

Divac, I. (1984). The neostriatum viewed orthogonally. In *Functions of the Basal Ganglia* (pp. 201-215). CIBA Foundation Symposium 107. London: Pitman.

Divac, I., & Diemer, N.H. (1980). The prefrontal system in the rat visualized by means of labelled deoxyglucose. Further evidence for functional heterogeneity of the neostriatum. *J. Comp. Neurol., 190,* 1-13.

Divac, I., Gade, A., & Wikmark, R.G.E. (1975). Taste aversion in rats with lesions in the frontal lobes: no evidence for interoceptive agnosia. *Physiol. Psychol., 3,* 43-46.

Divac, I., Öberg, R.G.E., & Rosenkilde, C.E. (1987). Patterned neural activity: Implications for neurology and neuropharmacology. In: J.S. Schnider and T.I. Lidsky (Eds.), *Basal Ganglia and Behavior: Sensory Aspects of Motor Functioning* (pp. 61-67). Hans Huber Publ., Bern.

Divac, I., Rosvold, H.E., & Swarcbart, M. (1967). Behavioral effects of selective ablation of the caudate nucleus. *J. Comp. Physiol. Psychol., 63,* 184-190.

Divac, I., Wikmark, R.G.E., & Gade, A. (1975). Spontaneous alternation in rats with lesions in the frontal lobes. An extension of the frontal lobe syndrome. *Physiol. Psychol., 3,* 39-42.

Frysz, W., & Stepien, I. (1983). Function of cat's caudate nucleus in tasks involving spatial discontiguity between location of cue and response. *Acta Neurobiol. Exp., 43,* 103-113.

Larsson, K., Öberg, R.G.E., & Divac, I. (1980). Frontal cortical ablations and sexual performance in male albino rats (Abstract). *Neurosci. Lett., Suppl. 5,* p. 319.

Öberg, R.G.E., & Divac, I. (1979). "Cognitive" functions of the neostriatum. In: I. Divac and R.G.E. Öberg (Eds.), *The Neostriatum* (pp. 291-313). Oxford: Pergamon.

Pisa, M. (1988a). Motor functions of the striatum in the rat: Critical role of the lateral region in tongue and forelimb reaching. *Neuroscience, 24,* 453-463.

Pisa, M. (1988b). Motor somatotopy in the striatum of rat: manipulation, biting and gait. *Behav. Brain Res., 27,* 21-35.

Rolls, E.T., & Williams, G.W. (1987). Sensory and movement-related neuronal activity in different regions of the primate striatum. In J.S. Schneider and T.I. Lidksy (Eds.), *Basal Ganglia and Behavior: Sensory Aspects of Motor Functioning* (pp. 37-59). Toronto: Hans Huber Publ.

Rosenkilde, C.E., & Divac, I. (1976) Time discrimination performance in cats with lesions in the prefrontal cortex and the caudate nucleus. *J. Comp. Physiol. Psychol., 90,* 343-352.

Ursin, H., & Divac, I. (1975). Emotional behavior in feral cats with ablations of the pre-frontal cortex and subsequent lesions in amygdala. *J. Comp. Physiol. Psychol., 88,* 36-39.

Whishaw, I.Q., O'Connor, W.T., & Dunnett, S.B. (1986). The contribution of motor cortex, nigrostriatal dopamine and caudate-putamen to skilled forelimb use in the rat. *Brain, 109,* 805-843.

Wikmark, R.G.E., Divac, I., & Weiss, R. (1973). Retention of spatial delayed alternation in rats with lesions in the frontal lobes. Implications for a comparative neuropsychology of the prefrontal system. *Brain Behav. Evol., 8,* 329-339.

16 Social Inhibition - Development and Implications

C. Philip Hwang and Anders G. Broberg
University of Göteborg

Social inhibition is a common condition. According to a recent estimate, 10 percent of all children are inhibited (Kagan, Reznick, & Snidman, 1987). They cling to their parents and only reluctantly venture into an unfamiliar room. Faced with strangers, they first freeze, fall silent, and stare at them. They seem visibly tense until they have had a chance to size up the new scene. Parents are likely to say that such children have always been on the timid side. "It's just his way", they might say of the child.

The present chapter deals with the issue of social inhibition. Our goal is threefold. First of all, we summarize recent psychobiological research on inhibited children. Secondly, we present some findings from "The Gothenburg Child Care Project", a study where inhibition, its links to other social behaviors, and the possible role of out-of-home care in altering the tendency to be either inhibited or outgoing, was examined. Finally, we discuss social inhibition and the possible clinical implications of this aspect of personality.

Definitions

In their definition of inhibition, Kagan, Reznick, and Snidman (1988) emphasized initial reactions to unfamiliar events, and especially unfamiliar people, noting that the typical inhibited child was slow to explore unfamiliar environments and more likely to withdraw from novel stimuli. Other researchers have used the terms shy (e.g., Daniels & Plomin, 1985; Hinde, Stevenson-Hinde, & Tamplin, 1985), or unsociable (e.g., Clarke-Stewart, Umeh, Snow, & Pedersen, 1980; Stevenson & Lamb, 1979) to refer to related behaviors although both these terms have broader connotations [shy = "self-conscious and uncomfortable in the presence of others" and sociable = "fond of the company of others", according to Hornby (1985)]. Plomin and Stocker (in press) have emphasized that emotionality, fearfulness, and shyness constitute some of the more important components of inhibition. In our own study (presented later in the chapter), it was obvious that many of the children rated as inhibited were indeed gregarious when given enough time to warm-up. Perhaps all shy children are inhibited, whereas some inhibited children are also shy or asocial. In the present chapter we will use the term inhibition instead of "shyness" or "asociability".

Inhibition - research findings

Ever since Thomas, Chess, and their colleagues began publishing data from the New York Longitudinal Study (Thomas, Chess, Birch, Herzig, & Korn, 1963; Thomas, Chess, & Birch, 1970; Thomas & Chess, 1977), individual differences in temperament have been invoked to explain children's reactions to novel circumstances and people. When challenged in this fashion, some young children become quiet, cease activity, or retreat to a familiar person. Other children of similar intellectual ability and social background do not change their ongoing behavior, and may even approach the unfamiliar object or person. The aim of the present longitudinal study was to examine these individual differences in behavioral tendencies, their links to other social behaviors, and the possible role of out-of-home care in altering the tendency to be either shy or outgoing. More specifically, we studied children's sociability in relation to unfamiliar adults, their fearfulness in different situations, their capacity to play with peers, their dependency on their mothers, and their adaptation to out-of-home care settings.

Garcia-Coll, Kagan, and Reznick (1984) have described children who retreat from new situations and cease their activity as "inhibited" and children who seem unafraid of novelty as "uninhibited". Other researchers have restricted their focus to children's interactions with people, excluding reactions to inanimate social stimuli (Beckwith, 1972; Bretherton, 1978; Clarke-Stewart et al., 1980; Stevenson et al., 1979; Thompson & Lamb, 1982, 1983). Within this tradition, "inhibited" and "uninhibited" behavior are seen as poles of a sociability continuum.

Researchers have found inhibition to be an important dimension of temperament that is at least partly hereditary (Buss & Plomin, 1984; Daniels et al.,, 1985; Plomin & Rowe, 1979; Plomin et al., in press), and stable over time (Bronson & Pankey, 1977; Caspi, Elder, & Bem, 1988; Hinde et al., 1985; Kagan & Moss, 1962; Kagan, Reznick, Clarke, Snidman, & Garcia-Coll, 1984; Kagan et al., 1987; Kagan, Reznick, Snidman, Gibbons, & Johnson, 1988). In addition, the behavioral profile has physiological correlates (Kagan et al., 1987; Kagan et al., 1988), and it may be related to the introversion-extroversion dimension in adulthood (Hinton & Craske, 1977; Kagan et al., 1987, 1988; Plomin et al., 1979).

Kagan and Moss (1962) found inhibition to the unfamiliar to be the only one of 15 infantile behaviors to predict later behavior. Children who were extremely inhibited until age 3 were easily dominated by their peers and were likely to withdraw from social interaction between 3 and 6 years of age. Between ages 6 and 10, they were still socially timid and, if boys, they avoided sports and other "masculine" activities as adolescents. Kagan and his colleagues have since published a number of reports on the effects of inhibition on children's behavior (Garcia-Coll et al., 1984; Kagan et al., 1984; Kagan et al., 1987; Kagan et al., 1988; Reznick, Kagan, Snidman, Gersten, Baak, & Rosenberg, 1986). According to Kagan et al. (1987), around 10 % of all children show marked inhibition in a variety of contexts between 2 and 5-1/2 years of age, with uninhibited behavior showing greater stability than inhibited behavior. Kagan et al. (1984, 1987, 1988) have suggested that this difference reflects a tendency for American parents to encourage their children to become outgoing and sociable, rather than shy and timid.

Both low levels of sociability and high levels of inhibition have been related to parental ratings of temperament, especially fearfulness. Thompson and Lamb (1982) and Reznick et al. (in press) found significant correlations of around .35 between observed sociability/inhibition and scores on the Fear-subscale of the IBQ.

Low levels of sociability and high levels of inhibition have both been associated with difficulties in interaction with peers. Bronson and Pankey (1977) found wariness to be the major source of individual differences in reactions to peers at the beginning of the second year. By the end of the second year, however, children's peer play styles reflected "the individual babies' cumulative evaluations of their accrued social experience", which were the only significant predictors of social orientation at 3 1/2 years of age. Hinde et al. (1985) reported that children who were rated as shy by their mothers at home likewise tended not to interact with peers when observed in the preschool. Garcia-Coll et al. (1984) reported that each of five children who had been inhibited in response to unfamiliar settings were also inhibited in interaction with peers, whereas none of 10 consistently uninhibited children were inhibited with peers. At 4 and 5-1/2 years of age, the formerly inhibited children showed longer latencies to initiate play and made fewer approaches to an unfamiliar child and they spent more time next to their mothers staring at the unfamiliar playmate in a laboratory setting (Reznick et al., 1986). When observed with familiar peers in kindergartens on two different occasions at 5-1/2 years of age, children rated as inhibited at 21 months of age spent more time alone looking at other children and less time in social interaction with peers. Quality of peer play has not been studied in relation to inhibition, however, although one would expect inhibited children to become less socially competent with peers as a result of their tendency to withdraw from formatively important social interaction with them.

Inhibition has also been related to dependency. In fact, Kagan et al. have used one aspect of dependency, namely, "prolonged clinging or remaining proximal to the mother" (Kagan et al., 1988) in the presence of an unfamiliar peer in a laboratory setting as one of the operational definitions of inhibition. Unfortunately, it is unclear whether the tendency of inhibited children to stay close to their mothers primarily reflects withdrawal from the unfamiliar peer or dependency on mothers. Hinde et al. (1985), however, found a substantial correlation ($r = .64$) between maternal ratings of shyness and dependency.

The results of studies that have related sociability to experiences of nonparental care are mixed. Maccoby and Feldman (1972), Tizard and Tizard (1971), and Clarke-Stewart et al. (1980) found sociability with strange adults to be negatively related to nonparental child care experiences, whereas Thompson et al. (1982) found no consistent associations between measures of nonparental care experience and sociability.

Finally, girls appear to be somewhat more sociable than boys (Clarke-Stewart et al., 1980, Thompson et al., 1983), as well as more likely than boys to be inhibited (Reznick, Gibbons, Johnson, & McDonough, in press; Rosenberg & Kagan, in press). In a recent review, Marks (1987) concluded that girls respond earlier, more often, and with a more intense fear of strange adults than boys do. Overall, therefore, girls appear to be both more sociable and more fearful - a pattern of findings consistent with other evidence that girls are more emotionally responsive than boys are (Block, 1976). Plomin et al. (1979), however, found neither sex- nor age-related differences in 13- to 37-month-old children's reactions to strange adults.

In sum, several lines of research indicate that inhibition is an enduring aspect of personality which, at least to a certain extent, is grounded in inborn ways of responding physiologically to the environment. In short, it is part of some children's basic temperament.

"The Gothenburg Child Care Project"

One aim of this study was to investigate the longitudinal stability of inhibition, to relate inhibition to the quality of peer play and to children's adjustment to out-of-home care. More specifically, we studied children's sociability in relation to unfamiliar adults, their fearfulness in different situations, their capacity to play with peers, their dependency on their mothers, and their adaptation to out-of-home care settings.

One hundred and forty-four first born children (72 girls) from Gothenburg (Sweden) participated in the study. The median age was 15.9 months at the time of the initial assessment. Names of children on the waiting lists for municipal child care facilities in Gothenburg, Sweden, were obtained from local authorities in all areas of the city. Because the number of available places in municipal day care was limited, only some were successful in being placed. One group consisted of children receiving center-based day care ($N = 53$).

Others were unable to be placed in centers but were either offered care in municipal family day care homes or the parents made private arrangements with daymothers (N = 33). A final group of children did not enter either centers or family day care facilities - they remained at home in the care of their parents (N = 59).

All families were visited in their homes by a member of the research staff. The research assistant rated the child's reactions when the assistant, as a strange adult, made increasingly intimate overtures to the child during the first 5 minutes of her/his visit. The parents were given a copy of Rothbart's (1981) Infant Behavior Questionnaire and instructed in how to observe their child and fill out the questionnaire during the next two weeks. A second visit was then arranged. On this occasion, the child was observed interacting at home for 30 minutes with a child (selected by the parents) of roughly the same age. The children in the alternative care group began out-of-home care within two weeks of the second home visit. Four to six weeks later, the out-of-home care facilities were visited by a member of the research staff and the child was observed for 30 minutes interacting with peers.

One year after the first interview, the families were visited again. At that time, the children's reactions to strange adults were again rated and the parents were asked about the children's adjustment to the current child care arrangements. On a second visit to the home, the children were observed playing alone during short separations from their mothers, and were then observed interacting for 30 minutes with familiar peers of the parents' choice. On a subsequent visit to the child-care facility (for children in the out-of-home care group), the children were again observed for 30 minutes interacting with peers.

Two years after the initial assessment, the children's reactions to strange adults were again measured and parents were asked to rate their children's adjustment to current child care arrangements. The children were also observed interacting with peers, both at home and in the alternative care settings.

Our index of inhibition included ratings of the children's behavior toward unfamiliar adults during the first five minutes of interaction, and measures of noninvolvement in peer play during the first 30 minutes of interaction with a peer. At 16 months, the measure also included parental ratings of children's fearfulness as measured by Rothbart's IBQ.

The main results of the assessments are summarized as follows:

Stability over time: Children who were inhibited at 16 months of age were more likely than were uninhibited children to be inhibited at 28 and 40 months. The composite measure of inhibition predicted subsequent sociability and peer noninvolvement at least as well as did earlier measures of sociability and peer noninvolvement themselves. This underscores the validity of our composite measure of inhibition.

Involvement in high-quality peer play: Uninhibited children were less likely to be uninvolved with peers in the home setting, while children rated as inhibited at 16 months were less involved with peers contemporaneously. In addition, inhibited children were less likely than uninhibited children to play at a qualitatively high level at both 28 and 40 months of age.

Ability to play alone: Children who had only experienced home care were better able to play alone in their mothers' absence than were children with one year of out-of-home care experience. Separate analysis of home and out-of-home care children revealed that inhibition at 16 months of age was significantly correlated with ability to play alone. In the home care group, inhibited children were less able to play alone in their mothers' absence than were uninhibited children, whereas the relation was reversed in the out-of-home care group.

Adjustment to out-of-home care settings: We found that inhibition was related to the quality of peer play in the out-of-home care setting in much the same way as to the quality of peer play at home, although the associations were somewhat weaker. Our prediction that inhibition (measured in the home setting) would predict adjustment to out-of-home care, as rated by teachers and parents, was supported. Inhibition at 16 months correlated significantly with ratings of adjustment to out-of-home care. However, the effect seemed to be transient since subsequent ratings of adjustment were unrelated to scores of inhibition.

Clinical comments

Social inhibition affects children's behavior in a variety of contexts and circumstances. In addition, it seems to be stable across the toddler and preschool years. Does this mean that inhibited children are at risk for psychological problems?

Findings from several studies suggest indeed this is the case. Recent work by Buss et al. (1984), Kagan et al. (1987), and Fox (1989), show that children who have a low threshold for arousal when faced with uncertainty, may evidence particular physical and physiological changes that may make them difficult to care for. It is possible that some parents find their infants' responses to novelty and mild stress aversive (Kagan et al., 1984). Consequently, they may react to their babies with insensitivity, lack of affection, and/or neglect.

Interestingly, Rubin (1990), has recently pointed out that these parental variables are all predictive of insecure attachment relationships at 12 and 18 months. In addition, the so called "C" baby, according to the classification of Ainsworth, Blehar, Waters, and Wall (1978), typically demonstrates wariness in the face of novelty, clingingness to and difficulty in separating from the parent, and reluctance to explore objects or to approach unfamiliar adults in unfamiliar settings.

The possible connection between insecure-resistent status in infancy and temperamental characteristics in early childhood has been suggested by Fox (1989) and Kagan (1984). A possible connection between inhibited temperament, insecure attachment, and later adjustment problems, could be as follows: a child's inhibited temperament affects the parent-child interaction and constitutes a risk factor for the development of an insecure parent-child attachment (type c-babies). The insecure attachment, in turn, constitutes a risk factor for later development of separation-anxiety disorders and other adjustment problems.

Other psychological difficulties that socially inhibited children exhibit more frequently are nightmares and unusual fears (Kagan, 1984). In addition, they seem to be more sensitive to parental reprimand, and they are generally more obedient to parental requests (Kagan, 1984). Although obedience may be regarded as something positive, inhibited children's acceptance of socialization may also in some cases, indirectly, promote later psychological disorder.

A possible pathway for such a development has been suggested by Kagan (1984). He argues that the combination of a close relationship to parents and more conflicted relationships with peers may draw the inhibited child toward adults and the adoption of adult values. If, however, the parents are extremely threatened by this characteristic and react to the child with hostility, the child may become very aggressive and disobedient. Such a profile then is the joint product of the child's temperament and the adult reactions.

What can be done? Is it possible to encourage a less fearful approach to unfamiliar people and situations? Several researchers have attempted to develop intervention procedures aimed at reducing withdrawn behavior (e.g., Strain & Kerr, 1981). Unfortunately, the results from these training efforts are ambiguous, thereby making any firm conclusions premature. Our own findings, that a supportive and stable out-of-home care settings may help inhibited children overcome their adjustment problems, is partial evidence that intervention attempts can be successful. However, it seems crucial that the environments are of high quality and that they provide a "good fit" to children with inhibited styles (Lerner, Lerner, & Zabski, 1985).

In sum, the evidence reviewed in this chapter suggests that inhibition can develop into a socioemotional problem. A shy and socially withdrawn child will have severe difficulties when confronted with unfamiliar people and environments, but may interact easily with familiar people. Although inhibition is a fairly stable dimension that is not systematically affected by ordinary life changes, inhibited children could be supported by providing environments for them that are stable, and by advancing their ability to develop positive relationships with others.

REFERENCES

Ainsworth, M.D., Blehar, M.C., Waters, E., & Wall, S. (1978). Patterns of attachment. Hillsdale, N:J Erlbaum.

Beckwith, L. (1972). Relationships between infants' social behavior and their mothers' behavior. *Child Develop., 43,* 397-411.

Block, J.H. (1976). Issues, problems, and pitfalls in assessing sex differences: a critical review of the psychology of sex differences. *Merrill-Palmer Quart., 22,* 283-308.

Bretherton, I. (1978). Making friends with one-year-olds: An experimental study of infant-stranger interaction. *Merrill-Palmer Quart., 24,* 29-51.

Bronson, G.W., & Pankey, W.B. (1977). On the distinction between fear and wariness. *Child Develop., 48,* 1167-1183.

Buss, A.H., & Plomin, R. (1984). *Temperament: Early Developing Personality Traits.* Hillsdale, N J: Erlbaum.

Caspi, A., Elder, G.H. Jr., & Bem, D.J. (1988). Moving away from the world: Life course patterns of shy children. *Develop. Psychol., 24,* 824-831.

Clarke-Stewart, K.A., Umeh, B.J., Snow, M.E., & Pederson, J.A. (1980). Development and prediction of children's sociability from 1 to 21/2 years. *Develop. Psychol., 16,* 290-302.

Daniels , D., & Plomin, R. (1985). Origins of individual differences in shyness. *Develop. Psychol., 21,* 118-121.

Fox, N.A. *Infant temperament and security of attachment: a new look.* Paper presented at the International Society for Behavioral Development, Jyväskylä, Finland, July, 1989.

Garcia-Coll, C.T., Kagan, J., & Reznick, J.S. (1984). Behavioral inhibition in young children. *Child Develop., 55,* 1005-1019.

Hinde, R.A., Stevenson-Hinde, J., & Tamplin, A. (1985). Characteristics of three- to four-year olds assessed at home and their interactions in preschool. *Develop. Psychol., 21,* 130-140.

Hinton, J.W., & Craske, B. (1977). Differential effects of test stress on the heart rates of extroverts and introverts. *Biolog. Psychol., 5,* 23-28.

Hornby, A.S. (Ed.) (1985). *Oxford Advanced Learners' Dictionary of Current English..* Oxford: Oxford University Press.

Kagan, J. (1984). *The Nature of the Child.* New York: Basic Books, Inc., Publishers.

Kagan, J., & Moss, H.A. (1962). *Birth to Maturity.* New York: Wiley.

Kagan, J., Reznick, J. S., & Snidman, N. (1987). The physiology and psychology of behavioral inhibition in children. *Child Develop., 58,* 1459-1473.

Kagan, J., Reznick, J. S., & Snidman, N. (1988). Biological bases of childhood shyness. *Science, 240,* 167-171.

Kagan, J., Reznick, J.S., Clarke, C., Snidman, N., & Garcia-Coll, C. (1984). Behavioral inhibition to the unfamiliar. *Child Develop., 55,* 2212-2225.

Kagan, J., Reznick, J.S., Snidman, N., Gibbons, J., & Johnson, M.O. (1988). Childhood derivatives of inhibition and lack of inhibition to the unfamiliar. *Child Develop., 59,* 1580-1589.

Lerner, S.V., Lerner, K.M., & Zabski, S. (1985). Temperament and elementary school children's actual and rated academic performance. *J. Child Psychol. Psychiat. 26,* 125-136.

Maccoby, E.E., & Feldman, S. (1972). Mother-attachment and stranger-reactions in the third year of life. *Monog. Soc. Res. Child Develop., 37, No. 1.*

Marks, I.M. (1987). *Fears, phobias and rituals.* New York: Oxford University Press.

Plomin, R, & Rowe, D.C. (1979). Genetic and environmental etiology of social behavior in infancy. *Develop. Psychol., 15,* 62-72.

Plomin, R, & Stocker, C. (in press). Behavioral genetics and emotionality. In S. Reznick (Ed.), *Perspectives on Behavioral Inhibition.* Chicago: University of Chicago Press.

Reznick, S., Gibbons, J., Johnson, M., & McDonough, P. (in press). Behavioral inhibition in a normative sample. In S. Reznick (Ed.), *Perspectives on Behavioral Inhibition.* Chicago: University of Chicago Press.

Reznick, J.S., Kagan, J., Snidman, N., Gersten, M., Baak, K., & Rosenberg, A. (1986). Inhibited and uninhibited behavior: A follow-up study. *Child Develop., 57,* 660-680.

Rosenberg, A. A., & Kagan, J. (in press). Physical and physiological correlates of behavioral inhibition. *Develop. Psychobiol.*

Rothbart, M.K. (1981). Measurement of temperament in infancy. *Child Develop., 52,* 569-578.

Rubin, K. (1990). Social withdrawal in childhood. Presentation at the MacArthur Symposium on Shyness, Social Withdrawal and Inhibition in Childhood, Langdoe Hall, Canada, July 15-17.

Stevenson, M.B., & Lamb, M.E. (1979). Effects of infant sociability and the caretaking environment on infant cognitive performance. *Child Develop., 50,* 340-349.

Strain, P.S., & Kerr, M.M. (1981) Modifying children's social withdrawal: Issues in assessment and clinical intervention. In: M. Hersen, R.M. Heisler, & P.M. Miller (Eds), *Progress in behavior modification* (Vol. 11, pp. 203-248). New York: Academic Press.

Thomas, A., & Chess, S. (1977). *Temperament and Development.* New York: Brunner/Mazel.

Thomas, A., Chess, S, & Birch, H.G. (1970). The origins of personality. *Scient. American, 223,* 102-109.

Thomas, A., Chess, S., Birch, H., Herzig, M.E., & Korn, S. (1963). *Behavioral Individuality in Early Childhood.* New York: New York University Press.

Thompson, R.A., & Lamb, M.E. (1982). Stranger sociability and its relationships to temperament and social experience during the second year. *Infant Behav. Develop., 5,* 277-287.

Thompson, R.A., & Lamb, M.E. (1983). Security of attachment and stranger sociability in infancy. *Develop. Psychol., 19,* 184-191.

Tizard, J., & Tizard, B. (1971). The social development of two-year old children in residential nurseries. In H.R. Schaffer (Ed.), *The Origins of Human Social Relations.* (pp. 147-163). New York: Academic Press.

17 Psychobiological Approaches in Clinical Practice

Sven G. Carlsson
University of Göteborg

INTRODUCTION

In this chapter I will discuss some concepts, problems, and attempts related to a clinical psychobiological tradition. This tradition is characterized by combinations of psychological and biomedical knowledge and methodology, whereby it differs from approaches where a disease is understood and managed mainly within either a psychological or a biomedical paradigm. The discussion will be limited to clinical problems related to somatic conditions with more or less direct evidence of a psychophysiological background. The most typical methodology is biofeedback, a technique where some physiological activity of an individual is recorded in such a way that he can observe the variations in the recorded activity "on line". This usage of biological parameters requires that they can be measured continuously and directly; typical methodologies are based upon EMG or ECG techniques, or recordings of electrical parameters or temperature of the skin. Neuroendocrine variables are, unfortunately, less useful for biofeedback purposes.

Specifically, I will describe two ways of applying biofeedback clinically. One is the most traditional: the physiological target is a process which is believed to have a pathophysiological significance for the condition, and the aim of the biofeedback training is to teach the patient to modify the activity until symptoms disappear or decrease. The theoretical model is derived from learning theory: the intended change in the neuromuscular activity is supposed to follow the laws of instrumental conditioning. One of the most well-researched examples is tension headache, where patients are taught how to control their tension responses in forehead and temporal muscles (Budzynski, Stoyva, Adler, & Mullaney, 1973).

The other, less traditional variety of biofeedback training, presupposes much more complicated psychobiological interactions during the biofeedback training process. Gary Schwartz (1984) was the first to offer an alternative theoretical model. He has presented a hierachically arranged system of possible changes during biofeedback training: his model includes several levels above the traditional conditioning phenomena, for instance, motivational, emotional, and attitude changes.

Pyschobiological models

Before we return to the two varieties of biofeedback training we shall discuss, briefly, some of the psychobiological models which have been used for understanding and treatment of somatic disorders. Figure 1 is an account of some of the "psychosomatic" interconnections which have been considered important for the etiology or the maintenance of a somatic condition. In different traditions, emphasis has been on one or the other of the relationships indicated in the figure.

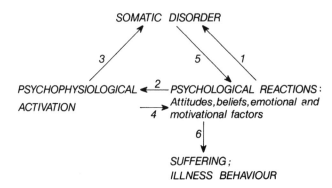

FIG. 17.1. Some psychobiological relationships in the etiology and maintenance of somatic disorders.

The psychoanalytically oriented psychosomatic medicine has been primarily concerned with the causative impact of emotional factors in the etiology of somatic disorders (connection 1 in the figure). According to one of the foremost proponents of this tradition, Franz Alexander, each of a specified number of disease states has a specific psychogenesis; essential hypertension occurs in individuals with an unresolved conflict related to the expression of aggressive impulses, and asthma patients suffer from a dependency conflict emanating from unfortunate early transactions with their mothers (Alexander, 1950). The claims were far-reaching; however, the evidence to justify them has remained limited.

One of the emerging theoretical problems with the psychoanalytically oriented view has been its emphasis upon an undirectional causation between, functionally and temporally, widely separate factors. There is no doubt a considerable amount of unknown land between the mother's inability to accept the early autonomy tendencies in her child and the child's asthma attacks several years later. The metaphor which has been presented to fill the gap conceives the asthma attack as a conflict-charged cry for the mother; however, such an explanation leaves the psychobiologist with a feeling of dissatisfaction.

Harold G. Wolff, an extremely influential researcher, represents a slightly different approach. He demonstrated interrelationships between emotional problems, psychophysiological activation, and symptoms (Wolf & Goodell, 1968). While most psychophysiologically oriented researchers have used standard laboratory stress conditions, Wolff mostly used individually relevant stimuli: based upon his knowledge of the individual patient's emotional problems, he activated them during a conversation with the patient, while recording some physiological parameter that he presumed important for the somatic symptom the patient was suffering from. When symptoms were aggravated after the activation of such idiosyncratic psychophysiological reactions, Wolff had demonstrated a plausible chain of causative links, represented in Figure 1 by connections 2 and 3. In this manner he could, for instance, demonstrate the functional relationships between emotional conflict, scalp muscle activation and headache symptoms (Simons, Day, Goodell, & Wolff, 1943).

Often, the total impact of a condition on the diseased individual is an effect not only of the somatic pathology itself but to a large degree of the cognitive and emotional systems which the individual builds around it (Fig. 1, connections 5 and 6). Such identification of non-organic factors which influence the severity of a disease has been one of the main themes in modern psychosomatic research. One of the first examples, and perhaps the best known, is chronic pain, where Wilbert Fordyce has advocated the importance of psychological factors for the severity of e.g., low back pain (Fordyce, 1976). Another example is the research by Dirks, Kinsman, and others in the "Denver Group". They have studied what they call "panic-fear" among asthma patients, demonstrating that a habitual tendency towards "panic-fear" seems to be an important emotional factor, with pronounced influence upon a number of illness behavior variables, for instance, the patient's need of medication (Dahlem, Kinsman, & Horton, 1979).

Tinnitus is a condition where a more or less constant sound is perceived, which has no external source. The etiology is poorly understood, and the medical treatment modalities which have been attempted have had limited success. It has been presumed that the severity of the condition reflects the intensity and duration of the perceived sound. However, this does not seem to be the case: the soundlevel perceived by patients who are profoundly disturbed can be just a few decibels, while others tolerate higher sound levels without the slightest problems. Hallam, after extensive exploration of the non-physical determinants of symptom impact in tinnitus patients, concludes that maintained suffering is due to negative beliefs and fears related to the symptoms and their underlying pathology. Hallam has proposed an explanatory model for tinnitus suffering, based upon habituation theory (Hallam, Rachman, & Hinchcliffe, 1984). Due to long-term habituation to the tinnitus sound, many afflicted individuals become less and less affected. However, the habituation process may be prevented, and the sound remains as a disturbing element in the patient's mind. Experimental habituation research has recognized two factors which contribute to lack of habituation: high arousal levels and significance of the stimulus to be habituated. The tinnitus patient becomes more disturbed, consequently, when in a state of more or less constant stress and when negative beliefs are attached to the sound. Hallam concludes that the adequate therapy should aim at promoting habituation by changing negative beliefs and reducing the stress level.

Biofeedback

In traditional psychophysiological therapy, like biofeedback training, emotional factors have been largely neglected. Biofeedback is targeted at the physiological overactivity believed to give rise to the symptom (Fig. 1, connection 3), and the aim of the treatment is to enable the patient to lower the activity: in tension headache, the pain symptoms are expected to decrease when the patient has acquired control over his head muscle tension reactions. We relied upon this rather simplistic learning model in our attempts to develop a biofeedback treatment method for mandibular dysfunction.

Mandibular dysfunction (or temporo-mandibular pain and dysfunction syndrome) is a syndrome with pain and tenderness in the masticatory muscles, and other signs of dysfunction from the masticatory apparatus. The condition is quite common, and is most often handled by dentists. While the most common etiological models have been purely somatic, the importance of psychophysiological muscle overactivity has been considered (Laskin, 1969). Our decision to attempt EMG biofeedback was inspired by the promising outcome of biofeedback treatment of tension headache, a condition with an etiology similar to what we hypothesized for a mandibular dysfunction. The biofeedback training, as we applied it, was based on the assumption that the mandibular dysfunction symptoms reflect an increased activity in the muscles of mastication, in the form of daytime and/or nighttime bruxing and clenching. During biofeedback training, the patient is first made aware of his muscular tension and is trained to lower it as securely as possible; the changes in muscular habits are seen as a result of an instrumental conditioning process, where the lowering of muscle tension, as displayed to the patient, is positively reinforced by the experience of success.

The clinical outcome was quite promising: most of the patients, even the chronic cases, were cured or significantly improved according to subjective as well as objective assessments (Carlsson, Gale, & Öhman, 1975, Carlsson, & Gale,1977). We were, however, disturbed by the lack of significant correlations between muscle tension change and clinical outcome.

While the correlation could be quite obvious in some of the patients (Carlsson, & Gale, 1978), most often the muscle tension parameters did not seem to relate to clinical effect. Because this observation seemed to violate our basic understanding of the treatment, we decided to explore it in further detail. We subjected 20 patients to a series of six biofeedback sessions, which is, in our experience, the minimum number of sessions necessary to reach clinical effect. Masseter muscle activity was recorded during all training sessions, and the EMG data were reduced in a number of ways, including trends within sessions and trends between sessions. At a group level, EMG levels changed in the expected direction, both within and between sessions; also, signs and symptoms of the disorder were significantly reduced. However, there were no meaningful correlations at all between clinical outcome and any EMG parameter we could think of (Dahlström, Carlsson, Gale, & Jansson, 1984). Obviously, we could not, from the data we collected, support the original theory of how the therapy works: patients showing a good outcome did or did not rank high in motor learning. We know, however, that patients were significantly improved clinically. We know, also, that the patients changed their tendencies to respond to stress with masseter muscle tension. Before and after the treatment period, we subjected patients and a group of control subjects to a sequence of experimental stressors, while recording masseter activity (Fig. 2). Before therapy, patients showed significantly more tension than controls; after training, the difference disappeared (Dahlström, Carlsson, Gale, & Jansson, 1985).

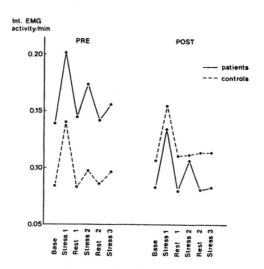

FIG. 17.2. Masseter area muscle tension levels during rest and experimental stress conditions, in mandibular dysfunction patients and control subjects, before and after biofeedback therapy.

So, there is evidence for a therapeutic change, and it is plausible that the normalization of the muscular overresponsivity to stress had contributed to the clinical improvement. What we do not know, however, are in which ways the biofeedback sessions worked.

One way to handle the situation is to question the theoretical model upon which the application was based. One of the roots of biofeedback is the learning-theoretical paradigm of instrumental learning: in EMG biofeedback training, clinical effects should mirror a specific, conditioned neuromuscular change. At the same time as the evidence for such a simplistic treatment model is lacking, clinicians begin to report other types of changes taking place during biofeedback training. Such changes, summarized by Schwartz as a basis for his theoretical reformulation of biofeedback effects (1984), concern the interaction between the biofeedback target physiology and psychological, cognitive, and emotional factors. They represent a methodological challenge, as psychological treatment process assessments have to be added to the physiological ones. We are presently involved in clinical research with biofeedback training of tinnitus patients where the treatment process is viewed in this broader mode (Erlandsson, Carlsson, & Svensson, 1989). Compared to the traditional biofeedback treatment approach, the role played by the physiological recording (masseter or forehead area EMG) has changed. The connection with the somatic symptom may be of direct importance in some patients (Fig. 1, connection 3); however, the EMG activity, through its interplay with psychological reactions (Figs. 1, 2 and 4), is now also used for identifying and modifying those cognitive and emotional factors which negatively affect the state of the patient. In this work we have utilized some of the experiences from our work with dental fear, where we have made a similar, "broader" use of biofeedback training.

A psychophysiological therapy for dental fear

We have taken a rather extensive interest in clinical and theoretical problems related to dental fear. The patients in our study groups all have a long-standing, phobic fear of dentistry, with avoidance patterns causing severe dental decay. The condition includes psycho-somatic as well as somato-psychic relationships: it is quite clear that the dental problems are due to psychological factors, and that they affect the individual both psychologically and socially.

If one wants to look upon dental fear as a psychophysiological condition, the situation is, of course, quite different compared to disorders like tension headache, where a presumed pathophysiology guides the researcher towards a specific physiological target. Psychophysiological phenomena are, however, obvious: the fearful dental patient complains of somatic disturbances in relation to dental visits or the mere thought of dental treatment. If psychophysiological recordings are made when exposing a fearful dental patient to video-taped dental scenes, pronounced activation can be observed (e.g., forehead area muscle tension).

The aims of the treatment modality we have developed for dental fear are to reduce the negative psychological and physiological concomitants of dentistry which have been developed by the treatment avoiding dental patient (Carlsson, Linde, & Öhman, 1980). For phobias in general, a methodology called Systematic Desensitization (SD) has been successively adopted: in SD, the patient is trained to eliminate responses of fear and anxiety to the phobic object by successively approaching it while in a state of calmness and relaxation. In order to focus the psychophysiological facets of the patient's reactions more effectively, we have included the continuous usage, during the therapy process, of a forehead EMG biofeedback recording. After an extensive interview, the patient is taught relaxation by means of a variety of progressive muscular relaxation training: the patient is given taped instructions of how to reduce muscle tension, and is required to train at home for a week or so. Because muscular relaxation is highly correlated to mental calmness, the technique has been widely used in fear and anxiety conditions. When at the next appointment the patient starts EMG biofeedback, the tension control is further enhanced. We then show to the patient video-taped dental scenes, which are to be viewed in a relaxed state: the patient is instructed to stop the scene with a remote control if he experiences tension and/or if he observes increased tension on the EMG meter. The scenes are arranged so that the more taxing ones (like the dentist using the drill) are attempted only when a number of less disturbing episodes (like the dentist making an examination) can be watched without subjective or objective tension. Throughout the whole treatment process, patient and therapist discuss thoughts, feelings, and questions which come to the patient's mind. After about eight sessions, the patient manages to view all the scenes without fear or tension and dental treatment is commenced. Around 80% of the patients are able to tolerate dental treatment (and are tolerated by their dentists), and can pursue a normal contact with dental care (Berggren, & Carlsson, 1984). The effects of biofeedback training (besides improving tension headache is some patients) involve emotional and cognitive factors: the patient's concepts and feelings concerning his fearfulness are mirrored in the observed EMG recording, whereby he is given an opportunity to reconstruct his previous fear-maintaining cognitions and feelings concerning dentistry, and his own inability to handle it.

CONCLUSION

The lack of convincing evidence for the traditional learning-theoretical model of biofeedback training has left us with a treatment modality with promising clinical effects but with largely unknown treatment mechanisms. There are two ways of handling this dilemma. One is to conclude that clinical biofeedback effects are of a non-specific "placebo" character. However, this is a meaningful conclusion only in relation to a definition of "specific" biofeedback factors. The alternative strategy is to widen the treatment mechanism model by including changes at cognitive, emotional, and motivational levels. The convergent evidence for the importance of such factors in somatic disorders may encourage the clinical biofeedback researcher to include them in his assessments of therapeutic changes. When he can relate such changes to what is going on between the patient and the biofeedback machine, he will promote the psychobiological understanding of somatic disorders. There are theoretical and, above all, methodological obstacles that have to be overcome. But, like Harold G. Wolff used to say: "Fixity of purpose required flexibility of method".

REFERENCES

Alexander, F. (1950). *Psychosomatic medicine: Its principles and applications.* New York: Norton Press.

Berggren, U., & Carlsson, S.G. (1984). A psychophysiological therapy for dental fear. *Behav. Res. Ther., 22,* 487-492.

Budzynski, T., Stoyva, J., Adler, C.S., & Mullaney, D.J. (1973). EMG biofeedback and tension headache: A controlled outcome study. *Psychosom. Med., 35,* 484-496.

Carlsson, S.G., Gale, E.N., & Öhman, A. (1975). Treatment of temporomandibular joint syndrome with biofeedback training. *J. Am. Dental Assoc., 91,* 602-605.

Carlsson, S.G., & Gale, E.N. (1978). Biofeedback for muscle pain associated with the temporomandibular joint. *J. Behav. Ther. Exp. Psychiatry, 7,* 383-385.

Carlsson, S.G., & Gale, E.N. (1977). Biofeedback in the treatment of long.term temporomandibular joint pain: An outcome study. *Biofeed. Self-Regul., 2,* 161-171.

Carlsson, S.G., Linde, A., & Öhman, A. (1980) Reduction of tension in fearful dental patients. *J. Am. Dental Assoc., 101,* 638-641.

Dahlem, N.W., Kinsman, R.A., & Horton, D.J. ((1979). Requests for as-needed medications by asthma patients. *J. Allerg. Clin. Immunol., 63,* 23-27.

Dahlström, L., Carlsson, S.G., Gale, E.N., & Jansson, T.G. (1984). Clinical and electromyographic effects of biofeedback training in mandibular dysfunction. *Biofeed. Self-Regul., 9,* 37-47.

Dahlström, L., Carlsson, S.G., Gale, R.N., & Jansson, T.G. (1985). Stress-induced muscular activity in mandibular dysfunction: Effects of biofeedback training. *J. Behav., Med., 8,* 191-200.

Erlandsson, S., Carlsson, S.G., & Svensson, S.G. (1989). Biofeedback treatment of tinnitus: A broadened approach. *Göteborg Psychol. Rep., 19,* No. 6.

Fordyce, W.E. (1976). *Behavioral Methods in Chronic Pain and Illness.* St. Louis: C.V. Mosby.

Hallam, R.S., Rachman, S., & Hinchcliffe, R. (1984). Psychological aspects of tinnitus. In S. Rachman (Ed,), *Contributions to Medical Psychology.* New York: Pergamon Press.

Laskin, D.M. (1969). Etiology of the pain-dysfunction syndrome. *J. Am. Dental Assoc., 79,* 147-153.

Schwartz, G.E. (1984). Toward a comprehensive theory of clinical biofeedback: A systems perspective. In C. Van Dyke, L. Temoshok, and L.S. Zegans (Eds.), *Emotions in Health and Illness: Applications to Clinical Practice.* Orlando, Grune & Stratton.

Simons, D.J., Day, E., Goodell, H., & Wolff, H.G. (1943). Experimental studies on headache: Muscles of the scalp and neck as sources of pain. *Res. Publ. Assoc. Nerv. Ment. Dis., 23,* 228-244.

Wolf, S., & Goodell, H. (Eds.) (1968). *Harold G. Wolff's Stress and Disease, 2nd Ed.* Springfield, IL: Thomas.

V Animal Psychobiology

Berend Olivier and Jan Mos
Department of Pharmacology
Duphar B.V.
Weesp

INTRODUCTION

Animal psychobiology is a rapidly developing area within neuroscience research. It comprises complex, multidisciplinary approaches towards the study of brain and behavior in its broadest sense. In this contribution, we will focus on some aspects of animal psychobiology, viz., neuropharmacology and behavior.

Neuropharmacology really began when people became aware of the relevance of chemical neurotransmission in the central nervous system, especially by the discovery of monoamines and their functional implications. Carlsson (1987) gives an excellent overview of the historical developments in this field, including functional and clinical perspectives. The emerging area of neuropsychopharmacology is closely linked to the serendipitous finding of antipsychotics and antidepressants and the subsequent elucidation of their presumed working mechanism. These kinds of discoveries also showed that monoamines appeared to play an important role in the modulation of e.g., mood, psychomotor activity, anxiety, aggression, sexual behavior, and food intake.

Instead of taking a very broad approach, the present contribution focuses on one monoamine, serotonin (5-HT) and especially on the 5-HT_{1A} agonists, whose functions and pathophysiology are still poorly understood, although impressive progress is accomplished in the involvement of 5-HT in the field of psychobiology, neuropharmacology, and behavior (see different books on 5-HT emerging in the last couple of years).

Serotonin-receptors

During the last decade, the number of 5-HT receptors has dramatically increased. At the moment, at least three main types of receptors are known, 5-HT_1, 5-HT_2, and 5-HT_3 receptors, whereas the 5-HT_4 receptor has been claimed (Dumuis, Sebben, & Bockaert, 1989). Several subtypes of the 5-HT_1 receptors are known, e.g., the 5-HT_{1A}, 5-HT_{1B}, 5-HT_{1C}, 5-HT_{1D}, and 5-HT_{1E} subtype (Leonhardt, Herrick-Davis, & Titeler, 1989), whereas the picture is less clear for the 5-HT_2 and 5-HT_3 receptors. The 5-HT_2 has been divided into

5-HT$_{2A}$ and 5-HT$_{2B}$ (Schmidt & Peroutka, 1989), but recent evidence suggests that the 5-HT$_{2A}$ could be equivalent to the 5-HT$_{1C}$ site (Van Wijngaarden, Tulp, & Soudijn, 1990). 5-HT$_3$ receptors have been subclassified into 5-HT$_{3A}$, 5-HT$_{3B}$, and 5-HT$_{3C}$, but these three subsites have not yet been delineated by receptor-binding experiments (Hoyer, Waeber, Karpf, Neijt, & Palacios, 1989; Kilpatrick, Jones, & Tyers, 1989). To date, three 5-HT receptors have been cloned, viz., the 5-HT$_{1A}$, 5-HT$_{1C}$, and 5-HT$_2$. Hartig (1989), on the basis of such cloning data, has proposed a new classification of 5-HT receptors. The three receptors mentioned belong to a G-protein receptor class, whereas 5-HT$_3$ receptors belong to a ligand-gated ion superfamily.

Most 5-HT receptor subtypes have been characterized by radioligand binding techniques in brain membranes. Before one considers such a "binding site" as a real "receptor", functional correlates of such a "binding site" should be delineated. Such functional correlates should be physiologically relevant. For instance, the 5-HT$_{1A}$ "receptor" has been correlated with adenylate cyclase stimulation, inhibition of hippocampal CA1 and raphe cells, canine basal artery contractions, induction of lower lip retraction, facilitation of ejaculation, hypothermia, and hypotension (cf. Peroutka, 1988). Similar correlates have been made for 5-HT$_{1B}$, 5-HT$_{1C}$, 5-HT$_{1D}$. 5-HT$_2$, and 5-HT$_3$, although the picture for 5-HT$_{1B}$, 5-HT$_{1D}$, and 5-HT$_3$ is far from clear. Moreover, some subtypes seem species-specific. Waeber, Schoeffter, Palacios, & Hoyer (1989a) and Waeber, Dietl, Hoyer, & Palacios (1989b) have clearly shown that the 5-HT$_{1B}$ receptor only occurs in myomorph rodents, whereas the 5-HT$_{1D}$ receptor seems to occur in the brain of the majority of vertebrate species, excluding rat and mouse. Whether the 5-HT$_{1D}$ and 5-HT$_{1B}$ sites fulfill a similar function (autoreceptor or postsynaptic receptor) in all species is unclear, but is suggested by a congruence of enrichment of both types in similar brain areas, especially the basal ganglia and substantia nigra (Waeber et al., 1989a, b).

5-HT$_{1A}$ receptors seem to occur in all vertebrate species (Waeber et al., 1989,a, b) and this strongly suggests an important role of this kind of receptor in phylogenetically old processes in the brain, like food intake, fear reactions, aggression, and sexual behavior. Because of the universality of the 5-HT$_{1A}$ receptor, experiments with agonists for this subtype form a good illustration os psychobiological research.

5-HT$_{1A}$ ligands, like 8-OH-DPAT, have been developed only recently and the serendipitous finding of the anxiolytic activity of buspirone (Goldberg & Finnerty, 1979) has led to a renewed interest of 5-HT in anxiety and subsequently in depression. Several new 5-HT$_{1A}$ agonists have been introduced since then, e.g., ipsapirone, gepirone, and flesinoxan, and the present paper focuses on their effects in animal models of 1) anxiety, 2) aggression, 3) feeding, and 4) sexual behavior. The receptor binding profile of 8-OH-DPAT, flesinoxan, buspirone, and ipsapirone is given in Table 1 in order to facilitate the description and discussion of the findings with these drugs.

1. Effects of 5-HT$_{1A}$ agonists in animal models of anxiety

Since the introduction of the DSM-III and its reversed version DSM-III R in the eighties, anxiety disorders have been divided into two classes: phobic disorders and anxiety states. Both classes consist of several subcategories. Phobic disorders consist of agoraphobia, with or without panic attacks, and simple and social phobia, whereas anxiety states comprise panic disorder (PD, generalized anxiety disorder (GAD), obsessive compulsive disorder (OCD), post-traumatic stress disorder (PTSD), and atypical anxiety disorder. This ordering suggests a clear demarcation of the different disorder, but there is a lot of dispute about this classification, and the boundaries between different disorders are sometimes extremely vague (cf. Den Boer, 1988 for an extensive discussion).

The human anxiety disorder classification should optimistically be mirrored in the field of anxiety research in animals. However, no clear differentiation in the various classes and categories has been made or is possible, and for some disorders, e.g., OCD, even no animal model has been developed, although some attempts have been made (Insel, 1988; Winslow

Table V.1 Receptor binding profile of 8-OH-DPAT, flesinoxan, buspirone, and ipsapirone. Affinities are expressed as pK_i-values ($-\log K_i$ in nM), which are average values of at least three dependent measurements.

	5-HT							Ach	Others					
	1A	1B	1C	1D	2	3	Uptake		α_1	α_2	$\beta 1,2$	D2	H_1	μ
8-OH-DPAT	8.6	5.8	5.1	6.0	<5	<5	6.3	5.7	5.6	6.5	<5	5.7	<5	<5
Flesinoxan	8.8	6.1	<5	6.8	5.4	<5	<5	<5	6.4	<5	<5	6.4	6.1	<5
Buspirone	7.8	5.5	5.4	<5	6.0	<5	<5	<5	6.2	<5	<5	7.4	6.0	5.2
Ipsapirone	8.3	5.5	<5	<5	5.6	<5	<5	<5	6.6	5.6	<5	6.4	5.9	<5

All four compounds had no or weak (pK_i <5) affinity for D1, κ, δ, BDZ, $GABA_A$, Gly, TRH, CCK_A, CCK_B receptors. Data from Van Wijngaarden et al. (1990).

& Insel, 1990). In most animal experiments, drugs are described as either anxiolytic or anxiogenic, and animal models are used which are aimed to detect anxiolytic/anxiogenic effects but do not differentiate in the kind of anxiety involved. It appears very difficult to assess what kind of anxiety is most clearly revealed in different animal tests.

There is, however, no *a priori* reason why animals should not experience different states of anxiety. The evidence from diverse animal tests definitely shows that animals may experience at least severe anxiety expressed on behavioral and physiological levels. Most animal tests of anxiety have been validated by means of pharmacology. Ideally, clinically effective anxiolytics, e.g., the benzodiazepines (BDZ), should exert a specific behavioral profile, not shared by other pharmacological classes, in a putative anxiolytic test. Most data on anxiolytic drugs have been gathered in animal tests from two main classes. The first group of tests is based on conflict behavior or conditioned fear. Examples include the Geller-Seifter test, the Vogel punished-licking test, and the four-plate test. The second group is based on the induction of anxiety or uncertainty by a novel environment; examples include light-dark exploration, open-field, elevated plus-maze, and social interaction test.

Benzodiazepines have clear anxiolytic actions in all these animal anxiety tests (cf. Hommer, Skolnick, & Paul, 1987), and there is also some evidence that nonbenzodiazepine drugs, which, however, act on the BDZ-GABA-A receptor, have a similar anxiolytic profile (Langer & Arbilla, 1988; Patel, Meiners, Salama, Malick, U'Pritchard, Giles, Goldberg, & Bare, 1988).

Recent evidence strongly suggests that certain serotonergic drugs also exert anxiolytic activity (Chopin & Briley, 1987; Dourish, 1987; Broekkamp, Berendsen, Jenck, & Van Delft, 1989). This evidence most strongly derives from buspirone, a 5-HT$_{1A}$ agonist, which is on the prescription market for generalized anxiety disorders. Buspirone and 5-HT$_{1A}$ agonists, e.g., 8-OH-DPAT, gepirone, ipsapirone, and flesinoxan, have anxiolytic activity in conflict tests, although negative findings have also been reported (Broekkamp et al., 1989; Dourish, 1987). Most studies have been performed in rats and, in general, the anxiolytic effects of 5-HT$_{1A}$ ligands are weak to moderate compared to those of BDZs. However, in pigeons, these drugs exert a very potent anticonflict effect (Barrett, Witkin, Mansbach, & Skolnick, 1986; Barrett, Gleeson, Nader, & Hoffman, 1989; Mansbach, Harrod, Hoffman, Nader, Lei, Witkin, & Barret, 1988), suggesting that the rat models are particularly sensitive towards BDZ, but less so towards 5-HT$_{1A}$ ligands.

A very sensitive animal model for anxiety seems to be separation-induced distress of young animals when separated from their mother (Gardner, 1985; Insel, Hill, & Mayor, 1986). A frequently used model to study the effects of anxiolytic drugs is separation-induced ultrasonic calls from rat pups. When rat pups, 9-10 days old, are separated from their mother and placed on a cold (18° C) or warm (37° C) plate, they display ultrasonic sounds in the range of 35-55 kHz. In the cold plate situation, which is more stressful than that of the warm plate, pups display a higher level of ultrasonic calls than in the warm plate situation (Fig. 1). Drugs were tested under both conditions (for experimental details, see Mos & Olivier, 1988, 1989). In this experimental setup, several benzodiazepines and 5-HT$_{1A}$ ligands were tested (Fig. 2).

All three BDZ-agonists reduce ultrasounds under both conditions. Typically, at the cold plate, there remains a certain minimum level of ultrasonic calling, whereas at the warm plate, ultrasounds are almost reduced to zero. This profile is different after treatment with 5-HT$_{1A}$ agonists, which reduce ultrasounds under both conditions similarly. Thus, 5-HT$_{1A}$ agonists have anxiolytic activity in this animal test, but there are differences from that induced by benzodiazepines. A role of 5-HT in anxiety processes is further supported by the clinical findings that specific 5-HT re-uptake blockers and antidepressants, e.g., fluvoxamine and fluoxetine (Goodman, Price, Rasmussen, Heninger, & Charney, 1989; Murphy, Zohar, Benkelfat, Pato, Pigott, & Insel, 1989), suppress panic attacks and OCD. Benzodiazepines (with the possible exception of alprazolam) and specific NA-uptake blockers and antidepressants, e.g., maprotiline and DMI, are not active in panic disorders and OCD (Den Boer & Westenberg, 1988; Murphy et al., 1989). Serotonergic re-uptake blockers (fluvoxamine, fluoxetine, zimeldine) are active in the pupvocalization test with a specific profile; they only decrease the vocalization under the cold stressful condition and have no effect on the warm plate condition (Mos et al., 1989). Maprotiline is inactive under both conditions. This indicates that differential profiles in the pupvocalization test may predict

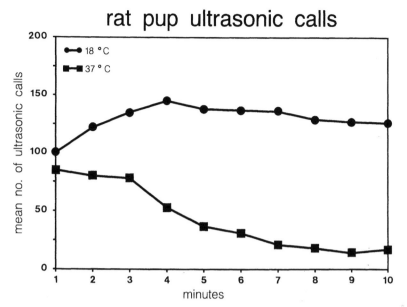

Fig. V.1. The mean number of ultrasonic calls of rat pups is shown in a 10 min test for two temperature conditions: a cold plate (18° C) and a warm plate (37° C).

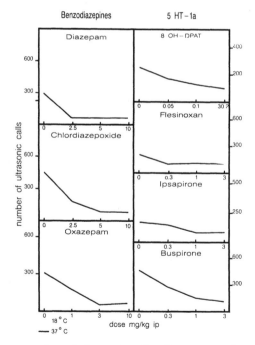

Fig. V.2. The mean number of ultrasonic calls of rat pups under a cold (18° C) and a warm (37° C) plate condition is shown after treatment with benzodiazepines (left column and 5-HT$_{1A}$ agonists (right column). The dose (mg/kg i.p.) is given 30 min before a five min test.

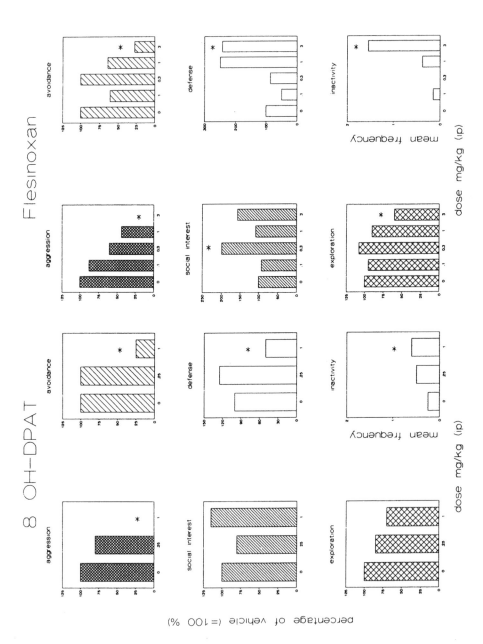

212

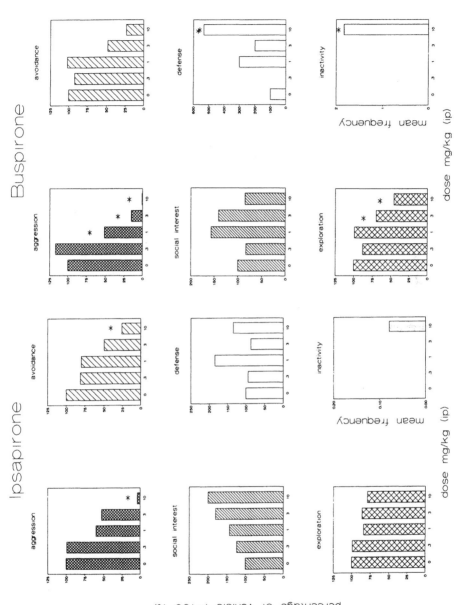

Fig. V.3. The effects of four 5-HT$_{1A}$ agonists on the mean frequency of 6 behavioral categories in the social interaction test in male mice. Fig.3A shows the effects of 8-OH-DPAT and flesinoxan, Fig. 3B the effects of ipsapirone and buspirone. Data are expressed as percentage of vehicle, except for inactivity which shows the real frequency.
** indicates significant differences from vehicle (* $p < 0.05$; ** $p < 0.01$).

whether a compound may have anti-panic and anti-OCD activity. If this hypothesis is true, it may be predicted that 5-HT$_{1A}$ agonists, because of their potent effects under cold conditions, may have such effects. As yet, preliminary data have not conclusively shown that 5-HT$_{1A}$ agonists - in particular, buspirone - are active as anti-panic or anti-OCD agents (Sheenan, Raj, Sheenan, & Soto, 1988). However, buspirone may not be the most explicit drug to test the hypothesis because of its potent dopamine-D$_2$ antagonistic properties. Experiments with more specific 5-HT$_{1A}$ agonists, e.g., gepirone, ipsapirone, and flesinoxan, may answer these questions better.

2. Effects of 5-HT$_{1A}$ ligands in animal models of aggression

Serotonin has been regularly implicated in the modulation of aggressive behavior (Miczek & Donat, 1989). However, regarding the various types of aggressive behavior distinguished (Brain, 1981; Moyer, 1968) and the suggested different neural substrates involved in these various types (cf. Adams, 1979), it seems unlikely that the neurotransmitter 5-HT is simply linked, in a very general (inhibitory) manner, to the expression of aggression. Moreover, taking into consideration the complexity in the localization of the 5-HT neuronal cell bodies, their afferent and efferent connections, and the different kinds of 5-HT receptors or binding sites, such a simple function of 5-HT in aggression becomes highly untenable. One avenue to unravel the role of (parts of) the serotonergic system in aggressive behavior is to study specific 5-HT ligands in specific aggression paradigms (Olivier, Mos, Schipper, Tulp, Van der Heyden, Berkelmans, & Bevan, 1987; Olivier, Mos, Tulp, Schipper, & Bevan, 1989a). In this case, specific 5-HT$_{1A}$ agonists were tested in three aggression tests, all representing different aspects of offensive aggression, viz., social interaction in male mice (Olivier, Mos, Van der Heyden, & Hartog, 1989b), resident-intruder aggression in male rats (Olivier, 1981), and maternal aggression in lactating female rats (Olivier, Mos & Van Oorschot, 1985; Olivier & Mos, 1986). In all paradigms, extensive behavioral recordings were made, but for comparative reasons, only behavioral categories are shown.

In social interaction in mice (Fig. 3), all 5-HT$_{1A}$ agonists decreased but not in a behaviorally specific manner (Olivier et al., 1989b). 8-OH-DPAT decreased aggression, but at the same time, decreased exploration and avoidance, and enhanced inactivity. Typically, such animals show enhanced defensive behavior when approached by the opponent. Buspirone and ipsapirone show a comparable profile, although somewhat milder, whereas flesinoxan resembles more the 8-OH-DPAT profile. Such a behavioral profile of 5-HT$_{1A}$ agonists suggests that the 5-HT$_{1A}$ receptor is not critically involved in the specific modulation of offensive aggression, although a role for this kind of receptor certainly cannot be excluded (Olivier et al., 1989b).

In the resident-intruder paradigm also, 5-HT$_{1A}$ agonists have no behaviorally specific profile (Fig. 4). In this paradigm, we have tested buspirone, 8-OH-DPAT, and ipsapirone. Buspirone has, in contrast to 8-OH-DPAT and ipsapirone, a much more sedative profile. Presumably, this is due to buspirone's dopamine-blocking (D$_2$) properties, which on itself induces sedation in this paradigm (Olivier, Van Aken, Jaaarsma, Van Oorschot, Zethof, & Bradford, 1984). However, the more specific 5-HT$_{1A}$ and any other receptor (cf. Olivier et al., 1989a), also show no specific inhibition of aggressive behavior (Fig. 4), thereby confirming the non-specific role of this type of receptor in the modulation of offensive aggression.

Buspirone

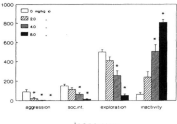

Ipsapirone

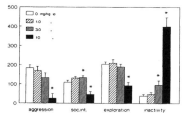

8 OH-DPAT

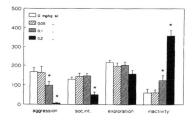

Fig. V.4. The effects of buspirone, ipsapirone, and 8-OH-DPAT on the duration (seconds) of four behavioral categories are shown in the resident-intruder paradigm. * p<0.05 compared to vehicle.

In maternal aggression, a paradigm using lactating female rats at 3 to 9 days postpartum (Mos et al., 1989; Olivier et al., 1986), 5-HT$_{1A}$ agonists again show no specific behavioral profile. For instance, 8-OH-DPAT has, up to 0.05 mg/kg (s.c.) no anti-aggressive effects, but doubling the dose (to 0.1 mg/kg) completely abolishes aggression (cf. Olivier et al., 1989a), presumably due to severe side effects as evidenced in the remaining behavior.

This paradigm suggests that 5-HT$_{1A}$ receptors are not specifically involved in the modulation of offensive aggression. This notion is further supported by the failure of 5-HT$_{1A}$ agonists to inhibit hypothalamically induced attacks and predatory attacks in rats (Olivier et al., 1989a).

Although it seem premature to exclude 5-HT$_{1A}$ receptors form playing a role in aggressive (offensive) behavior, the data gathered so far (cf. Olivier et al., 1989a; Olivier, & Mos, 1989) do not support a specific role in this behavior. In contrast, 5-HT$_{1B}$ receptors seem to play a critical and specific role in offensive aggression (Olivier et al., 1987, 1989a). Whether or not the 5-HT$_{1A}$ receptor (and others) interacts with the 5-HT$_{1B}$ receptor in the specific modulation of offensive aggression is unclear and will need further investigation.

3. Effects of 5-HT$_{1A}$ ligands in feeding

Until the finding of the specific 5-HT$_{1A}$ receptor ligand 8-OH-DPAT (Arvidsson, Hacksell, Nilsson, Hjort, Carlsson, Lindberg, Sanchez, & Wikström, 1981), brain 5-HT was generally accepted to have an inhibitory role in food intake (Blundell & Latham, 1982). However, administration of the potent 5-HT$_{1A}$ agonist 8-OH-DPAT to satiated rats caused marked hyperphagia (Dourish, Hutson, & Curzon, 1985a). The hyperphagia, but not the serotonergic behavioral syndrome which appeared at higher doses (Dourish, Hutson, & Curzon, 1985b), could be attributed to activation of serotonergic somatodendritic autoreceptors located in the raphe nuclei (Bendotti & Samanin, 1986; Dourish , Hutson, & Curzon,1986; Hutson, Dourish, & Curzon, 1988).

Further studies have indicated that 8-OH-DPAT stimulates food consumption of a highly palatable diet both in non-deprived and partially satiated rats (Cooper, 1988). 8-OH-DPAT was able to reverse the decreased intake of palatable food induced by fenfluramine, but not that induced by the β-carboline FG7142, an inverse agonist of BDZ-receptors (Cooper, 1988). Other 5-HT$_{1A}$ ligands, such as buspirone, ipsapirone (Dourish et al., 1986), gepirone (Gilbert & Dourish, 1987), and MDL 72832 (Neill & Cooper, 1988) also increase food intake.

Although some reports (Fletcher, 1987; Montgomery, Willner, & Muscat, 1988) shed some doubts on the specificity of the enhanced feeding responses after 5-HT$_{1A}$ agonists, several other reports have now confirmed that 8-OH-DPAT enhances appetite for different types of food and fluids (Cooper, Fryer, & Neill, 1988; Dourish, Cooper, Gilbert, Goughlan, & Iversen, 1988). The latter authors suggest that 8-OH-DPAT specifically stimulates appetite by counteracting a tonic serotonergic inhibition of feeding. However, several authors report contrasting findings on the "intake-releasing" properties of 5-HT$_{1A}$ ligands. Chaouloff, Serrurrier, Mérino, Laude, & Elghozi, (1988) found that 8-OH-DPAT enhanced intake of pellets, but not powder in young rats, but in contrast did not enhance it in adult rats. Aulakh, Wozniak, Haas, Hill, Zohar, & Murphy, (1988) even reported dose-dependent decreases in food-intake (pellets) at doses at which others clearly showed hyperphagia.

We studied the effects of 8-OH-DPAT and flesinoxan in rats. For flesinoxan, male rats (N=10 per dose) were deprived of food for 24 hours before the test; 8-OH-DPAT was tested in undeprived (N=14 per dose). Animals were acquainted with this procedure which had been performed before, however, without drug testing. Flesinoxan (i.p.) was administered 30 min before the food was returned. After 60 and 120 min, the amount of food consumed (grams) was measured. Kruskal-Wallis analysis was applied to test overall drug effects followed by Mann-Whitney U tests. Figure 5 shows the effects of 5-HT$_{1A}$ ligands. 8-OH-DPAT had no effect in doses of up to 0.2 mg/kg in undeprived rats, whereas flesinoxan dose-dependently decreased food intake, most notably at 60 min. Typically, under circumstances where many reports have described food intake enhancement (Dourish et al., 1985a, b; Gilbert et al., 1987), 8-OH-DPAT did not increase food intake; neither was it possible to find a stimulating effect of flesinoxan in 24 hr food deprived rats. The latter effect may be explained by similar findings after 8-OH-DPAT; a given dose may enhance food intake in free-feeding rats and decrease food intake in fasting rats (Dourish et al., 1985a, b; Bendotti & Samanin, 1987).

Because a recent study (Kalkman, Siegl, & Fozard, 1989) found a stimulatory effect of flesinoxan on food intake in free-feeding rats, an apparent conclusion is that 5-HT$_{1A}$ agonists may stimulate food intake only under certain conditions. This notion is further corroborated by findings of Chaouloff et al. (1988), who found age-dependent effects on the stimulatory effects of 8-OH-DPAT on either solid or powdered food intake. Moreover, they suggest (in Chaouloff et al., 1988) that 8-OH-DPAT treatment at the onset of darkness, the period during which rats normally consume most of their daily food intake, does not further elicit hyperphagia.

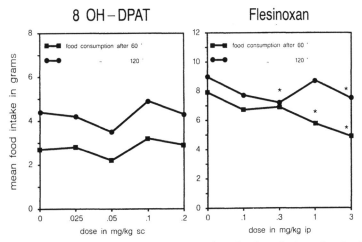

Fig. V.5. The effects of 8-OH-DPAT (mg/kg s.c.) on food intake in undeprived rats and the effects of flesinoxan (mg/kg i.p.) on food intake in 24 hr food deprived rats. Food intake was measured 60 and 120 min after returning the food to the animals. * depicts a significant difference (p<0.05) from vehicle.

Our experiments were performed in the early dark phase, when normally most intake occurs. It is, therefore, conceivable that we have missed, by using the natural and physiologically most salient experimental conditions, the hyperphagic effects of 5-HT$_{1A}$ agonists. A simple review of the experimental conditions used in almost all studies reported on the hyperphagic effects of 5-HT$_{1A}$ ligands, shows, indeed, that most studies are performed in the light part of the cycle, i.e., under normally low feeding conditions. Further work is clearly needed to understand whether a (population of) 5-HT$_{1A}$ receptor(s) is physiologically involved in the regulation of feeding behavior.

4. Effects of 5-HT$_{1A}$ ligands on sexual behavior of male rats

In an early publication (Ahlenius, Larsson, Svensson, Hjort, Carlsson, Lindberg, Wikström, Sanchez, Arvidsson, Hacksell, & Nilsson, 1981), the prosexual effects of 8-OH-DPAT on male sexual behavior were first reported. Subsequent studies have shown that other 5-HT$_{1A}$ agonists may share similar properties, e.g., ipsapirone (Glaser, Dompert, Schuurman, Spencer, & Traber, 1987), buspirone, and gepirone, and even beneficial effects in male humans after buspirone treatment have been reported (Othmer & Othmer, 1987).

We (Olivier & Mos, 1988) were not able to find clear prosexual effects after 8-OH-DPAT, although some indications were present. We explained this result by the level of sexual experience; very experienced male rats were used which had been selected for high levels of sexual performance. In order to study the effect of sexual experience on the level of facilitation of male sexual behavior, we manipulated the level of sexual performance of male rats and tested various 5-HT$_{1A}$ ligands on their prosexual effects.

A. The effects of 8-OH-DPAT in sexually experienced rats

Male Wistar rats were selected on their level of sexual performance in 4 subsequent weekly tests of 15 min in which the males were tested against estrous females. The selection-criterion was that males should, on an average, achieve at least two ejaculations in each subsequent test. Such rats (400-500 g) were housed singly, under a reversed day-night regimen (night 7.00-19.00 hrs). Female Wistar rats (200-350 g) were housed in groups of three under the same conditions. Thirty-six hours before testing, the females were injected s.c. with 50 μg estradiol benzoate and screened for receptivity. 8-OH-DPAT (0, 0.1, 0.2, and 0.4 mg/kg s.c.) were given to the male 30 min before testing. The sexual activity was measured in a testing cage (80x60x50 cm) in which the male was placed 30 min before testing, which started with the introduction of an estrous female into the cage. The ejaculation, intromission, and mount frequencies and the ejaculation, mount, and intromission latencies were measured per ejaculation series. Each animal was tested at one week intervals with vehicle or 8-OH-DPAT randomized according to a Greek-Latin square design. The duration of the test was 30 min. 8-OH-DPAT (Fig. 6) had no significant effect on any measure of male sexual behavior (shown here by the mean number of ejaculations), although a slight stimulating tendency was present. Presumably, by using sexually experienced rats, a ceiling effect was present which could not be overcome by further stimulating. However, it has been described that some animals immediately ejaculate after 8-OH-DPAT (Ahlenius et al., 1981) but this was certainly not the case with these animals which had average ejaculation latencies of 2 to 3 min. Apparently, some other factors may play a role, although the baseline levels of sexual behavior certainly influence the prosexual effects of 5-HT$_{1A}$ agonists. In order to further study the effects of baseline levels, some experiments on "average" sexually performing and "naive" sexually rats were performed.

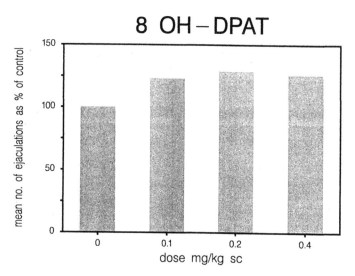

Fig. V.6. The mean number of ejaculation in percentage of vehicle (control) is shown in sexually experienced rats after treatment with 8-OH-DPAT (mg/kg s.c.).

B. The effects of 5-HT$_{1A}$ agonists in "average" performing rats

From a group of 100 male rats, the best and the poorest performers in sexual behavior tests (selected on the number of ejaculations in a 15 min test) were chosen for other experiments and the remaining 40 males were used for the present experiment. All 40 selected males had been tested five times before the present experiments. One vehicle group and 3 doses of a drug were tested on the same day. Each group consisted of 10 animals. At weekly intervals, the animals were tested and treatment was randomized over the groups. All drugs were given 30 min before the test of 15 min in which the same parameters were measured as in Experiment A. Figure 7 summarizes the data.

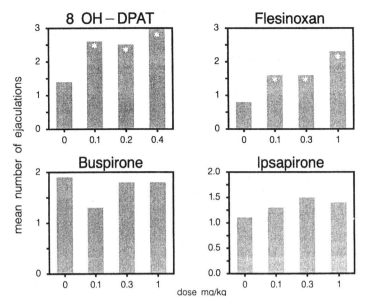

Fig. V.7. The effects of four 5-HT$_{1A}$ agonists are shown on the mean number of ejaculations in sexually "naive" rats. Doses are give i.p. (flesinoxan, buspirone, and ipsapirone) or s.c. (8-OH-DPAT) 30 min before a test of 15 min with an estrous female. In the cases of buspirone and ipsapirone, two experiments have been performed, one with a low dose range (0.1-1 mg/kg) and one with a high dose range (3-10 mg/kg).
* indicates a significant difference (p<0.05) from vehicle.

8-OH-DPAT increased the number of ejaculations dose-dependently, and decreased the latency to the first ejaculation. The frequency of mounts and intromissions also decreased significantly. This pattern, of faster and more ejaculations with less mounts and intromissions, is typical of the prosexual effects of 8-OH-DPAT. Flesinoxan has a, more or less, comparable profile but is clearly less potent than 8-OH-DPAT. Buspirone failed to facilitate sexual behavior at the selected doses, whereas ipsapirone had a small facilitating effect, viz., a decrease in the number of intromissions to reach ejaculation. The doses used of buspirone and ipsapirone (0.1-1 mg/kg) were most likely too low, as suggested from data in the literature (Glaser et al., 1988).

C. The effects of 5-HT$_{1A}$ agonists on sexually naive rats

Since the prosexual effects of 5-HT$_{1A}$ agonists seem to be critically dependent upon the baseline levels of sexual performance, we also tested such drugs on sexually naive male rats which had not had previous experience with sexual behavior. Usually, such rats do not reach ejaculation in a short duration tests (15 min). Facilitatory effects on sexual behavior in such rats may result in an enhanced number of ejaculations and in a reduced latency to mounts and/or intromissions.

Groups of 60 naive male rats were used for each test. Procedures of drug administration, testing, scoring, and statistics were similar to those described in A and B. Figure 8 shows the effects of 8-OH-DPAT, flesinoxan, buspirone, and ipsapirone on the sexual behavior of naive male rats in a 15 min test with an estrous female.

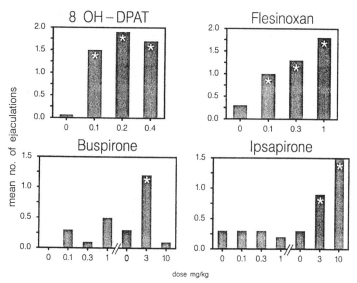

Fig. V.8. The effects of four 5-HT_{1A}-agonists is shown on the mean number of ejaculations in sexually "naive" rats. Doses are given i.p. (flesinoxan, buspirone and ipsapirone) or s.c. (8-OH-DPAT) 30 min before a test of 15 min with an estrous female. In the case of buspirone and ipsapirone two experiments have been performed, one with a low dose range (0.1-1 mg/kg) and one with a high dose range (3-10 mg/kg). * indicates a significant difference (p<0.05) from vehicle.

All 5-HT_{1A} agonists facilitated sexual behavior, although their potency differed dramatically. 8-OH-DPAT and flesinoxan already exerted prosexual effects at approximately 0.1 mg/kg, whereas buspirone and ipsapirone were active at doses of 1 mg/kg or higher, especially when taking into account the effects on the number of ejaculations. Buspirone had a quite narrow dose range for these prosexual effects; at 10 mg/kg, sexual behavior was even decreased, presumably due to a sedating effects at this dose.

DISCUSSION

The present contribution shows the impact of activation of one serotonin receptor type, the 5-HT_{1A}, on several types of behavior. Apparently, depending upon the external situation, such a single receptor modulates several different behaviors with different functions, e.g., feeding, anxiety, sexual behavior, and with less specificity, aggressive behavior. Moreover, the 5-HT_{1A} receptor is, at the least, also involved in the temperature regulation and 'the serotonergic syndrome'. Some 5-HT_{1A} agonists lower blood pressure (Wouters, Tulp, & Bevan, 1988; Ramage & Fozard, 1987) via a central mechanism. Enhancement of 5-HT neurotransmission in the brain of rats induces a complex behavioral pattern, the so-called 5-HT syndrome (Jacobs, 1976), consisting of a number of components: wet-dog shakes, lower lip retraction, flat body posture, hindlimb abduction, spread paws, arched back, forepaw treading, head weaving, pilo-erection, and tremor. Some distinct components could be linked to specific (subtypes of) serotonin receptors, e.g., wet-dog shakes to 5-HT_2 (Lucki, Nobler, & Frazer, 1984) and lower lip retraction to 5-HT_{1A} receptor activator (Berendsen, Jenck, & Broenkkamp, 1989; Molewijk, Van der Heyden, & Olivier, 1987). All 5-HT_{1A} agonists, indeed, induced lower lip retraction and hypothermia. Both effects could be antagonized by the 5-HT_{1A} antagonist (±)-pindolol, suggesting that these effects are specifically modulated by 5-HT_{1A} receptors.

This illustrates the complexity of interventions in one neurotransmitter system - in this case, even only a part of it - on the possible behavioral outcome. Therefore, an integrative approach should be taken, using multidisciplinary psychobiological techniques from all areas in pharmacology, neuro(bio)chemistry, and behavior.

In the case of the 5-HT$_{1A}$ receptor, we have studied several aspects of its function by using specific 5-HT$_{1A}$ agonists. Because no specific 5-HT$_{1A}$ antagonists are available, it is difficult to delineate the specificity of the effects observed after using 5-HT1A agonists, but the receptor-specificity of some of the 5-HT$_{1A}$ agonists used is so high (at least a factor 100 affinity distance from other receptors) that, certainly at low dosages, hardly any doubt exists about the specific involvement of the 5-HT$_{1A}$ sites. Evidence also stems from studies using drug discrimination techniques.

By training an animal to press either one of two levers, depending on the drug or saline given, a very sensitive procedure can be used to study the "in vivo" characteristics of drugs. 5-HT$_{1A}$ ligands induce a cue which is very specific to activation of the 5-HT$_{1A}$ receptor, as evidenced by cross-generalization of all 5-HT$_{1A}$ agonists towards each other (Glennon, 1986; Mansbach & Barrett, 1987; Spencer & Traber, 1987; Tricklebank, Neill, Kidd, & Fozard, 1987; Ybema, Slangen, Olivier, & Mos, 1990; Nader, Hoffman, Gleeson, & Barret, 1989; Barrett et al., 1989). Moreover, the 5-HT$_{1A}$ cue could be antagonized by 5-HT$_{1A}$ antagonists, e.g., pindolol or propranolol, and partly by the partial 5-HT$_{1A}$ antagonist BMY 7378 (Nader et al., 1989). Several other 5-HT agonists and antagonists, e.g., the 5-HT$_{1B}$ agonist TFMPP (Cunningham, Callahan, & Appel, 1987; Glennon, 1986, Nader et al., 1989) did neither substitute for, nor antagonize the 5-HT$_{1A}$ cue (Tricklebank et al., 1987). All these data strongly suggest the importance and relevance of the 5-HT$_{1A}$ receptor in the modulation of several behavioral states.

Table 2 summarizes the effects of four 5-HT$_{1A}$ agonists on a number of parameters apparently modulated by the 5-HT$_{1A}$ receptor. In various animal models for anxiety and depression, 5-HT$_{1A}$ agonists show activity, although their pattern of activity deviates from that of the benzodiazepines.

5-HT$_{1A}$ agonists enhance intake behavior (feeding and drinking) in rats, although not under all conditions. Sexual behavior of male rats is stimulated by 5-HT$_{1A}$ agonists, especially when the baseline level or sexual experience is low to moderate. Aggressive behavior is decreased by 5-HT$_{1A}$ agonists, but the available evidence points to nonspecific influences (Olivier et al., 1989a, b), indicating that the 5-HT$_{1A}$ receptor may not be directly involved in the modulation of offensive aggression. Considerable evidence exists, however, that 5-HT$_{1A}$ agonists reduce defensive and flight behavior (Blanchard, Rodgers, Hendrie, & Hori, 1988; Rodgers & Shepherd, 1989).

Comparing the receptor affinities of the different drugs for the 5-HT$_{1A}$ sites with their behavioral effects, it is evident that 8-OH-DPAT and flesinoxan are the most potent drugs. Not only are their affinities high, but these drugs are presumably, also full agonists, whereas buspirone and ipsapirone are partial agonists. It remains to be established whether partial or full agonism also plays an important in clinical effects. It seems that the 5-HT$_{1A}$ receptor contributes a special influence in each behavioral state and has no general facilitatory role. The influence of the 5-HT$_{1A}$ receptor seems to depend on the internal and external environment as evidenced in, e.g., anxiety, depression, and feeding. It would be interesting to modulate the environment, e.g., the hunger state of an animal, to see whether this would influence, for instance, the drug-discrimination qualities of a drug.

Psychobiology faces the intriguing task of unraveling the puzzles of brain, mind, and behavior. In view of the complexities, only multidisciplinary approaches may enlarge our horizon, and guard against naive and too simple unitary concepts.

Table V.2. Overview of the effect of 5-HT$_{1A}$ agonists in various behaviors and physiological parameters.

	Anxiety	Depression	Feeding	Sexual Behavior	Aggressive Behavior		Blood Pressure	Lower Lip Retraction	Temperature (rat)	Drug Discrimination
					Offense	Defense				
8-OH-DPAT	↓↓	↓↓↓	↑↑↑	↑↑↑	↓	?	↓	↑↑↑	↓↓↓	5-HT$_{1A}$ cue
Flesinoxan	↓↓↓	↓↓	?(↑)	↑↑	↓↓	?	↓	↑↑	↓↓	5-HT$_{1A}$ cue
Ipsapirone	↓↓	↓	↑	↑	↓	↓	?	↑	↓	5-HT$_{1A}$ cue
Buspirone	↓↓	↓	↑	↑	↓	↓	?	↑	↓	5-HT$_{1A}$ cue

↓ decrease; ↑ increase

↓↓↓ very strong effects; ↓↓ strong effects; ↓ moderate effects; ? not known

ACKNOWLEDGEMENTS

We thank Marijke Mulder and Ruud von Oorschot for their excellent technical contributions.

REFERENCES

Adams, D.B. (1979). Brain mechanisms for offense, defense and submission. *Behav. Brain Sci., 2,* 201-241.

Ahlenius, S., Larsson, K., Svensson, L., Hjorth, S., Carlsson, A., Lindberg, P., Wikström, H., Sanchez, D., Arvidsson, L.E., Hacksell, U., & Nilsson, J.L.G. (1981). Effects of a new type of 5-HT receptor agonist on male rat sexual behavior. *Pharmacol. Biochem. Behav., 15,* 785-792.

Arvidsson, L.E., Hacksell, U., Nilsson, J.L.G., Hjorth, S., Carlsson, A., Lindberg, P., Sanchez, D., & Wikström, H. (1981). 8-Hydroxy-2-(di-n-propylamino) tetralin, a new centrally acting 5-hydroxytryptamine receptor agonist. *J. Med. Chem., 24,* 921-923.

Aulakh, C.S., Wozniak, K.M., Haas, M., Hill, J.L., Zohar, J., & Murphy, D.L. (1988). Food intake, neuroendocrine and temperature effects of 8-OH-DPAT in the rat. *Eur. J. Pharmacol., 146,* 253-259.

Barrett, J.E., Gleeson, S., Nader, M.A., & Hoffman, S.M. (1989). Anticonflict effects of the 5-HT$_{1A}$ compound flesinoxan. *J. Psychopharmacol., 3,* 64-69.

Barrett, J.E., Witkin, J.M., Mansbach, R.S., Skolnick, P., & Weisman, B.A. (1986). Behavioral studies with anxiolytic drugs. III. Antipunishment actions of buspirone do not involve benzodiazepine receptor mechanisms. *J. Pharmacol. Exp. Ther., 238,* 1009-1013.

Bendotti, C., & Samanin, R. (1986). 8-hydroxy-2-(di-n-propylamino) tetralin (8-OH-DPAT) elicit eating in free-feeding rats by acting on central serotonin neurons. *Eur. J. Pharmacol., 121,* 147-150.

Bendotti, C., & Samanin, R. (1987). The role of putative 5-HT$_{1A}$ and 5-HT$_{1B}$ receptors in the control of feeding in rats. *Life Sci., 41,* 635-642.

Berendsen, H.H.G., Jenck, F., & Broekkamp, C.L.E. (1989). Selective activation of 5-HT$_{1A}$ receptors induces lower lip retraction in the rat. *Pharmacol. Biochem. Behav., 33,* 821-827.

Blanchard, D.C., Rodgers, R.J., Hendrie, C.A., & Hori, K. (1988). "Taming" of wild rats (Rattus rattus) by 5-HT$_{1A}$ agonists buspirone and ipsapirone. *Pharmacol. Biochem. Behav., 31,* 269-278.

Blundell, J.E., & Latham, C.J. (1982). Behavioral pharmacology of feeding. In T. Silverstone (Ed.), *Drugs and Appetite* (pp. 41-80). London: Academic Press.

Brain, P.F. (1981). Differentiating types of attack and defense in rodents. In P.F. Brain and D. Benton (Eds.), *Multidisciplinary Approaches to Aggression Research* (pp. 53-77). Elsevier/North Holland Biomedical Press.

Broekkamp, C.L.E., Berendsen, H.H.G., Jenck, F., & Van Delft, A.M.L. (1989). Animal models for anxiety and response to serotonergic drugs. *Psychopathology, 22 (S1),* 2-12.

Carlsson, A. (1987). Monoamines of the central nervous system: a historical perspective. In H.Y. Meltzer (Ed.), *Psychopharmacology: The Third Generation of Progress* (pp. 39-48). New York: Raven Press.

Chaouloff, F., Serrurrier, B., Mérino, D., Laude, D., & Elghozi, J.L. (1988). Feeding responses to a high dose of 8-OH-DPAT in young and adult rats: influence of food texture. *Eur. J. Pharmacol., 151,* 267-273.

Chopin, P., & Briley, M. (1987). Animal models of anxiety: the effects of compounds that modify 5-HT neurotransmission. *Tr. Pharmacol. Sci., 8,* 383-388.

Cooper, S.J. (1988). Palatability-induced food consumption is stimulated by 8-hydroxy-2-(di-n-propylamino) tetralin (8-OH-DPAT). In C.T. Dourish, S. Ahlenius, and P.H. Hutson (Eds.), *Brain 5-HT$_{1A}$ Receptors* (pp. 233-242). Chichester, England: Ellis Horwood.

Cooper, S.J., Fryer, M.J., & Neill, J.C. (1988). Specific effect of putative $5\text{-}HT_{1A}$ agonists, 8-OH-DPAT and gepirone, to increase hypertonic saline consumption in the rat: evidence against a general hyperdipsic action. *Physiol. Behav., 43,* 533-537.

Cunningham, K.A., Callahan, P.M., & Appel, J.B. (1987). Discriminative stimulus properties of 8-hydroxy-2-(di-n-propylamino) tetralin (8-OH-DPAT): implications for understanding the actions of novel anxiolytics. *Eur. J. Pharmacol., 138,* 29-36.

Den Boer, J.A. (1988). *Serotonergic mechanisms in anxiety disorders. An inquiry into serotonin function in panic disorder* (pp. 1-232). Ph.D. Thesis, Utrecht.

Den Boer, J.A., & Westenberg, H.G.M. (1988). Effects of serotonin and noradrenalin uptake inhibitors in panic disorders: a double blind comparative study with maprotiline and fluvoxamine. *Int. Clin. Psychopharmacol., 3,* 59-74.

Dourish, C.T. (1987). Brain $5\text{-}HT_{1A}$ receptors and anxiety. In C.T. Dourish, S. Ahlenius, P.H. Hutson (Eds.), *Brain $5\text{-}HT_{1A}$ Receptors* (pp. 261-277). Chichester, England: Ellis Horwood.

Dourish, C.T., Cooper, S.J., Gilbert, F., Goughland, J., & Iversen, S.D. (1988). The $5\text{-}HT_{1A}$-agonist 8-OH-DPAT increases consumption of palatable wet mash and liquid diets in the rat. *Psychopharmacology, 94,* 58-63.

Dourish, C.T., Hutson, P.H., & Curzon, G. (1985a). Low doses of the putative serotonin agonist 8-hydroxy-2-(di-n-propylamino) tetralin (8-OH-DPAT) elicit feeding in the rat. *Psychopharmacology, 86,* 197-204.

Dourish, C.T., Hutson, P.H., & Curzon, G. (1985b). Characteristics of feeding induced by the serotonin agonist 8-hydroxy-2-(di-n-propylamino) tetralin (8-OH-DPAT). *Brain Res. Bull., 15,* 377-384.

Dourish, C.T., Hutson, P.H., & Curzon, G. (1986). Putative anxiolytics 8-OH-DPAT, buspirone and TVX Q7821 are agonists at $5\text{-}HT_{1A}$ autoreceptors in the raphe nuclei. *Tr. Pharmacol. Sci., 7,* 212-214.

Dumuis, A., Sebben, M. & Bockaert, J. (1989). The gastrointestinal prokinetic benzamide derivatives are agonists at the nonclassical 5-HT receptor ($5\text{-}HT_4$) positively coupled to adenylate cyclase in neurons. *Naunyn-Schmiedeberg's Arch. Pharmacol., 340,* 403-410.

Fletcher, P.J. (1987). 8-OH-DPAT elicits gnawing, and eating of solid but not liquid foods. *Psychopharmacology, 92,* 192-195.

Gardner, C.R. (1985). Inhibition of ultrasonic distress vocalizations in rat pups by chlordiazepoxide and diazepam. *Drug Dev. Res., 5,* 185-193.

Gilbert, F., & Dourish, C.T. (1987). Effects of the novel anxiolytics gepirone, buspirone and ipsapirone on free feeding and on feeding induced by 8-OH-DPAT. *Psychopharmacology, 93,* 349-352.

Glaser, T., Dompert, W.U., Schuurman, T., Spencer, D.G., Jr., & Traber, J. (1988). Differential pharmacology of the novel $5\text{-}HT_{1A}$ receptor ligands 8-OH-DPAT, BAY R 1531 and ipsapirone. In C.T. Dourish, S. Ahlenius, and P.H. Hutson (Eds.), *Brain $5\text{-}HT_{1A}$ Receptors* (pp. 106-119). Chichester, England: Ellis Horwood.

Glennon, R.A. (1986). Discriminative stimulus properties of the $5\text{-}HT_{1A}$ agonist 8-hydroxy-2-(di-n-propylamino) tetralin (8-OH-DPAT). *Pharmacol. Biochem. Behav., 25,* 135-139.

Goldberg, H.L., & Finnerty, R.J. (1979). The comparative efficacy of buspirone and diazepam in the treatment of anxiety. *Am. J. Psychiatry, 136,* 1184-1187.

Goodman, W.K., Price, L.H., Rasmussen, S.A., Heninger, G.R., & Charney, D.S. (1989). Fluvoxamine as an antiobsessional agent. *Psychopharmacol. Bull., 25,* 31-35.

Hartig, P.R. (1989). Molecular biology of 5-HT receptors. *Tr. Pharmacol. Sci., 10,* 64-69.

Hommer, D.W., Skolnick, P., & Paul, S.M. (1987). The benzodiazepine/GABA receptor complex and anxiety. In H.Y. Meltzer (Ed.), *Psychopharmacology: The Third Generation of Progress* (pp. 977-983). New York: Raven Press.

Hoyer, D., Waeber, C., Karpf, A., Neijt, H., & Palacios, J.M. (1989). [3H]-ICS 205-930 labels $5\text{-}HT_3$ recognition sites in membranes of cat and rabbit vagus nerve and superior cervical ganglion. *Naunyn-Schmiedeberg's Arch. Pharmacol., 340,* 396-402.

Hutson, P.H., Dourish, C.T., & Curzon, G. (1988). 8-hydroxy-2-(di-n-propylamino) tetralin (8-OH-DPAT)-induced hyperphagia: neurochemical and pharmacological evidence for an involvement of 5-hydroxytryptamine somatodendritic autoreceptors. In C.T. Dourish, S. Ahlenius, and P.H. Hutson (Eds.), Brain 5-HT$_{1A}$ Receptors (pp. 211-232). Chichester, England: Ellis Horwood.

Insel, T.R. (1988). Obsessive-Compulsive Disorder: a neuroethological perspective. Psychopharmacol. Bull., 24, 365-369.

Insel, T.R., Hill, J.L., & Mayor, R.B. (1986). Rat pup ultrasonic isolation calls: possible mediation by the benzodiazepine receptor complex. Pharmacol. Biochem. Behav., 24, 1263-1267.

Jacobs, B.L. (1976). An animal model for studying central serotonergic synapses. Life Sci., 19, 777-786.

Kalkman, H.O., Siegl, H., & Fozard, J.R. (1989). Comparison of the behavioral responses to 8-OH-DPAT and flesinoxan. Abst. Int. Symp. Serotonin, Florence, p. 45.

Kilpatrick, G.J., Jones, B.J., & Tyers, M.B. (1989). Binding of the 5-HT$_3$ ligand [^{3}H]GR65630, to rat area postrema, vagus nerve and the brains of several species. Eur. J. Pharmacol., 159, 157-164.

Langer, S.Z., & Arbilla, S. (1988). Imidazopyridines as a tool for the characterisation of benzodiazepine receptors: a proposal for a pharmacological classification as omega receptor subtypes. Pharmacol. Biochem. Behav., 29, 763-766.

Leonhardt, S., Herrick-Davis, K., & Titeler, M. (1989). Detection of a novel serotonin receptor subtype (5-HT$_{1E}$) in human brain: interaction with a GTP-binding protein. J. Neurochem., 53, 465-471.

Lucki, I., Nobler, M.S., & Frazer, A. (1984). Differential actions of serotonin antagonists on two behavioural models of serotonin receptor activation in the rat. J. Pharmacol. Exp. Ther., 228, 133-139.

Mansbach, R.S., & Barrett, J.E. (1987). Discriminative stimulus properties of buspirone in the pigeon. J. Pharmacol. Exp. Ther., 240, 364-359.

Mansbach, R.S., Harrod, C., Hoffman, S.M., Nader, M.A., Lei, Z., Witkin, J.M., & Barrett, J.E. (1988). Behavioural studies with anxiolytic drugs. V. Behavioral and in vivo neurochemical analysis in pigeons of drugs that increase punished responding. J. Pharmacol. Exp. Ther., 246, 114-120.

Miczek, K.A., & Donat, P. (1989). Brain 5-HT system and inhibition of aggressive behaviour. In P. Bevan, A.R. Cools, and T. Archer (Eds.), Behavioural Pharmacology of 5-HT (pp. 117-144). Hillsdale, NJ: Lawrence Erlbaum.

Molewijk, H.E., Van der Heyden, J.A.M., & Olivier, B. (1989). Lower lip retraction is selectively mediated by activation of the 5-HT$_{1A}$ receptor. Eur. J. Neurosci., 64 (S2), 23.

Montgomery, A.M.J., Willner, P., & Muscat, R. (1988). Behavioural specificity of 8-OH-DPAT induced feeding. Psychopharmacology, 94, 110.114.

Mos, J., & Olivier, B. (1988). Ultrasonic vocalizations by rat-pups as an animal model for anxiolytic activity: effects of antidepressants. In B. Olivier, and J. Mos (Eds.), Depression, Anxiety and Aggression. Preclinical and Clinical Interfaces (pp. 85-93). Houten: Medidact.

Mos, J., & Olivier, B. (1989). Ultrasonic vocalizations by rat-pups as an animal model for anxiolytic activity: effects of serotonergic drugs. In P. Bevan, A.R. Cools, and T. Archer (Eds.), Behavioural Pharmacology of 5-HT (pp. 363-366). Hillsdale, NJ: Lawrence Erlbaum.

Mos, J., Olivier, B., Van Oorschot, R., Van Aken, H., & Zethof, T. (1989). Experimental and ethological aspects of maternal aggression in rats: five years of observations. In R.J. Blanchard, P.F. Brain, D.C. Blanchard, and S. Parmigiami (Eds.), Ethoexperimental Approaches to the Study of Behaviour (pp. 385-399). Dordrecht: Kluwer Acad. Publ.

Moyer, K.E. (1968). Kinds of aggression and their physiological basis. Comm. Behav. Biol., 2, 65-87.

Murphy, D.L., Zohar, J., Benkelfat, C., Pato, M.T., Pigott, T.A., & Insel, T.R. (1989). Obsessive-Compulsive Disorder as a 5-HT subsystem-related behavioural disorder. Br. J. Psychiatry, 155 (S8), 15-24.

Nader, M.A., Hoffman, S., Gleeson, S., Barrett, J.E. (1989). Further characterisation of the discriminative stimulus effects of buspirone using monoamine agonists and antagonists in the pigeon. *Behav. Pharmacol., 1,* 57-67.

Neill, J.C., & Cooper, S.J. (1988). MDL 72832, a selective 5-HT$_{1A}$ receptor ligand, stereoselectively increases food intake. Eur. J. Pharmacol., 151, 329-332.

Olivier, B. (1981). Selective anti-aggressive properties of DU 27725: ethological analysis of intermale and territorial aggression in the male rat. *Pharmacol. Biochem. Behav., 14 (S1),* 61-77.

Olivier, B., & Mos, J. (1986). A female aggression paradigm for use in psychopharmacology: maternal agonistic behaviour in rats. In P.F. Brain and J. Martin Ramirez (Eds.), *Cross-Disciplinary Studies on Aggression* (pp. 73-111). Seville: Univ. Seville Press.

Olivier, B., & Mos, J. (1988). Effects of psychotropic drugs on sexual behaviour in male rats. In B. Olivier and J. Mos (Eds.), *Depression, Anxiety and Aggression. Preclinical and Clinical Interfaces* (pp. 121-133). Houten: Medidact.

Olivier, B., & Mos, J. (1989). Serotonergic and benzodiazepine modulation of agonistic behaviour: ethopharmacological analysis. *Biotemas, 2,* 1-48.

Olivier, B., Mos, J., Schipper, J., Tulp, M.Th.M., Van der Heyden, J.A.M., Berkelmans, B., & Bevan, P. (1987). Serotonergic modulation of agonistic behaviour. In B. Olivier, J. Brain, and P.F. Brain (Eds.), *Ethopharmacology of Agonistic Behavior in Animals and Humans* (pp. 162-186). Dordrecht: Martinus Nijhoff.

Olivier, B., Mos, J., Tulp, M.Th.M., Schipper, J., Bevan, P. (1989a). Modulatory action of serotonin in aggressive behaviour. In P. Bevan, A.R. Cools, and T. Archer (Eds.), *Behavioural Pharmacology of 5-HT* (pp. 89-115). Hillsdale, NJ: Lawrence Erlbaum.

Olivier, B., Mos, J., & Van Oorschot, R. (1985). Maternal aggression in rats: effects of chlordiazepoxide and fluprazine. *Psychopharmacology, 86,* 68-76.

Olivier, B., Mos, J., Van der Heyden, J.A.M., & Hartog, J. (1989b). Serotonergic modulation of social interaction in male mice. *Psychopharmacology, 97,* 154-156.

Olivier, B., Van Aken, H., Jaarsma, I., Van Oorschot, R., Zethof, T., & Bradford, L.D. (1984). Behavioural effects of psychoactive drugs on agonistic behaviour of male territorial rats (resident-intruder paradigm). In K.A. Miczek, M.R. Kruk, and B. Olivier (Eds.), *Ethopharmacological Aggression Research* (pp. 137-156). New York: Alan R. Liss.

Othmer, E., & Othmer, S.C. (1987). Effects of buspirone on sexual dysfunction in patients with generalized anxiety disorders. J. Clin. Psychiatry, 48, 201-203.

Patel, J.B., Meiners, B.A., Salama, A.I., Malick, J.B., U'Pritchard, D.C., Giles, R.E., Goldberg, M.E., & Bare, T.M. (1988). Preclinical studies with pyrazolopyridine non-benzodiazepine anxiolytics: ICI 190, 662. *Pharmacol. Biochem. Behav., 29,* 775-779.

Peroutka, S.J. (1988). Functional correlates of central 5-HT binding sites. In N.N. Osborne and M. Hamon (Eds.), *Neuronal Serotonin* (pp. 423-447). London: John Wiley & Sons Ltd.

Ramage, A.G., & Fozard, J.R. (1987). Evidence that the putative 5-HT$_{1A}$ receptor agonists, 8-OH-DPAT and ipsapirone, have a central hypotensive action that differs from that of clonidine in anaesthetized cats. *Eur. J. Pharmacol., 138,* 179-191.

Rodgers, R.J., Shepherd, J.K. (1989). Prevention of the analgesic consequences of social defeat in male mice by 5-HT$_{1A}$ anxiolytics, buspirone, gepirone and ipsapirone. Psychopharmacology, 99, 274-380.

Schmidt, A.W., & Peroutka, S.J. (1989). 5-Hydroxytryptamine receptor "families". *The FASEB J., 3,* 2242-2249.

Sheehan, D.V., Raj, A.B., Sheehan, K.H., & Soto, S. (1988). The relative efficacy of buspirone, imipramine and placebo in panic disorder: a preliminary report. *Pharmacol. Biochem. Behav., 29,* 815-817.

Spencer, D.G., & Traber, J. (1987). The interoceptive discriminative stimuli induced by the novel putative anxiolytic TVX Q7821: behavioural evidence for the specific involvement of serotonin 5-HT$_{1A}$ receptors. *Psychopharmacology, 91,* 25-29.

Tricklebank, M.D., Neill, J., Kidd, E.J., & Fozard, J.R. (1987). Mediation of the discriminative stimulus properties of 8-hydroxy-2-(di-n-propylamino) tetralin ((8-OH-DPAT) by the putative 5-HT$_{1A}$ receptor. *Eur. J. Pharmacol., 133,* 47-56.

226

Van Wijngaarden, I., Tulp, M.Th.M., & Soudijn, W. (1990). The concept of selectivity in 5-HT receptor research. *Eur. J. Mol. Pharmacol.,* in press.

Waeber, C., Dietl, M.M., Hoyer, D., & Palacios, J.M. (1989). 5-HT_1 receptors in the vertebrate brain. Regional distribution examined by autoradiography. *Naunyn-Schmiedeberg's Arch. Pharmacol., 340,* 486-494.

Waeber, C., Schoeffter, P., Palacios, J.M., & Hoyer, D. (1989a). 5-HT_{1D} receptors in guinea pig and pigeon brain: Radioligand binding and biochemical studies. *Naunyn-Schmiedeberg's Arch. Pharmacol., 340,* 479-485.

Winslow, J.T., & Insel, T.R. (1990). Neuroethological models of Obsessive-Compulsive Disorder. In press.

Wouters, W., Tulp, M.Th.M., & Bevan, P. (1988). Flesinoxan lowers blood pressure and heart rate in cats via 5-HT_{1A} receptors. *Eur. J. Pharmacol., 149,* 213-223.

Ybema, C.E., Slangen, J.L., Olivier, B., & Mos, J. (1990). Discriminative stimulus properties of flesinoxan. *Pharmacol. Biochem. Behav., 35,* 781-784.

18 Studies on the Effects of Ethyl Alcohol on Rodent Behavior in Diverse Tests

Paul F. Brain and
University College of Swansea

Mansour A. Al-Hazmi
King Abdul-Aziz University

INTRODUCTION

Attempts to develop an armamentarium of behavior-changing drugs have involved varied tests situations. The initial approaches systematically used on a large scale were termed "psychopharmacology" and employed rapid, "easy-to-perform" tests, based on methodologies supplied by experimental psychology. These included techniques such as open field testing, intracranial self-stimulation, operant conditioning in a Skinner box, shuttle-box behavior or performance in a Hebb-Williams maze. In many situations, subjects are trained to a criterion of performance *before* drug treatments are commenced (see Sanger & Blackman, 1984). Such tests facilitate rapid screening of the array of compounds developed by chemists to assess their potential behavioral potency. These techniques supply 'behavioral profiles' such that materials can be classified as belonging to particular drug types. The perceived advantages of these tests are that they can be carried out quickly by relatively untrained individuals (or quantified automatically) and the generated data facilitate statistical analysis. Their *now* recognized disadvantage is a failure to deal with the fact that drugs influence behavior in diverse ways (see Brain, 1989). 'Direct' (motivational) and 'indirect' (via intraspecific signals or perception of such signals) effects are rarely distinguished, necessitating multiple 'control' groups. Further, the generated 'behavioral profile' is necessarily restricted (potentially important behavioral effects may go unrecorded). Some workers have questioned whether activities recorded in highly artificial and "impoverished" (intentionally so to reduce the range of potential responses) situations by psychopharmacologists have any relevance to 'normal' behavior (where the organism has a range of options).

Responding to some of the problems listed above, workers (e.g., Silverman, 1965, 1978; Mackintosh, 1970; Chance, Mackintosh, & Dixon, 1973; Flannelly & Lore, 1977; Miczek, 1978; Olivier & van Dalen, 1982; Poshivalov, 1982; Blanchard, Blanchard & Flannelly, 1985; Rodgers & Randall, 1985; Blanchard & Blanchard, 1987; Brain, McAllister, & Walmsley, 1989) have strongly advocated applying ethological considerations to assess drug actions on behavior. Attempts are made to provide test situations which conform more closely to the animal's natural environment. Further, styles of inclusive behavioral recording are used providing more extensive behavioral descriptions than is attempted in psychopharmacology. This tendency has been facilitated by using videotape recorders and microprocessors. The disadvantages of these 'ethopharmacological' or 'pharmacoethological' techniques are that they are time-consuming, analysis is mathematically-complex and they require skilled operatives, who have to maintain their performance level over extended periods. There, however, indications that this approach has advantages over traditional 'psychopharmacological' methodologies, amplified by the recent output of papers in this area (Poshivalov, 1981; Smoothy, Bowden, & Berry, 1982; Dixon, 1982; Olivier & van Dalen, 1982; Rodgers & Hendrie, 1983; Benton, Smoothy, & Brain, 1985; Brain, Smoothy, & Benton, 1985a; Smoothy, Brain, Berry, & Haug, 1986; Blanchard & Blanchard, 1987; Brain, McAllister, & Walmsley, 1987; Olivier, Mos, Heyden, Zethof, Aken, & Oorschot, 1987).

In spite of the debate, systematic comparisons of more than two behavioral test situations in drug research are rare. Workers generally indicate differences between their results and those of others using alternative strategies. This paper sets out to systematically examine the predictive powers of a number of traditional and novel methods in behavioral pharmacology. The tests selected range from extremely simplistic measures allied to operant behavior to ethologically-inspired situations. It appears that only systematic studies (where subjects are extremely comparable *and* the observer constant) are likely to answer questions about the relative utilities of different tests. These questions are important because drug studies require techniques which are rapid but do not lack predictive power.

The Tests

The most overtly simple test employed here is the *tube restraint* situation developed by Wagner, Beauving, and Hutchinson (1979; 1980). This technique when applied to psychopharmacological studies is much influenced by sedation (Smoothy, 1985), as only biting a target is recorded.

Another test advocated for drug studies, is the *platform* (holeboard) test. This situation provides a simple method of measuring the effects of drugs on a mouse's response to a novel environment. Kaesermann (1986) used the test to study different elements of behavior such as locomotion, grooming, attend, immobility, or rearing.

Another model examined the influence of drug administrations on the individual activity of *nest-building behavior*. It was thought that this might provide useful information on drug effects on a variety of behavioral elements (Schneider & Chenoweth, 1970; Moschovakis et al., 1978) as a permanent record of behavior (the nest) is generated by the animal.

Many studies on social behavior have concerned the acute or chronic effects of drugs on interacting rodents. More complex tests have considered activities variously known as *intermale fighting* (Moyer, 1968), *social conflict* (Brain, 1977) or *territorial behavior* (Flannelly & Lore, 1977; Dixon, 1978; Mackintosh, 1981; Blanchard & Blanchard, 1987). The staged encounters with *standard opponents* in the present studies looked at female as well as male behavior. They also examined the impact of experimental techniques in drug research.

This paper essentially compares and contrasts varied tests to determine what information one receives from particular situations in studies of ethyl alcohol. It was hoped to identify the strengths and weaknesses of particular methodologies and to offer guidelines for the selection of tests for specific purposes. The experiments described here seemed likely to settle some of the claims and counterclaims of individuals who work with behavior and this ubiquitous drug.

Ethyl alcohol

As with most other pharmacological agents, alcohol's actions depend on the dose administered or consumed, the species and/or strain of animal taking it, the route of administration, the time interval between application and behavioral assessment, and the test animal's previous experience of the compound. Ethanol is mainly metabolized in the liver but it is thought that up to 20% of a given dose may be transformed extra-hepatically. Three main enzyme systems metabolize ethanol *in vitro* in liver preparations, namely alcohol dehydrogenase (ADH), catalase, and the so-called microsomal ethanol-oxidizing system (MEOS) (Badawy, 1986). The three enzyme systems differ in many respects, including their subcellular compartmentalization and cofactor requirements. The major enzyme systems probably involved in ethanol metabolism *in vivo* are also ADH, catalase and MEOS. The abundance of catalase and the widespread occurrence of peroxidative reactions in the body suggest that catalase has an important and active role in ethanol oxidation *in vivo*.

Ethanol was obtained from B.D.H. Chemicals (Poole, U.K.), as a 99.7% pure material (Spectroscopic grade). Solutions (in saline) were arranged to produce doses of 0.125; 0.25; 0.5; 1.0, and 2.0 g of ethanol per kg body weight. The doses were selected on the basis of studies on the effects of alcohol on social aggression in mice of the Swiss-Webster line (Smoothy, 1982). Some studies, using this test have reported a biphasic action of alcohol on 'aggression', with low doses (0.3-0.8 g/kg) increasing and higher doses suppressing attack (Krsiak, 1976; Miczek & Barry, 1977). Smoothy, Bowden, and Barry (1982) failed to find low dose potentiation but higher doses of ethanol do suppress attack behavior before sedation is obvious (Berry & Smoothy, 1986).

Effects of ethanol on tube restraint-induced attack by Swiss mice

Most studies assessing the effects of alcohol on rat or mouse aggression have focused on a range of doses that suppress inter-male (Krsiak & Borgesova, 1973; Lagerspetz, 1980; Lagerspetz & Ekqvist, 1978) or shock-induced fighting. Alcohol is often perceived to exert its aggression-modifying effects primarily by acting directly or indirectly on the neural mechanisms that presumedly control aggressive behavior. In the tube restraint paradigm, males given alcohol exhibited a dose-dependent suppression of biting frequency (Smoothy & Berry, 1984a). The present study re-investigated the use of this last paradigm to study the impact of systemically-applied alcohol.

Methods

Seventy-two naive group-housed Swiss male and 72 female counterpart mice were allocated at 8 weeks of age to categories (N=12) injected i.p. with one of the standard doses of ethanol. Twenty minutes after injection, each test animal was placed for 10 minutes into the tube restraint apparatus (see Brain et al., 1983) and a variety of measures were obtained.

Results

Kruskal-Wallis tests revealed a significant overall drug effect of *latency to first bite* in both males (h=9.67, P<0.05) and females (h=10.16, P<0.05). There was also a significant overall drug effect on *biting frequency* in both males (h=22.86, P<0.0001) and females (h=18.23, P<0.003). *Post hoc* paired comparisons with saline treated controls using Mann-Whitney 'U' tests showed significant suppressive effects c.f. controls on bite frequencies in males at doses of 0.5 (U=29.0, P<0.01= and 2.0 g/kg (U=18.0, P<0.001). In females, the 1.0 (U=29.0, P<0.01= and 2.0 g/kg (U=14.0, P<0.004) doses were effective. Latencies to first bite were significantly increased at the highest dose in both male (U=28.5, P<0.01) and female (U=35.0, P<0.05) mice. *Proportions* of both males and females biting the target c.f.

controls were also suppressed by the highest dose of ethanol (both P<0.05, Fisher's test). Male and female controls showed similar bite frequencies.

Discussion

The substantial suppression of biting at the highest (2.0 g/kg) dose used suggests nonspecific CNS depression. However, ethological studies using males of the same line in a social conflict situation have shown that forms of activity requiring high degrees of motor coordination (e.g., bipedal postures such as 'rearing') are still observed at this dose (Smoothy & Berry, 1984b). Direct observation of animals within the tube revealed (as in Smoothy & Berry, 1984a) that mice receiving this dose still vigorously attempted to escape from the apparatus.

It is unclear whether the high dose reduced 'aggression' or simply produced difficulties in biting. Cats given comparable doses of ethyl alcohol have difficulty biting anaesthetized rats (MacDonnell & Ehmer, 1969). In the present study, initial bites at the target occurred as quickly as in controls at the 0.125; 0.25, and 1.0 g/kg doses in males and the 0.25 g/kg dose in females. Over the course of the trial, however, biting becomes less frequent than in the saline-treated animals. Motor impairment thus seems an unlikely explanation of the decline.

Earlier studies found that alcohol does not potentiate biting 'attack' in rats and mice in a tube restraint situation ay any dose (Tramill et al., 1980; Smoothy & Berry, 1984b). In the present study, there was also no significant increase in biting at low doses of ethanol as in the 'typical' biphasic effect of ethanol on social conflict in male mice (Miczek & O'Donnell, 1980). Other studies on the 'intermale' test have also provided little support for a low-dose potentiation of attack (Bertilson et al., 1977; Smoothy & Berry, 1983).

Effects of doses of alcohol on murine exploration in the platform apparatus

As experience influences performance in the tube restraint situation, a simpler behavioral situation for naive animals might reveal whether alcohol influences a basic behavioral attribute which modifies attack rather than "aggression" *per se*. This could account for variance in repeated tests on the effects of alcohol on behavior. For example, aggressive mice show less exploratory activity and an altered activity and an altered sensitivity to psychoactive drugs compared with "normal" mice (Valzelli, 1971). The present study consequently examined the effects of ethanol on performance of mice in the platform test.

Methods

Sixty male and 60 female 8-week-old group-housed Swiss mice were allocated to categories (N=12) injected i.p. with saline vehicle or one of the standard doses of ethanol. Twenty minutes later, they were placed for 10 minutes on the platform situation (similar to that used by File & Wardill) consisting of a 12cm high platform measuring 15 x 33cm with 24, 13mm diameter holes arranged in 6 equally-spaced rows of 6. Measures of a) latency to first place nose in a hole, b) number of different holes examined, c) total holes examined, d) total time spent immobile, e) number of extended stretches, f) number of rears, g) number of bouts of grooming or washing were obtained. The raw data were subjected to two-way analysis of variance with sex and dose of ethanol as the major variables. The data was subsequently subjected to paired comparisons (Student's t-test).

Results

Few of the measures showed significant interactions between dose of ethanol administered and sex of the subject but a significant interaction was evident in numbers of *extended stretches* (d.f. 5, 108, F=25.72, P<0.001) as the males showed a more marked reduction with increasing dose of ethanol than did their female counterparts. Alcohol dose had an overall effect on *latency to explore the first hole* (d.f. 5, 108, F=40.45, P<0.0001). The doses of 1.0 and 2.0 g/kg produced significant (all P<0.005) increases in latency to explore the first hole in male (t=4.7 and t=8.85, respectively, and female (t=4.28 and t=7.4, respectively) mice. The sex of subject significantly influenced the *number of head-dips* and the *number of different holes investigated* (d.f. 5, 108, F=119.65, P<0.0001 and d.f. 5, 108, F=112.09, =<0.0001, respectively). There was a significant influence of ethanol treatment on numbers of head-dips (d.f. 5, 108, F=2.341, P<0.05), but there was no significant influence of this drug on numbers of different holes investigated. The male subjects showed, however, significantly fewer head-dips at doses of 0.125, 1.0, and 2.0 g/kg of ethanol (t=5.24, t=4.56, and t=6.22, respectively, all P<0.005) than their female counterparts. The drug (d.f. 5, 108, F=5.4, P<0.02) and subject's sex (d.f. 5, 108, F=5.93, P<0.02) significantly changed the *time subjects spent immobile.* Alcohol at doses of 0.05, 1.0, and 2.0 g/kg significantly increased the time spent immobile in males (t=2.61, P<0.05; t=3.2, P<0.025; and t=2.96, P<0.025, respectively). In female mice, only the highest dose significantly increased (t=3.21, P<0.01) this measure. Drug treatment also produced significant variance in the *numbers of extended stretches* (d.f. 5, 108, F=8.85, P<0.0001) and male mice showed significant decreases in this attribute at doses of 0.25, 0.5, 1.0, and 2.0 g/kg of ethanol (t=3.39; t=6.32; t=6.43; t=6.43, respectively, all P<0.005). Female mice did not show significant changes on this measure. Sex had a significant influence (d.f. 5, 108, F=25.72, P<0.0001) on numbers of extended stretches. There was an overall influence of dose of ethanol on *numbers of rears* (d.f., 108, F016.47; P<0.0001). Ethanol doses of 0.25, 1.0, and 2.0 g/kg significantly decreased the number of rears shown by female mice c.f. controls (t=3.88; t=4.47; and t=5.05, respectively, all P<0.005). *Grooming* was not significantly influenced by ethanol dose or sex in the present study.

Discussion

A wide variety of effects of ethanol on varied measures of exploratory activity in the platform test are thus evident. Major effects of alcohol in mice were evident in terms of the number of head-dips, the latency to head-dipping, and the number of holes explored. Strikingly, the number of head-dips *decreased* with increasing alcohol dose. In contrast, 0.5 and 1.0 g/kg of alcohol reportedly *increase* head-dipping in mice (Joyce, Steele, & Summerfield, 1972). This diametrically opposite result might be related to differences in the equipment (the apparatus used by Joyce et al. had four holes, whereas the apparatus used in the present study had 24 holes) or conditions of testing. Different periods of exposure to the apparatus were also used; in the present study, the subjects were placed for 10 minutes onto the apparatus whereas Joyce et al. tested for 3 minutes.

The number of "extended stretches" (claimed to be an index of tentative exploratory behavior by Kaesermann, 1986), were clearly reduced by alcohol treatment, with males showing the more striking decline with increasing dose. Immobility has long been regarded as an expression of suppressed exploratory motivation. This attribute was significantly increased by alcohol treatment although rather different patterns were evident in males and females.

Few published studies concern the effects of alcohol on exploration. File and Wardill (1975), in a notable exception, reported that alcohol increased the frequency and duration of head-dips. This result also differs from present findings but File and Wardill placed the subjects *underneath* the holes, whereas in the present study, the subjects were placed on the top of the apparatus.

One may conclude that when ethanol is administered to naive mice before they are placed in a novel situation (as is employed in most aggression tests), there is generally an increase in the latency to move and a relative increase in immobility. Exploratory behavior is also somewhat reduced. It is quite possible that some effects of alcohol observed in 'aggression tests' are due to changes in the exploration of novel situations rather than direct actions on aggressiveness *per se.*

Effects of alcohol on nest-building in male and female mice

Female rats show a decline in daily nesting activity after repeated feeding of a liquid diet containing ethanol (Bond, 1979). Indeed, these subjects are less active and spend more time lying on the nest, supporting the view that large doses of ethanol depress activity (including that of nest-building in rats) (Pohorecky, 1977; Bond, 1979). The present study assessed the effects of repeated injections of alcohol on nest-building in virgin mice of both sexes. This method of administration gives a more reliable dose of ethanol although it is more stressful.

Methods

Fifty male and 50 female mice (eight weeks old at the beginning of the experiment) were individually housed in type M1 cages. On the first day of the experiment, 15 g of shredded paper was placed in each subject's food hopper, twenty minutes after categories of animals (N=10) had been injected i.p. with one of the standard doses of ethanol. Nest-building behavior was directly observed for 15 minutes in each test animal, assessing a) latency to first pull in material, b) the total time spent in nest-building, and c) the number of times the test animal returned to the hopper to collect material, and d) the total time spent immobile. After 24 hours, the resultant nest was weighed and the style determined. The old nest was removed, the sawdust replaced, new material provided, and the procedure repeated over 14 consecutive days.

Results

Behavioral observations

Kruskal-Wallis tests revealed no significant effects over test days in either males or females in terms of *latency to first pull nesting material,* the *total time spent in nest-building* or the *number of occasions the test animal returned to collect nest materials.* The only significant change over days occurred at the highest dose of ethanol where the time spent *immobile* by female mice showed marked variance.

Kruskal-Wallis tests showed marked differences (all P<0.001) over the doses of ethanol (irrespective of test day) in terms of latency to first pull nesting material, the total time spent nest-building, the total number of occasions the test animal returned to collect nest materials or the total time spent immobile in male (h=112.033; h=147.733; h=122.401, respectively), and female (h=135.991; h=138.608; h=135.238.548, respectively) subjects.

There were no significant effects of the lowest dose of alcohol used on any measure of behavior in either sex. Male mice showed highly significant increases in latencies to first pull material into the cage at doses of 0.5, 1.0, and 2.0 g/kg (U=7997.5, P<0.01; U=6435.5, P<0.001, and U=3920.0, P<0.001, respectively). Females also showed significant increases on this measure at these doses (U=7995.0, P<0.01; U=7829.0, P<0.01, and U=3024.0, P<0.001, respectively). Ethanol, especially at higher doses, greatly increased the time spent immobile in both sexes. Male mice given 0.5, 1.0, or 2.0 g/kg of ethanol showed significantly extended periods of immobility (U=4636.5 and U=3672.0, respectively, all P<0.001). Females also significantly increased the time they spent immobile at doses of 1.0 and 2.0 g/kg of ethanol (U=6995.5 and U=1163.0, respectively). Alcohol significantly decreased the time allocated to nest building activity in males at doses of 0.5, 1.0, and 2.0 g/kg (U=8374.0, P<0.05; U=6294.5, P<0.001 and U=3920.0, P<0.001, respectively). Female mice also showed significant declines on this measure at doses of 0.5, 1.0 and 2.0 g/kg (U=7507.5; U=7495.5; U=3122.0, respectively, all P<0.001). 0.5, 1.0, and 2.0 g/kg of ethanol significantly reduced the number of times males returned to collect nest materials (U=7917.5, P<0.01; U=6468.5, P<0.001 and U=3920.0, P<0.001, respectively). Females also showed significant declines on this measure at these doses (U=7445.0, U=7510.5, and U=3085.5, respectively, all P<0.001).

Measures of materials used

The effects of ethanol doses on the two sexes over the test days were assessed statistically using analysis of variance. There was no significant sex difference with respect to the amount of nesting material used (F sex = 0.01, d.f. 1, 90, n.s.) nor was there any significant difference in the amount of material used on different days of the experiment F day = 1.45, d.f. 13, 1170, n.s.). There was, however, a highly significant treatment effect (F dose 0 5.07, d.f. 4, 90, P<0.001). The interaction between sex and dose was not significant (F sex x dose = 1.06, d.f. 4, 90, n.s.) but the sex by day interaction was significant (F sex x day = 1.79, d.f. 1, 1170, P<0.05). This last effect seems due to female mice pulling more nesting material on the first three days than their male counterparts. In addition, the interaction between drug dose and days was highly significant (F dose x days = 5.44, d.f. 4, 1170, P<0.001). This seems due to the progressive reduction in the amount of nest material used over test days at the highest dose of ethanol. The three way interaction sex by dose by day was significant (F sex x dose x day = 1, 1170, P<0.001). This may be due to the dramatically different response of females given the highest ethanol dose on the first 3 days of testing compared with their male counterparts.

Further analyses using the simple main effects test confirmed that ethanol produced significant effects from days 4 to 14. There was significant variance over days in animals given the control treatment (d.f. 13, 1170, F=3.75, P<0.001), 0.25 g/kg (d.f. 13, 1170, F=2.17, P<0.01) or 2.0 g/kg (d.f. 13, 1170, F=14.6, P<0.001), but not in subjects given 0.5 or 1.0 g/kg of ethanol.

An arbitrary rating scale score for the different types of nest made by animals with each type of nest being allocated a value as follows: Plate = 1, Bowl = 2, Covered = 3, and Double = 4. Animals not building a nest received a score of 0. The scale (Rajendram, Brain, Parmigiani, & Mainardi, 1987) was such that constructions of greater complexity received progressively higher arbitrary (unweighted) 'scores'. The resulting data was subjected to two-way analysis of variance with sex and dose of ethanol as the major variables. The data was subsequently subjected to paired comparisons (using Student's t test) to give the direction of changes between the control and the drug doses. None of the measures were significantly influenced by the subject's sex (F sex = 0.1, d.f. 1, 90, n.s.). Drug treatment produced significant variance in scores for the type of nest that was built (F dose = 4.7, d.f. 4, 90, P<0.002), seemingly because ethanol (especially at the higher doses), resulted in females building 'plate' style nests rather than 'bowl' or 'covered' nests and males showing a greater tendency to build no nest. Alcohol at the highest dose, significantly decreased the style score of nest construction in male and female subjects (t=3.09, P<0.01 and t=2.53, P<0.05, respectively) but there was no interaction (F dose x sex = 2.3, d.f. 4, 90, n.s.) between dose of ethanol and sex on this measure.

Discussion

The failure to find significant sex differences here in terms of the weight of nest material used and type of construction generated, supports Brain's and Rajendram's (1986) findings on the same strain of mice. Initially, both sexes pulled much nest material into the cage at the lowest dose of ethanol but less material was used with increasing ethanol doses. Although both sexes showed a progressive decline over days in time allocated to nest-building, this effect was most clear with the highest dose of ethanol. This decline in nest-building in all subjects seems simply due to animals spending more time immobile when treated with higher doses.

As higher ethanol doses also increased the latency to first pulling in nest materials in male and female mice whereas control subjects and counterparts given low doses of ethanol showed vigorous responses to nesting material, the highest doses of ethanol may simply sedate the subjects. As the complexity of types of nest built by male and female subjects declined with dose of ethanol, one can maintain that the drug reduces the motivation for nest building.

When supplied with paper, mice in this study generally immediately started pulling material into the cage, carrying it around and fraying the strips, confirming observations by Rajendram (1984). The fact that this activity decreases in both sexes with increasing ethanol doses may have a number of possible explanations. Firstly, it is well-known that large doses of ethanol depress activity (Pohorecky, 1977; Smoothy, Bowden, & Berry, 1982; Smoothy et al., 1986; Brain, 1986). Thus, subjects treated with high doses of ethanol may simply be less active and less likely to show nest-building. Secondly, Bond (1979) recorded that alcohol-treated female rat spent more time lying in the nest and Ritchie (1980) noted that ethanol causes vasodilation and increases heat loss. Consequently, ethanol potentially facilitates huddling in the nest (in an attempt to keep warm), an activity which could also account for the reduction in nest-building.

Effects of alcohol on social behavior in naive male and female mice

Males and females, having very different baselines of 'fear'-related behavior, may respond differently to ethanol. Consequently, the effects of alcohol were reassessed on the behavior of naive mice and female mice, tested in dyadic encounters with male 'standard opponents'.

Methods

Sixty male and 60 female individually-housed Swiss mice were used in this study. Twenty minutes before behavioral encounters, categories of isolates (N=10) were injected i.p. with one of the standard doses of ethanol (see Brain, McAllister, & Walmsley, 1989). Twenty minutes later, a docile 'standard opponent' was introduced into the test animal's home cage and the social encounter was videotaped for 10 minutes (see Brain, McAllister, & Walmsley, 1989). The tape was later analysed in terms of latency to attack and the times (in seconds) allocated to social investigation, non-social investigation, attack, threat, defence, displacement behavior, and number of attacks. The individual categories are described in Brain et al. (op cit).

Results

Kruskal-Wallis tests showed significant variance over doses in male subjects in terms of *latency to attack* (h=13.371, P<0.05=, *time spent in non-social investigation* (h=29.212, P<0.001), time spent in attack (h=11.874, P<0.05), *time spent in threat* (h=13.775, P=0.05, and the *number of attacks* (h=19.830, P<0.01). Kruskal-Wallis tests also revealed significant differences in female subjects, in terms of the time they allocated to *social investigation, non-social investigation*, and *threat* (h=10.54; h=12.207; h=9.98, respectively, all P<0.05).

Although latency to attack by the males increased with increasing ethanol dose, paired Mann-Whitney 'U' test comparisons with controls only reached significance for the highest dose (U=16.5, P<0.01). Male and female subjects given this dose of ethanol allocated significantly more time to non-social investigation (U=13.0, P<0.01 and U=24.0, P<0.05, respectively), but showed reduced times allocated to social investigation in male or female subjects. Ethanol significantly decreased the time allocated to attack by males, especially at the highest dose (U=16.5, P<0.01). Significant decreases in numbers of attacks by male mice at doses of 0.5, 1.0, and 2.0 g/kg of ethanol (U=21.0, P<0.05; U=22.5, P<0.05; and U=16.5, P<0.01, respectively) were noted. Significant suppression of threat in both male and female subjects (U=18.0 and U=21.5, respectively, both P<0.05) was only seen after the highest dose. Alcohol had no effect on the time allocated to defensive or displacement behavior in either sex.

Discussion

The dose-dependent suppression of agonistic behavior in mice with increasing alcohol dose, differs from studies where biphasic actions of the drug have been claimed. This discrepancy may be due to variations in methodologies, such as the strain of mice employed and the use of naive animals rather than aggressive, sociable, or timid individuals (these different types of animal vary considerably in their response to ethanol (see Krsiak, 1976; Miczek & Barry, 1977). However, Mos, Olivier and Aken (1987), and Olivier et al. (1987), have also reported that the only decrease in aggression in male mice is seen at the 2.0 g/kg dose whereas lesser doses do *not* increase this measure. Although female rodents show less aggression than males, there was no evidence that alcohol stimulated threat or attack in the present situation in female mice, a finding in agreement with Smoothy et al. (1986).

The present study produced no significant increases in the time allocated to defensive behavior after treatment. Comparisons with similar studies also revealed marked differences in the dose of ethanol required to generate a significant increase in defensive behavior. Smoothy, Bowden, and Berry (1982) reported that 1.0 g/kg of ethanol was sufficient to significantly increase the time spent in flight behavior and also to elevate the frequency of defensive-escape postures. Krsiak (1976) found that timid-defensive-escape activities were altered by 2.4 g/kg of ethanol with the dose increasing flight behavior in aggressive males and reducing defensive upright postures in timid mice. The failure of ethanol to influence timidity to individual housed mice in the present study, supports the majority of negative results in the extensive literature that examines the claimed tension-reducing effects of ethanol in animals (reviewed by Cappell & Herman, 1972). In spite of their high baseline of defensive behavior in the present study, females did not show any effects of alcohol on this attribute.

As sociable activities were unchanged by lower doses of ethanol tested in either males or females, the present results confirm and extend previous studies which, although they used a narrower dose range of ethanol, found reduced or unchanged amounts of social investigation after treatment (Chance, Mackintosh, & Dixon, 1973; Smoothy & Berry, 1983; Smoothy, Bowden, & Berry, 1982). They contrast with Krsiak (1976) who recorded that 0.4 g/kg of ethanol *reduced* the incidence of sniffing and following of partners by sociable isolates. The present subjects may not properly be described as "sociable".

As non-social behavior was increased by the highest dose of ethanol in both sexes, this augmentation seems simply a consequence of the increased available time in animals whose social interactions were suppressed by the drug.

GENERAL DISCUSSION

A synopsis of the effects of ethyl alcohol on behavior in the different test situations is provided in Table 1.

None of the tests with ethanol provided convincing evidence of a biphasic effect on behavior. *All* studies confirmed that the highest dose used *reduced* various aspects of activity. It is difficult, however, to evaluate the reduction of biting in the tube test. The platform and nest-building tests simply suggest that the drug reduces activity but the standard opponent test confirms that the treated animals are simply less prone to interact with conspecifics than are control treated counterparts.

Some specific comments on the tests may be appropriate:

TABLE 18.1

Synopsis of the Behavioral Effects of Ethyl Alcohol in the Different Tests			
Tube Restraint	Elevated Platform	Nest Building	Standard Opponent
No evidence of a biphasic effect—highest dose reduces all biting activities.	In males, increasing doses progressively reduce head-dipping and extended stretches. In females, rears are suppressed.	All acutely assessed measures of nest building are reduced by higher doses - no evidence of low dose potentiation.	No biphasic action. The highest dose reduced offense and social investigation in males. Social investigation and threat reduced in females.

The Tube Test

Although the tube test is superficially easy to use it can be objected to on the following grounds: a) the generated restraint 'stress' does not seem entirely ethically acceptable if it is not an integral part of the experiment; b) interpretation of the results is difficult—it is certainly *not* possible to simply equate biting with aggression (see Brain, 1986): and c) the test does not seem very sensitive. In some cases, no significant effects are apparent—although actions are evident in other tests. Only the highest doses used seem to significantly changer performance in this test.

The Elevated Platform

This is a "simple-to-use" test in which relatively mild impositions are made on the test animals. It is suggested that latency to move, total number of head-dips, extended stretches, and rears would provide all the relevant information on exploratory behavior. A major difficulty of the test is, however, that the range of potential behavior is not very wide. All measures used here are *very* susceptible to sedative actions of drugs. For many preliminary rapid screening protocols, the method seems to have advantages over open field and maze tests.

The Nest-Building Test

This test measures a seemingly compulsive natural response which is rather repetitive. It is clear that such behavior *must* be assessed over several days and only the acutely-assessed measures of behavior consistently show significant drug effects (often the animals seem to make up any initial decrements in the behavior). As ethanol (and other drugs) influenced activity in males and these animals are less behaviorally variable than their female counterparts, it seems profitable to consider only using animals of this sex in such tests.

The 'Standard Opponent' Test

This paradigm still seems of great interest as the activities generated can be assessed in male and female subjects (with rather different baselines). Also a wide range of elements can be quantified and such tests are sensitive and amenable to analysis using videotape and microprocessor-assisted techniques. It is also very difficult to deal adequately with encounters involving more than two animals or more complex social relationships. Using the present tests, effects were apparent at high or low doses of drug treatment but it seems essential to use males *and* females to exploit the range of behaviors. Tests involving more naturalistic test situations are currently being developed.

REFERENCES

Badawy, A.A.B. (1986). Alcohol as a psychopharmacological agent. In P.F. Brain (Ed.), *Alcohol and Aggression* (pp. 55-83). London: Croom Helm.

Benton, D., Smoothy, R., & Brain, P.F. (1985). Comparisons of the influence of morphine sulfate, morphine-3-glucuronide and tifluadom on social encounters in mice. *Physiol. Behav., 35,* 689-693.

Berry, M.S., & Smoothy, R. (1986). A critical evaluation of claimed relationships between alcohol intake and aggression in infra-human animals. In P.F. Brain (Ed.), *Alcohol and Aggression* (pp. 84-137). London: Croom Helm.

Bertilson, H.S., Mead, J.D., Morgret, M.K., & Dengerink, H.A. (1977). Measurement of mouse squeals for 23 hours as evidence of long-term effects of alcohol on aggression in pairs of mice. *Physiol. Rep.,* 41, 247-250.

Blanchard, R.J., & Blanchard, D.C. (1987). The relationship between ethanol consumption and aggression: Studies using ethological models. In B. Olivier, J. Mos, and P.F. Brain (Eds.), *Ethopharmacology of Agonistic Behaviour in Humans and Animals* (pp. 144-160). Dordrecht: Martinus Nijhoff.

Blanchard, R.J., Blanchard, D.C., & Flannelly, K.J. (1985). Social stress, mortality and aggression in colonies and burrowing habitats. *Behav. Proc., 11,* 209-213.

Bond, N.W. (1979). Effects of postnatal alcohol exposure on maternal nesting behaviour in the rat. *Physiol. Behav., 7,* 396-398.

Brain, P.F. (1977). *Hormones and Aggression, Vol. 1: Annual Research Reviews.* Montreal: Eden Press.

Brain, P.F. (1986). The nature of aggression. In P.F. Brain (Ed.), *Alcohol and Aggression* (pp. 1-18). London: Croom Helm.

Brain, P.F. (1989). An ethoexperimental approach to behaviour endocrinology. In R.J. Blanchard, P.F. Brain, D.C. Blanchard, and S. Parmigiani (Eds.), *Ethoexperimental Approaches to the Analysis of Behavior* (pp. 539-557). Dordrecht: Kluwer Academic.

Brain, P.F., & Rajendram, E.A. (1986). Nest-building in rodents: A brief cross-species review. In P.F. Brain and J.M. Ramirez (Eds.), *Cross-Disciplinary Studies on Aggression* (pp. 157-182). Sevilla: Publications de la Universidad de Sevilla.

Brain, P.F., Al-Maliki, S., Parmigiani, S., & Hammour, H.A. (1983). Studies on tube restraint-induced attack on a metal target by laboratory mice. *Behav. Proc., 8,* 277-287.

Brain, P.F., Smoothy, R., & Benton, D. (1985). An ethological analysis of the effects of tifluadom on social encounters in male albino mice. *Pharmacol. Biochem. Behav., 23,* 979-985.

Brain, P.F., McAllister, K.H., & Walmsley, S. (1989). Drug effects on social behavior: methods in ethopharmacology. In A.A. Boulton, G.B. Baker, and A.J. Greenshaw (Eds.), *Neuromethods Volume 13: Psychopharmacology* (pp. 689-739). Clifton, NJ: Humana Press Inc.

Cappell, H., & Herman, C.P. (1972). Alcohol and tension reduction: A review. *Quart. J. Stud. Alcohol, 33,* 33-64.

Chance, M.R.A., Mackintosh, J.H., & Dixon, A.K. (1973). The effects of ethyl alcohol on social encounters between mice. *J. Alcoholism, 8,* 90-93.

Dixon, A.K. (1978). Rodent social behaviour in relation to biomedical research. In W.H. Weike (Ed.), *Das Tier im Experiment* (pp. 128-146).

Dixon, A.K. (1982). A possible olfactory component in the effects of diazepam on social behaviour of mice. *Psychopharmacology, 77,* 246-252.

File, S.E., & Wardill, A.G. (1975). Validity of head-dipping as a measure of exploration in a modified hole-board. *Psychopharmacology, 44,* 53-59.

Flannelly, K., & Lore, R. (1977). The influence of females upon aggression in domesticated male rats (Rattus norvegicus). *Anim. Behav., 25,* 654-659.

Joyce, D., Steele, J.W., & Summerfield, A. (1972). Chronic injection of nicotine modifies the behaviour of mice after ethanol. *Br. J. Pharmacol., 45,* 164-165.

Kaesermann, H.P. (1986). Stretched attend posture, a non-social form of ambivalence, is sensitive to a conflict-reducing drug action. *Psychopharmacology, 89,* 31-37.

Krsiak, K.M. (1976). Effect of ethanol on aggression and timidity in mice. *Psychopharmacology, 51,* 75-80.

Krsiak, M., & Borgesova, M. (1973). Effect of alcohol on behaviour of pairs of rats. *Psychopharmacology, 32,* 201-209.

Lagerspetz, K.M.J. (1980). Failure to induce aggression in mice with ethyl alcohol. In K. Eriksson, K. Kiianmaa, and J.D. Sinclair (Eds.), *Animal Models in Alcohol Research* (pp. 329-335). London: Academic Press.

Lagerspetz, K.M.J., & Ekqvist, K. (1978). Failure to induce aggression in inhibited and in genetically non-aggressive mice through injections of ethyl alcohol. *Aggr. Behav., 4,* 105-113.

MacDonnell, M.F., & Ehmer, M. (1969). Some effects of ethanol on aggressive behaviour in cats. *Quart. J. Stud. Alcohol, 30,* 312-319.

Mackintosh, J.H. (1970). Territory formation by laboratory mice. *Anim. Behav.,18,* 177-183.

Mackintosh, J.H. (1981). Behaviour of the house mouse. In R.J. Berry (Ed.), *The Biology of the House Mouse* (pp. 337-365). London: Academic Press.

Miczek, K.A. (1978). Δ^9-tetrahydrocannabinol: Antiaggressive effects in mice, rats and squirrel monkeys. *Science (N.Y.), 199,* 1459-1461.

Miczek, K.A., & Barry, H. (1976). Pharmacology of sex and aggression. In: S.D. Glick and J. Goldfarb (Eds.), *Behavioral Pharmacology* (pp. 176-257). St. Louis: C.V. Mosby Inc.

Miczek, K.A., & O'Donnell, J.M. (1980). Alcohol and chlordiazepoxide increase suppressed aggression in mice. *Psychopharmacology, 69,* 39-44.

Mos, J., Olivier, B., & Aken, V.H. (1987). Differential effects of psychoactive drugs on male rats housed in a colony. In J.M. Ramirez (Ed.), *Research on Aggression* (p. 88). Sevilla: Publicaciones de la Universidad de Sevilla.

Moschovakis, A., Liakopoulos, D., Armaganidis, A., Kapsambelis, V., Papanikalou, G., & Petroulakis, G. (1978). Cannabis interferes with nest-building behavior in mice. *Psychopharmacology, 58,* 181-183.

Moyer, K.E. (1968). Kinds of aggression and their physiological basis. *Comm. Behav. Biol., 2,* 65-87.

Olivier, B., & van Dalen, D. (1982). Social behavior in rats and mice: An ethologically based model for differentiating psychoactive drugs. *Aggr. Behav., 8,* 163-168.

Olivier, B., Mos, J., Heyden, V.D.J.A., Zethof, T., Aken, V.H., and Oorschot, V.R. (1987). Effects of DU 28853, a new serenic drug, in several experimental models for aggression. In J.M. Ramirez (Ed.), *Research on Aggression* (p. 93). Sevilla: Publicaciones de la Universidad de Sevilla.

Pohorecky, L.A. (1977). Biphasic action of ethanol. *Biobehav. Rev., 1,* 231-240.

Poshivalov, V.P. (1981). Pharmaco-ethological analysis of social behavior in isolated mice. *Pharmacol. Biochem. Behav., 14,* 53-59.

Poshivalov, V.P. (1982). Ethological analysis of neuropeptides and psychotropic drugs: Effects on intraspecies aggression and sociability of isolated mice. *Aggr. Behav., 8,* 355-369.

Rajendram, E.A. (1984). *The factors affecting nest-building behaviour in different species of rodents.* University of Wales: Ph.D. Dissertation.

Rajendram, E.A., Brain, P.F., Parmigiani, S., & Mainardi, M. (1987). Effects of ambient temperature on nest construction in four species of rodent. *Boll. Zool., 54,* 75-81.

Ritchie, J.M. (1980). The aliphatic alcohols. In L.S. Goodman and A. Gilman (Eds.), *The Pharmacological Basis of Therapeutics* (pp. 376-390). New York: MacMillan.

Rodgers, R.J., & Hendrie, C.A. (1983). Social conflict activates status-dependent analgesic or hyperalgesic mechanisms in male mice: Effects of naloxone on nociception and behavior. *Physiol. Behav., 30,* 775-780.

Rodgers, R.J., & Randall, J.I. (1985). Situation-dependent opioid or non-opioid analgesia intruder mice. In F. Le Moll (Ed.), *Multidisciplinary Approaches to Conflict and Appeasement in Animals and Man* (p. 72). Parma: Istituto di Zoologia.

Sanger, D.J., & Blackman, D.E. (Eds.) (1984). *Aspects of Psycho-pharmacology*. London: Methuen.

Schneider, C.W., & Chenoweth, M.B. (1970). Effects of hallucinogenic and other drugs on nest-building behaviour in mice. *Nature (London), 225,* 1262-1263.

Silverman, A.P. (1965). Ethological and statistical analysis of drug effects on the social behaviour of laboratory rats. *Br. J. Pharmacol., 24,* 579-590.

Silverman, A.P. (1978). Analysis of behaviour in social situations. In A.P. Silverman (Ed.), *Animal Behaviour in the Laboratory* (pp. 1294-1338). London: Chapman and Hall Ltd.

Smoothy, R. (1985). *Videotape analysis of the effects of alcohol on aggressive and other behaviour in laboratory mice (Mus musculus).* University of Wales: Ph.D. Dissertation.

Smoothy, R., & Berry, M.S. (1983). Effects of ethanol on behaviour of aggressive mice from two different strains: A comparison of simple and complex behavioral assessments. *Pharmacol. Biochem. Behav., 9,* 645-653.

Smoothy, R., & Berry, M.S. (1984a). Effects of ethanol on murine aggression assessed by biting of an inanimate target. *Psychopharmacology, 83,* 268-271.

Smoothy, R., & Berry, M.S. (1984b). Alcohol increases both locomotion and immobility in mice: An ethological analysis of spontaneous motor activity. *Psychopharmacology, 83,* 272-276.

Smoothy, R., Bowden, N.J., & Berry, M.S. (1982). Ethanol and social behaviour in naive Swiss mice. *Aggr. Behav., 8,* 204-207.

Smoothy, R., Brain, P.F., Berry, M.S., & Haug, M. (1986). Alcohol and social behavior in group-housed female mice. *Physiol. Behav., 37,* 689-694.

Tramill, J.L., Turner, P.E., Sisemore, D.A., & Davis, S.F. (1980). Hungry drunk and not real mad: The effects of alcohol injections on aggressive responding. *Bull. Psychonom. Soc., 15,* 339-341.

Valzelli, L. (1971). Further aspects of the exploratory behaviour in aggressive mice. *Psychopharmacologia (Berlin), 19,* 91-94.

Wagner, G.C., Beauving, L.J., & Hutchinson, R.R. (1979). Androgen dependency of aggressive target-biting and paired fighting in male mice. *Physiol. Behav., 22,* 43-46.

Wagner, G.C., Beauving, L.J., & Hutchinson, R.R. (1980). The effects of gonadal hormone manipulations on aggressive target biting in mice. *Aggr. Behav., 6,* 1-7.

19 Neurotransmitters in Affective Disorders

Gerard J. Marek and Lewis S. Seiden
University of Chicago

Monoaminergic neurotransmitters have played an important role in directing research strategies regarding understanding the etiology, pathogenesis and treatment of affective disorders. In this chapter, we will briefly describe the current classification of affective disorders, the role that monoaminergic neurotransmitters may play in the etiology and/or pathogenesis of affective disorders, and both clinical and preclinical evidence suggesting a role for monoaminergic neurotransmitters in the therapeutic effects and antidepressant treatments.

Classification of affective disorders

One of the major distinctions in the affective disorders is between unipolar and bipolar illness. Unipolar disorder is characterized by episodes of major depression while bipolar disorder is characterized by episodes of both major depression and mania. Criteria have been developed to aid in reliable diagnoses of major depression. These criteria include a dysphoric mood, loss of interest or pleasure in usual activities, change in appetite or weigh, sleep disturbances, psychomotor agitation or retardation, loss of energy, feelings of worthlessness, excessive and inappropriate guilt, decreased ability to concentrate or make decisions, and recurrent thoughts of death progressing to suicide ideation or suicide attempts. These symptoms must be present nearly every day for a period of at least two weeks to constitute a diagnosis of an episode of a major depressive illness. Other types of affective disorders are: cyclothymic disorder, which include numerous periods of both depression and hypomania over a two year period not severe enough to be classified as a bipolar disorder; dysthymic disorder, which is a chronic mood disorder over a two year period not severe enough to be classified as major depression; and seasonal affective disorder, which is characterized by the regular appearance of depression between October-November and full remission between February-April (Diagnostic and Statistical Manual of Mental Disorders, 3rd Ed.-Rev.; DSM-III-R). The focus of the current chapter will be on major depression.

Neurotransmitters in etiology and/or pathogenesis of affective disorders

Norepinephrine (NE) has long been implicated in playing a role in the etiology or pathogenesis of depression leading to the catecholamine (CA) theory of depression, which posits that depression is a result of decreased noradrenergic neurotransmission (Bunney & Davis, 1965; Schildkraut & Kety, 1967). The CA theory of depression is based in large part on the therapeutic effectiveness of tricyclic antidepressants (TCA) and monoamine oxidase (MAO) inhibitors which enhance synaptic availability of norepinephrine and block the hypothermic effects of reserpine. The latter finding was of particular interest as reserpine is known to deplete monoamines, and was associated with episodes of depression when used as an antihypertensive in the 1950's. The phenomenon of reserpine-induced depression, however, is not as clear cut as was originally thought (Goodwin, Ebert, & Bunney, 1972). A recent exciting development links DNA markers on tip of the short arm of chromosome 11 to bipolar illness in the Amish community of Lancaster county in Pennsylvania (Egeland, Gerhard, Pauls, Sussex, Kidd, Allen, Hostetter, & Housman, 1987). This marker is located in the same region as the gene coding for tyrosine hydroxylase. Another intriguing observation is the finding of an increased number of beta-adrenergic receptors in the frontal cortex of suicide victims (Mann, Stanley, McBride, & McEwen, 1986).

Serotonin (5-hydroxytryptamine; 5-HT) has also been implicated in playing a role in the etiology or pathogenesis of depression leading to the indolamine theory of depression, which posits that depression is a result of decreased serotonergic neurotransmission (Lapin & Oxenkrug,1969). One of the major lines of evidence supporting this theory is that baseline levels of 5-hydroxyindole acetic acid (5-HIAA; the deaminated 5-HT metabolite) tend to be significantly lower in depressed patients than in controls in approximately 50% of cerebrospinal fluid (CSF) monoamine metabolite studies (see Post, Ballenger, & Goodwin, 1980). Regarding the negative CSF studies, the work of Åsberg, Thoren, Traskman, Bertilsson, & Ringberger (1976) and Åsberg, Bertilsson, Mårtensson, Scalia-Tomba, Thoren, & Traskman-Bendz, (1984) has suggested that a heterogeneous patient population as well as methodological problems in collecting and analyzing CSF 5-HIAA may have contributed to those studies reporting no significant differences between depressed subjects vs controls. A second line of evidence supporting the indolamine theory is the observation of increased numbers of 5-HT-2 receptors in the frontal cortices of suicide victims (Stanley & Mann, 1983; Mann et al., 1986; Arora & Meltzer,1989). This finding is of considerable interest given that many antidepressant drugs decrease either the number and/or the responsiveness of the 5-HT-2 receptor system (Peroutka & Snyder, 1980a; Sanders-Bush, Breeding, Knoth, & Tsutsumi, 1989).

Other neurotransmitters that have been investigated for their role in affective disorders include GABA, dopamine, acetylcholine and histamine. Especially interesting are the ties between affective disorders to both the hypothalamic-pituitary-thyroid and the hypothalamic-pituitary -adrenal axis (see Nemeroff, 1989). Space precludes discussing these other neurotransmitter systems.

Mechanism of action of antidepressant drugs

There are several key facts which are important to keep in mind when attempting to formulate a mechanism to account for the therapeutic effects of antidepressant drugs. The first key observation is that antidepressants do not act immediately to relieve depression. One to three weeks of daily antidepressant drug administration is required to demonstrate a clinical effect (Oswald, Brezinova, & Dunleavy, 1972). This time lag poses a problem for the catecholamine and indoleamine theories of depression. These theories were built to a large degree on the known pharmacology of TCAs and MAOIs (inhibition of monoamine uptake and MAO, respectively). This 1-2 week time lag before therapeutic effects of antidepressant drugs are observed is a problem because the pharmacologic action of TCAs and MAOIs occur within minutes of administration.

An additional challenge to the catecholamine and indoleamine theories of depression was the emergence during the 1970's of drugs such as trazodone, mianserin and iprindole whose pharmacological actions did not include significant inhibition of monoamine uptake or MAO. The clinical activity of these drugs suggested that either the classical catecholamine and indoleamine theories of depression needed modification or that other mechanisms of antidepressant action may exist.

Several lines of work during the past two decades have continued to provide support for the relevance of 5-HT and NE's role in the therapeutic effects of antidepressant drugs. First, MAO-A inhibitors (e.g., clorgyline) that selectively block the degradation of 5-HT and NE are clinically active as antidepressants (Lipper, Murphy, Slater, & Buchsbaum, 1979) while MAO-B inhibitors that selectively block phenylethylamine deamination do not have a therapeutic effect at doses selective for inhibition of MAO-B (Mendis, Pare, Sandler, Glover, & Stern, 1981). Second, drugs which have been designed to inhibit NE uptake (e.g., maprotiline) or 5-HT uptake (e.g., fluoxetine, sertraline, and zimelidine, etc.), but which do not have many of the other concomitant side effects of TCAs that limit patient compliance, have demonstrated antidepressant action in the clinics (Coccaro & Siever, 1985; Åsberg, Eriksson, Mårtensson, Traskman-Bendz, & Wagner, 1986). Third, parachlorophenylalanine (PCPA), a tryptophan hydroxylase inhibitor, reverses the therapeutic effect of imipramine and tranylcypromine in patients (Shopsin, Gershon, Goldstein, Friedman, & Wilks, 1975; Shopsin, Friedman, & Gershon, 1976). Fourth, a tryptophan-free diet has recently been shown to result in remission of depressed patients while receiving concurrent antidepressant drugs (Delgado, Charney, Price, Goodman, Aghajanian, & Heninger, 1989).

Through the mid-1970's, preclinical research on antidepressant drugs focused on presynaptic events. Glowinski and Axelrod (1964) first found that the TCAs inhibited uptake of [3H]norepinephrine into brain nerve endings and suggested that this might be the mechanism of their antidepressant drug action. The TCAs were later shown to inhibit 5-HT uptake as well as NE uptake (Carlsson, Corrodi, Fuxe, & Hökfelt, 1969). Monoamine oxidase inhibitors were the first type of drug found to be useful for the treatment of depression during the 1950's when clinicians noticed that an experimental anti-tuberculosis drug, iproniazid, improved the mood of depressed patients. Shortly thereafter, the ability of iproniazid to inhibit MAO was recognized and other drugs designed to inhibit MAO were also found to be useful in the treatment of depression (Baldessarini, 1980).

By the early 1970's, a second generation of antidepressants were in development, exemplified by drugs such as mianserin and trazodone. Mianserin does not inhibit MAO or block in vivo measures of monoamine uptake (Pinder & van Delft,1983). One of the leading candidates to explain the drug's therapeutic action is its antagonist action at the presynaptic alpha-2 adrenoceptor, thereby increasing NE release (Pinder et al. 1983). Mianserin's blockade of central 5-HT receptors has also been postulated to account for the drugs therapeutic effects (Maj, Sowinska, Baran, Gancarczyk, & Rawlow, 1978; Ogren, Fuxe, Agnati, Gustafsson, Jonsson, & Holm,1979), although at first glance this would seem contradictory to the classical indoleamine theory of depression. One of the candidates to explain the action of trazodone that has been most discussed over recent years is specific blockade of 5-HT uptake. Trazodone does inhibit 5-HT uptake in vitro (Stefani, Fadda, Medda, & Gessa, 1976) but it is no more potent than desipramine (Riblet, Gatewood, & Mayol, 1979). Trazodone is known to have approximately a 100-fold selectivity for the 5-HT-2 relative to the 5-HT-1 binding site (Luttinger, Freedman, Hamel, Ward, & Perrone, 1985), similar to mianserin (Peroutka & Snyder, 1980b; Leysen, Awouterts, Kennis, Laduron, Vandenberk, & Janssen, 1981). Trazodone exerts effects on operant behavior that is consistent with antagonism of 5-HT-2 receptors (Hingtgen, Hendrie, & Aprison, 1984). Other in vivo work suggests that blockade of 5-HT uptake by trazodone is relatively unimportant in explaining the pharmacological effects of this drug, as compared to blockade of 5-HT-2 receptors by trazodone (Fuller, Snoddy, & Cohen, 1984). Relating the pharmacological actions of these atypical antidepressants in a logical fashion to the NE and 5-HT systems constituted one challenge for the classical catecholamine and indoleamine theories of depression. The second challenge for the classical theories involved explaining how the acute actions of TCAs and MAOIs were related to the therapeutic effects which are not demonstrated until after several weeks of daily drug administration.

An answer to the second challenge began to emerge in the mid-1970's with the development of ligand binding techniques. First, Vetulani and Sulser (1975) demonstrated a desensitization of the beta-adrenergic cyclic AMP-generating system in the limbic forebrain of rats that are treated daily for 4-8 weeks with the TCA desipramine and the atypical antidepressant iprindole as well as 8 daily electroconvulsive shock (ECS) sessions. Then, Banerjee, Kung, Riggi, & Chanda, (1977) showed that 6 weeks of daily desipramine, iprindole, and doxepin administration in rats decreased the number, but not the affinity of beta-adrenergic receptors. Since that time, decreased number of beta-adrenergic receptors or decreased sensitivity to NE has been observed following daily administration of virtually all antidepressant drugs and ECS (Sulser, 1983). The current view is that prolonged stimulation of the beta-adrenergic receptor results in its eventual down-regulation and/or desensitization. Co-administration of propranolol (Wolfe, Harden, Sporn, & Molinoff, 1978; Scott & Crews, 1986) or lesions of NE nerve terminals using 6-hydroxydopamine or N-(2-chloroethyl)-N-ethyl-2-bromobenzylamine (DSP4) (Schweitzer, Schwartz, Friedhoff, 1979; Hall, Ross, & Sallemark, 1984) block the down-regulation of beta-adrenergic receptors following daily tricyclic antidepressant drug administration. Thus, prolonged agonist action at the beta-adrenergic receptor that is indirectly induced by antidepressant drug administration appears to down-regulate the receptor.

The second most consistent change seen in receptor binding experiments following chronic antidepressant drug administration is a decreased number of 5-HT-2 receptors (Peroutka & Snyder,1980a; 1980b). A decreased number of 5-HT-2 binding sites has also been observed following administration of 5-HT-2 antagonists that are also antidepressants (e.g., mianserin, trazodone and ritanserin; Blackshear, & Sanders-Bush, 1982; Riblet & Taylor, 1981; Leysen, Van Gompel, Gommeren, & Woestenborghs, 1986). This effect is paradoxical in that chronic administration of antagonists usually results in an increased number or sensitization of receptors. This down-regulation of 5-HT-2 receptors, however, does not appear to be related to continued binding of the antagonist to the receptor (Sanders-Bush,.Breeding, & Roznoski, 1987), but appears to be related to antagonist action at the 5-HT-2 receptor (Leysen et al., 1986). Not all antidepressant drugs have been demonstrated to decrease the number of 5-HT-2 receptors, but it has recently been demonstrated that antidepressant drugs may cause a desensitization of the 5-HT-2 receptor mediated phosphoinositide turnover without altering 5-HT-2 receptor binding (Sanders-Bush et al.,1989). In addition, some 5-HT antagonists (e.g., methysergide and metergoline) that are probably not antidepressant drugs are able to decrease 5-HT-2 receptor number (Blackshear et al., 1982). This finding may not necessarily rule out the relevance of 5-HT-2 receptor regulation for antidepressant effects when given the effects of these 5-HT antagonists at other 5-HT receptors (see below). The 5-HT-2 receptor down-regulation has been observed following a single dose of some, but not all, 5-HT antagonists (Blackshear et al., 1982). However, daily administration of the antidepressant and 5-HT-2 antagonist mianserin results in a greater effect than a single dose. ECS has also been observed to increase the number of 5-HT-2 binding sites, but it is unknown how this change in receptor binding relates to the signal transduction through the 5-HT-2 receptor. In addition, ECS results in many other effects, such as down-regulation of beta-adrenergic receptors, which may be related to the mechanism of antidepressant drug effect.

A second line of evidence implicating 5-HT in the effects of chronic antidepressant treatment began with the observation that TCAs, 5-HT uptake inhibitors, MAOIs, and the putative antidepressant and 5-HT-1A agonist gepirone (deMontigny & Aghajanian,1978; Chaput , deMontigny, & Blier, 1986; Blier, deMontigny, & Azzaro, 1986; Blier & deMontigny,1987) all enhance the inhibitory effects of 5-HT in the hippocampus when administered daily, although the mechanism by which each drug achieves this effect is different.

Not only may 5-HT and NE be involved in the acute and chronic effects of antidepressant drugs, but there appear to be important interactions between the two transmitter systems. A reciprocal innervation is present between 5-HT and NE containing neurons (Pickel, Joh, & Reis, 1977; Baraban & Aghajanian, 1981). Alpha-1 adrenergic antagonists are known to suppress dorsal raphe cell firing (Baraban & Aghajanian,1980). Studies suggesting that 5,7-DHT lesions increase the number of beta-adrenergic receptors (Stockmeier, Martino, & Kellar, 1985) and block the antidepressant-induced down-regulation of beta-adrenergic receptors (Brunello, Barbaccia, Chuang, & Costa, 1982; Janowsky, Okada, Manier, Applegate, Sulser, & Steranka, 1982) have been subsequently found to be complicated by the fact that a portion of specific 3H-dihydroalprenolol (3H-DHA) binding is to a 5-HT-1 receptor. Nevertheless, functional interrelationships are present between serotonergic and noradrenergic neurons. The beta adrenergic agonist isoproterenol increases the number of 5-HT-2 receptors (Scott & Crews, 1985) in vitro. Beta-adrenergic agonists are known to increase central 5-HT turnover (Waldmeier, 1981). In addition, noradrenergic denervation prevents the TCA-induced sensitization of hippocampal neurons to 5-HT (Gravel & deMontigny, 1987). Noradrenergic denervation has also been shown to potentiate 5-HT-2 mediated head twitches; this potentiation was related to up-regulation of beta-adrenergic receptors without a change in the Bmax or Kd of the 5-HT-2 receptor (Eison, Yocca, Gianutsos, 1988).

Use of behavioral antidepressant drug screens: Mechanism of drug action

One strategy in studying the mechanism of antidepressant action is to use behavioral antidepressant drug screens. The use of behavioral screens employing rodents allows a much greater degree of experimental invasiveness. Many antidepressant screens involve acute effects, whereas it has previously been pointed out that clinical effects are not immediately apparent. Thus, if one employs behavioral antidepressant drug screens to investigate the mechanism of antidepressant drug action, one assumes that the acute effects might be related to the delayed clinical effects.

The differential-reinforcement-of-low rate 72-sec (DRL 72-s) schedule of water reinforcement appears to be an animal screen that may be useful in studying the mechanism of antidepressant drug action because antidepressant drugs affect DRL 72-s scheduled behavior in a unique manner unlike other classes of drugs. TCAs, MAOIs, atypical antidepressants (e.g., trazodone, mianserin and iprindole), and electroconvulsive shock increase the reinforcement rate and decrease the response rate of rats performing on DRL schedules (McGuire & Seiden, 1980; O'Donnell & Seiden, 1982; 1983; Seiden, Dahms, & Shaughnessy, 1985). In these same studies, antipsychotics, antihistamines, anticholinergics, psychomotor stimulants, benzodiazepines, barbiturates, ethanol and narcotic analgesics did not test as antidepressant drugs. Hence, the DRL 72-s schedule appears to be a more specific test for the detection of antidepressant drugs than other screens, such as antagonism of the behavioral effects of reserpine or tetrabenazine (Sulser,.Bickel, & Brodie, 1964; Howard, Coroko, & Cooper, 1981), antagonism of muricidal behavior in rats, and the forced swim test (Porsolt, Le Pichon, & Jalfre, 1977). These other screens are plagued by the problem of anticholinergics, antihistamines and psychomotor stimulants testing as antidepressant drugs (Howard et al.,1981; Ueki, 1982; Einsenstein, Iorid, & Clody, 1982; Porsolt, 1981; Luttinger et al.,1985). In addition, many of the behavioral antidepressant drug screens are unable to detect atypical antidepressants such as trazodone, mianserin or iprindole as antidepressants (Ayd & Settle,1982; Brogden, Heel, Speight, & Avery, 1978; Luttinger et al.,1985). The only area of controversy concerning the specificity of the DRL schedule as an antidepressant drug screen involves the effect of antipsychotics. When investigators have used food as a reinforcer, antidepressant-like effects of chlorpromazine and haloperidol have been observed (Howard and Pollard, 1984; Britton & Koob, 1989). This observation is complicated by the known weak efficacy of antipsychotics in treating depression (Paykel, Price, Gillan, Palmai, & Chesser, 1968). In addition, the antidepressants that possess psychomotor stimulant properties such as bupropion and nomifensine (Seiden et al.,1985; O'Donnell et al., 1983) have a profile of activity like psychomotor stimulants, not antidepressant drugs.

NE clearly plays a role in the effects of some known and putative antidepressant drugs on the DRL 72-s schedule. Oxaprotiline, a derivative of the tetracyclic antidepressant maprotiline derivative, tests as an antidepressant on the DRL 72-s schedule. The (+)-enantiomer is responsible for the antidepressant-like effects of oxaprotiline on the DRL schedule (Marek, Li, & Seiden, 1988). Both the (+)- and (-)-enantiomers have a similar potency at several neurotransmitter receptors but differ dramatically in their potency at blocking NE uptake with the (+)-enantiomer being several orders of magnitude more potent (Delini-Stula, Hauser, Baumann, Olpe, Waldmeier, & Storni, 1982; Mishra, Gillepsie, Lovell, Robson, & Sulser, 1982; Waldmeier, Bauman, Hauser, Maitre, & Storni, 1982). Both the reinforcing-rate increasing and the response-rate decreasing effects of DMI are blocked by a relatively low dose of the beta-adrenergic antagonists propranolol (1 mg/kg; O'Donnell, 1987). O'Donnell (1987) has also shown that the beta-2 adrenergic agonist and putative antidepressant clenbuterol tests as an antidepressant drug on the DRL 72-s schedule with an ED50 of 0.03 mg/kg. Propranolol is able to completely block the effects of clenbuterol and the DRL schedule at 1 mg/kg and has an ED50 of 0.01 mg/kg. Further, down-regulation of beta-2 adrenergic receptors by daily treatment with clenbuterol produces a tolerance to the antidepressant-like effects of acute clenbuterol administration (personal communication, J. O'Donnell). These results are consistent with the hypothesis that stimulation of beta-adrenergic receptors is involved the therapeutic effects of antidepressant drugs (Sulser, 1983).

5-HT also clearly plays a role in the effects of some known antidepressant drugs on the DRL 72-s schedule. The antidepressant and MAO-A inhibitor clorgyline, which blocks degradation of both 5-HT and NE, tests as an antidepressant on the DRL 72-s schedule (Marek & Seiden,1988a). The MAO-B inhibitor deprenyl, which does not block degradation of either 5-HT or NE and does not appear to be an antidepressant at doses specific for MAO-B inhibition, does not test as an antidepressant on the DRL 72-s schedule at doses specific for inhibition of MAO-B (Marek et al., 1988a). The reinforcement-rate increasing effects of clorgyline are blocked by both methysergide, a 5-HT-1 and 5-HT-2 antagonist, as well as by specific 5-HT depletions following ivt infusion of 5,7-dihydroxytryptamine (Marek, Li, & Seiden, 1989a). Furthermore, the reinforcement-rate increasing effects of the TCAs imipramine, desipramine and the 5-HT uptake inhibitor fluoxetine are partially blocked by methysergide (Marek & Seiden,1988b; Marek et al.,1989a). These effects are consistent with the hypothesis that enhancing serotonergic neurotransmission may be important for the clinical effects of some antidepressants (Carlsson, 1984).

The effects of some of the atypical antidepressants with respect to the serotonergic system have been difficult to relate to possible mechanisms of antidepressant action because one of the most potent effects of trazodone and mianserin is blockade of 5-HT receptors. Recent work on the DRL schedule suggests that the action of trazodone and mianserin could be related to their moderately selective blockade of 5-HT-2 relative to 5-HT-1 receptors. Several years ago it was noted that drugs with an approximately 1000-fold selectivity for the 5-HT-2 site relative to the 5-HT-1 site, like ketanserin and pipamperone (Leysen et al.,1981), exert robust antidepressant-like effects on the DRL 72-s schedule while drugs with approximately an equal affinity at the 5-HT-2 and 5-HT-1 sites, like methysergide and metergoline, do not increase the reinforcement rate (Marek & Seiden, 1986; Marek et al., 1988b). It was subsequently shown that the selectivity of a more extensive series of 5-HT antagonists (including trazodone and mianserin) for the 5-HT-2 relative to the 5-HT-1 receptor is positively correlated with the maximal increase in the reinforcement rate on the DRL 72-s schedule (Marek et al., 1988b; Marek, Li, & Seiden, 1989b). In addition, the selective 5-HT-2 antagonist and putative antidepressant ritanserin (Hoppenbrowers, Gelders, & Van den Bussche, 1986) tests as an antidepressant on the DRL 72-s schedule (Marek et al.,1989b).

Other work suggests that trazodone and mianserin have antidepressant activity on the DRL 72-s schedule by blocking 5-HT-2 receptors. First, the often mentioned 5-HT uptake inhibiting property of trazodone is unlikely to account for the effects of trazodone on the DRL 72-s schedule as this property is quite weak in comparison to the ability of trazodone to block 5-HT-2 receptors (Fuller et al.,1984). In addition, the 5-HT antagonist methysergide blocks the reinforcement-rate increasing effects of the 5-HT uptake inhibitors fluoxetine and imipramine but not trazodone. Second, the formation of the 5-HT-1B/1C agonist and trazodone metabolite m-CPP is unlikely to mediate the effects of trazodone on the DRL 72-s schedule because m-CPP does not mimic the reinforcement-rate increasing effect of trazodone In contrast, inhibition of trazodone metabolism with SKF 525 A markedly enhances both the reinforcement-rate increasing and response-rate decreasing effects of trazodone (Li, Marek, Hand, & Seiden, 1990). Third, the antidepressant-like effects of trazodone are not mediated by the either alpha-1 adrenergic or alpha-2 adrenergic antagonist activity since none of these drugs: prazosin, phentolamine, phenoxybenzamine, yohimine or idazoxan increase the reinforcement rate as do antidepressant drugs (Marek et al.,1989b; Marek & Seiden,1988b; Sanger,1988). Fourth, (+)-mianserin is more potent on DRL behavior than (-)-mianserin, while the metabolites desmethylmianserin and 8-hydroxymianserin show no dose-related antidepressant-like effects. This profile of activity for the mianserin enantiomers and metabolites is most consistent with the hypothesis that the antidepressant activity of mianserin is related to antagonist activity at the 5-HT-2 rather than at the histamine-1 or the alpha-2 adrenergic receptors (manuscript in preparation).

The work cited above suggests that antagonist action at 5-HT-2 receptors may be important for the effects of atypical antidepressants on the DRL 72-s schedule. In addition, agonist action at a 5-HT-1 receptor appears to be important for the effects of TCAs and MAOIs on DRL behavior since both methysergide and specific 5-HT depletions block the effects of TCAs and MAOIs (Marek et al., 1988b; Marek et al.,1989a). If such reciprocal relationships between 5-HT-2 and 5-HT-1 receptors is important for therapeutic effects of some antidepressant drugs, it is easily seen why drugs that selectively block the 5-HT-2 receptor are antidepressants while drugs that block both 5-HT-1 and 5-HT-2 receptors (e.g., methysergide and metergoline) are not antidepressants. This reciprocal relationship between 5-HT receptors has been further defined by the observation that 5-HT-1A agonists, including the putative antidepressants buspirone and gepirone, increase the reinforcement rate as do known antidepressant drugs (Marek et al.,1989a; Hand, Marek, Jolly, & Seiden, 1989).

There is additional evidence of opposing effects between 5-HT-1 and 5-HT-2 receptors. The selective 5-HT-2 antagonist ketanserin enhances the inhibitory effect of iontophoretically applied 5-HT in the prefrontal cortex and ventral lateral geniculate nucleus of the rat (Lakoski & Aghajanian, 1985). Stimulation of 5-HT-2 receptors on guinea pig somatosensory cortical pyramidal neurons results in depolarization whereas stimulation of 5-HT-1A receptors on pyramidal cells in the same area resulted in a hyperpolarizing response (Davies, Deicz, Prince, & Peroutka, 1987). A similar pattern of opposing roles for 5-HT-1A and 5-HT-2 receptors exists for thermoregulation (Gudelsky, Koening, & Meltzer, 1986) and has been proposed for the biphasic effect of 5-HTP in the Vogel conflict model (Hjorth, Söderpalm, & Engel, 1987). Antidepressant drugs are felt to achieve their clinical effects in part either as indirect 5-HT agonists (Carlsson,1984; Asberg et al.,1986), or as 5-HT receptor antagonists (Fuxe et al.,1979; Maj, Palider, & Rawlow, 1980). Both hypotheses may be true, and heterogeneity of 5-HT receptor type and function may resolve the apparent clinical paradox just as heterogeneity of 5-HT receptor type and function appears to resolve the paradox of similar antidepressant-like effects of indirect 5-HT-1A agonists and 5-HT-2 antagonists on DRL 72-s behavior.

Both NE and 5-HT appear to be important in mediating the effects of antidepressant drugs on the DRL 72-s schedule, although it is still unclear as to the exact relationship between these two neurotransmitter systems with respect to DRL behavior. For some drugs like the TCAs, both systems may be important as both methysergide and low doses of propranolol block the reinforcement rate increasing effect of desipramine (Marek et al., 1988b; O'Donnell, 1987). For some drugs like the MAOIs the serotonergic system may be more important because methysergide and 5,7-DHT lesions block the reinforcement rate increasing effect of clorgyline while propranolol is without effect (Marek et al.,1989a). For other drugs like the putative antidepressant clenbuterol, propranolol is able to block the antidepressant-like effects while methysergide is ineffective (O'Donnell,1987; Marek et al., 1988b).

In summary, 5-HT and NE are neurotransmitters that have been suspected to be involved in the etiology, pathogenesis and treatment of depression for over 25 years. Most effective antidepressant treatments are thought to be mediated via either or both of these neurotransmitter systems. Understanding how these systems are related remains an important strategy to unraveling both the pathogenesis and treatment of depression.

ACKNOWLEDGEMENTS

We wish to thank Linh-Chi Nguyen and Barbara X. Knight for their help in preparing this manuscript.

REFERENCES

Arora, R.C., & Meltzer, H.Y. (1989). Serotonergic measures in the brains of suicide victims: 5-HT-2 binding sites in the frontal cortex of suicide victims and control subjects. *Am. J. Psychiatry, 146*, 730-736.

Asberg, M., Thoren, P., Traskman, L., Bertilsson, L., & Ringberger, V. (1976). 'Serotonin depression' - A biochemical subgroup within affective disorders. *Science, 191*, 478-480.

Asberg, M., Bertilsson, L., Martensson, B., Scalia-Tomba, G.-P., Thoren, P., & Traskman-Bendz, L. (1984). CSF monoamine metabolites in melancholia. *Acta Psychiat. Scand., 69*, 201-219.

Asberg, M., Eriksson, B., Martensson, B., Traskman-Bendz, L., & Wagner, A. (1986). Therapeutic effects of serotonin uptake inhibitors in depression. *J. Clin. Psychiat., 46 (4, Suppl)*, 23-35.

Ayd, F.J., & Settle, E.C. (1982). Trazodone: A novel, broad-spectrum antidepressant. In *Mod. Prob. Pharmacopsychiat., 18*, 49-69. Karger, Basel.

Baldessarini, R.J. (1980). Drugs and the treatment of psychiatric disorders. In: *Goodman's and Gilman's The Pharmacological Basis of Therapeutics* (Ed. A.F. Goodman, L.S. Goodman, & A. Gilman), 6th edition, pp. 391-447, Macmillan, New York.

Banerjee, S.P., Kung, L.S., Riggi, S.J., & Chanda, S.K. (1977). Development of ß-adrenergic receptor subsensitivity by antidepressants. *Nature, 268*, 455-456.

Baraban, J.M., & Aghajanian, G.K. (1980). Suppression of firing activity of 5-HT neurons in the dorsal raphe by alpha-adrenergic antagonists. *Neuropharmacology, 19*, 355-363.

Baraban, J.M., & Aghajanian, G.K. (1981). Noradrenergic innervation of serotonergic neurons in the dorsal raphe: Demonstration by electron microscopic autoradiography. *Brain Res., 204*, 1-11.

Blackshear, M.A., Friedman, R.L., and Sanders-Bush, E. (1983). Acute and chronic effects of serotonin (5-HT) antagonists on serotonin binding sites. *Naunyn-Schmiedeberg's Arch. Pharmacol., 324*, 125-129.

Blackshear, M.A., & Sanders-Bush, E. (1982). Serotonin receptor sensitivity after acute and chronic treatment with mianserin. *J. Pharmacol. Exp. Ther., 221*, 303.

Blier, P., deMontigny, C., & Azzaro, A.J. (1986). Effect of repeated amiflamine administration on serotonergic and noradrenergic neurotransmission: Electrophysiological studies in the rat CNS. *Naunyn-Schmiedeberg's Arch. Pharmacol., 334*, 253-260.

Blier, P., & deMontigny, C. (1987). Modification of 5-HT neuron properties by sustained administration of the 5-HT-1A agonist gepirone: Electrophysiological studies in the rat brain. *Synapse, 1*, 470-480.

Britton, K.T., & Koob, G.F. (1989). Effects of corticotropin releasing factor, desipramine and haloperidol on a DRL schedule of reinforcement. *Pharmacol. Biochem. & Behav., 32*, 967-970.

Brogden, R.N., Heel, R.C., Speight, T.M., & Avery, G.S. (1978). Mianserin: A review of its pharmacological properties and therapeutic efficacy in depressive illness. *Drugs, 16*, 273-301.

Brunello, N., Barbaccia, M.L., Chuang, D.-M., & Costa, E. (1982). Down-regulation of ß-adrenergic receptors following repeated injections of desmethylimipramine: Permissive role of serotonergic axons. *Neuropharmacology, 21*, 1145-1149.

Bunney, W.E., & Davis, J.M. (1965). Norepinephrine in depressive reactions. *Arch. Gen. Psychiat., 13*, 483-494.

Carlsson, A. (1984). Current theories on the mode of action of antidepressant drugs. In: *Frontiers in Biochemical and Pharmacological Research in Depression.* (Eds. E. Usdin, M. Asberg, L. Bertilsson, & F. Suoqvist), *Vol.39, Advances in Biochemical Psychopharmacology,* pp. 213-221, Raven Press, New York.

Carlsson, A., Corrodi, H., Fuxe, K., & Hokfelt, T. (1969). Effect of antidepressant drugs on the depletion of intraneuronal brain 5-hydroxytryptamine stores caused by 4-methyl-alpha-ethyl-meta- tyramine. *Eur. J. Pharmacol., 5,* 357-366.

Chaput, Y., deMontigny, C., & Blier, P. (1986). Effects of a selective 5-HT reuptake blocker, citalopram, on the sensitivity of 5-HT autoreceptors: Electrophysiological studies in the rat brain. *Naunyn-Schmiedeberg's Arch. Pharmacol., 333,* 342-348.

Coccaro, E.F., & Siever, L.J. (1985). Second generation antidepressants: A comparative review. *J. Clin. Pharmacol., 25,* 241-260.

Davies, M.F., Deicz, R.A., Prince, D.A., & Peroutka, S.J. (1987). Physiological and pharmacological characterization of the responses of guinea pig neocortical neurons to serotonin. *Soc. Neurosci. Abst., 3,* 801.

Delgado, P.L., Charney, D.S., Price, L.H., Goodman, W.K., Aghajanian, G.K., & Heninger, G.R. (1989). Behavioral effects of acute tryptophan depletion in psychiatric patients and healthy subjects. *Soc. for Neurosci. Abst.., 15,* 412.

Delini-Stula, A., Hauser, K., Baumann, P., Olpe, H.-R., Waldmeier, P., & Storni, A. (1982). Stereospecificity of behavioral and biochemical responses to oxaprotiline - a new antidepressant. In *Typical and Atypical Antidepressants: Molecular Mechanisms,* E. Costa and G. Racagni (Eds.), Raven Press, New York, pp. 265-275.

deMontigny, C. & Aghajanian, G.K. (1978). Tricyclic antidepressants: Long-term treatment increases responsivity of rat forebrain neurons to serotonin. *Science, 202,* 1303-1306.

Diagnostic and Statistical Manual of Mental Disorders (Third Edition-Revised), DSM-III-R, American Psychiatric Association, Washington, DC, 1987.

Egeland, J.A., Gerhard, D.S., Pauls, D.L., Sussex, J.N., Kidd, K.K., Allen, C.R., Hostetter, A.M., & Housman, D.E. (1987). Bipolar affective disorders linked to DNA markers on chromosome 11. *Nature, 325,* 783-787.

Einsenstein, N., Iorid, L.C., & Clody, D.E. (1982). Role of serotonin in the blockade of muricidal behavior by tricyclic antidepressants. *Pharmacol. Biochem. & Behav., 17,* 847-849.

Eison, A.S., Yocca, F.D., & Gianutsos, G. (1988). Noradrenergic denervation alters serotonin-2-mediated behavior but not serotonin-2 number in rats: Modulatory role of beta adrenergic receptors. *J. Pharmacol. Exp. Ther., 246,* 571-577.

Fuller, R.W., Snoddy,H.D., & Cohen, M.L. (1984). Interactions of trazodone with serotonin neurons and receptors. *Neuropharmacology, 23,* 539-544.

Glowinski, J., & Axelrod, J. (1964). Inhibition of uptake of tritiated- noradrenaline in the intact rat by imipramine and structurally related compounds. *Nature, 204,* 1318-1319.

Goodwin, F.K., Ebert, M.H., & Bunney, W.E. (1972). Mental effects of reserpine in man: A review. In R.I. Shader (Ed.), *Psychiatric Complications of Medical Drugs.* Raven Press, New York, pp. 73-101.

Gravel,P., & deMontigny, C. (1987). Noradrenergic denervation prevents sensitization of rat forebrain neurons to serotonin by tricyclic antidepressant treatment. *Synapse, 1,* 233-239.

Gudelsky, G.A., Koenig, J.I., & Meltzer, H.Y. (1986). Thermoregulatory responses to serotonin (5-HT) receptor stimulation in the rat: Evidence for opposing roles of 5-HT-2 and 5-HT-1A receptors. *Neuropharmacology, 25,* 1307-1313.

Hall, H., Ross, S.B., & Sallemark, M. (1984). Effect of destruction of central noradrenergic and serotonergic nerve terminals by systemic neurotoxins on the long-term effects of antidepressants on ß-adrenoceptors and 5-HT-2 binding sites in the rat cerebral cortex. *J. Neural Transm., 59,* 9-23.

Hand, T.H., Marek, G.J., Jolly, D., & Seiden, L.S. (1989). Antidepressant-like effects of the 5-HT-1A agonists buspirone,gepirone, 8-OH-DPAT, and 5-MeODMT in rats on the DRL 72 sec schedule: Differential blockade by methysergide and purported 5-HT-1A antagonists. *Soc. Neurosci. Abst.., 15,* 1282.

Hingtgen, J.N., Hendrie, H.C., & Aprison, M.H. (1984). Post-synaptic serotonergic blockade following chronic antidepressive treatment with trazodone in an animal model of depression. *Pharmacol. Biochem. Behav., 20,* 425-428.

Hoppenbrowers, M.L., Gelders, Y., & Vanden Bussche, G. (1986). Ritanserin (R55667) an original thymosthenic. *Boll. Chim. Farm., 125,* 136S-147S.

Hjorth, S., Soderpalm, B., & Engel, J.A. (1987). Biphasic effect of L-5-HTP in the Vogel conflict model. *Psychopharmacology, 92,* 96-99.

Howard, J.L., Coroko, F.E., & Cooper, B.R. (1981) Empirical behavioral models of depression, with emphasis on tetrabenazine antagonism. In *Antidepressants: Neurochemical, Behavioral, and Clinical Perspectives,* (S.J. Enna, J.B. Malick and E. Richelson, eds.), pp.107-120, Raven Press, New York.

Howard, J.L., & Pollard, G.T. (1984). Effects of imipramine, bupropion, chlorpromazine, and clozapine on differential-reinforcement of low rate (DRL) > 72-s and > 36-s schedules in rat. *Drug Dev. Res., 4,* 607-616.

Janowsky, A., Okada, F., Manier, D.H., Applegate, C.D., Sulser, F., & Steranka, L.R. (1982). Role of serotonergic input in the regulation of the ß-adrenergic receptor-coupled adenylate cylcase system. *Science, 218,* 900-901.

Lakoski, J.M. & Aghajanian, G.K. (1985). Effects of ketanserin on neuronal responses to serotonin in the prefrontal cortex, lateral geniculate and dorsal raphe nucleus. *Neuropharmacology, 24,* 265-273.

Lapin, I.P. and Oxenkrug, G.F. (1969). Intensification of the central serotonergic processes as a possible determinant of the thymoleptic effect. *The Lancet,* 132-136.

Leysen, J.E., Awouters, F., Kennis, L., Laduron, P.M., Vandenberk, J., & Janssen, P.A.J. (1981). Receptor binding profile of R 41 468, a novel antagonist at 5-HT2 receptors. *Life Sci., 28,* 1015-1022.

Leysen, J.E., Van Gompel, P.V., Gommeren, W., Woestenborghs, R., & Janssen, P.A.J. (1986). Down regulation of serotonin-S-2 receptor sites in rat brain by chronic treatment with the serotonin-S-2 antagonists: ritanserin and setoperone. *Psychopharmacology, 88,* 434-444.

Li, A.A., Marek, G.J., Hand, T., & Seiden, L.S. (1990). Antidepressant-like effects of trazodone on a behavioral screen is mediated by trazodone, not the metabolite m-chlorophenylpiperazine. *Eur. J. Pharmacol.,* in press.

Lipper, S., Murphy, D.L., Slater, S., & Buchsbaum, M.S. (1979) Comparative behavioral effects of clorgyline and pargyline in man: a preliminary evaluation. Psychopharmacology, 62, 123-128.

Luttinger, D., Freedman, M., Hamel, L., Ward, S.J., & Perrone, M. (1985). The effects of serotonin antagonists in a behavioral despair procedure in mice. *Eur. J. Pharmacol., 107,* 53-58.

Maj, J., Palider, W., & Rawlow, A. (1980). Trazodone, a central serotonin antagonist and agonist. *J. Neural Transm., 44,* 237-248.

Maj, J., Sowinska, H., Baran, L., Gancarczyk, L., & Rawlow, A. (1978). The central antiserotonergic action of mianserin. *Psychopharmacology, 59,* 79-84.

Mann, J.J., Stanley, M., McBride, A., & McEwen, B.S. (1986). Increased serotonin-2 and ß-adrenergic receptor binding in the frontal cortices of suicide victims. *Arch. Gen. Psychiatry, 43,* 954-959.

Marek, G.J., & Seiden, L.S. (1986). The effect of ketanserin and pipamperone on a differential-reinforcement-of-low-rate 72-sec schedule. *Federation Proceedings,45,* 680.

Marek, G.J., & Seiden, L.S. (1988a). Selective inhibition of MAO-A but not MAO-B results in antidepressant-like effects on DRL 72-s behavior. *Psychopharmacology, 96,* 153-160.

Marek, G.J., & Seiden, L.S. (1988b). Effects of selective 5-HT and non-selective 5-HT antagonists on the differential-reinforcement- of-low-rate 72-sec schedule. *J. Pharmacol. Exp. Ther., 244,* 650-658.

Marek, G.J., Li, A., & Seiden, L.S. (1988). Antidepressant-like effects of (+)-oxaprotiline on a behavioral screen. *Eur. J. Pharmacol., 157,* 183-188.

Marek,G.J., Li, A., & Seiden, L.S. (1989a). Evidence for involvement of 5-hydroxytryptamine-1 receptors in antidepressant-like drug effects on differential-reinforcement-of-low-rate 72-s behavior. *J. Pharmacol. Exp. Ther., 250*, 60-71.

Marek, G.J., Li, A., & Seiden, L.S. (1989b). Selective 5-hydroxytryptamine-2 antagonists have antidepressant-like effects on differential-reinforcement- of-low-rate 72-s schedule. *J. Pharmacol. Exp. Ther., 250*, 52-59.

McGuire, P.S., & Seiden, L.S. (1980). The effects of tricyclic antidepressants on performance under a differential-reinforcement-of-low-rates schedule in rats. *J. Pharmacol. Exp. Ther., 214*, 635-641.

Mendis, N., Pare, C.M.B., Sandler, M., Glover, V., & Stern, G.M. (1981). Is the failure of (-) deprenyl, a selective monoamine oxidase B inhibitor, to alleviate depression related to freedom from the cheese effect. *Psychopharmacology, 73*, 87-90.

Mishra, R., Gillepsie, D.D., Lovell, R.A., Robson, R.D., & Sulser, F. (1982). Oxaprotiline: Induction of central noradrenergic subsensitivity by its (+)-enantiomer. *Life Sci., 30*, 1747-1755.

Nemeroff, C.B. (1989). Clinical significance of psychoneuroendocrinology in psychiatry: Focus on the thyroid and adrenal. *J. Clin. Psychiatry , 50:5 (Suppl)*, 13-22.

O'Donnell, J.M., & Seiden, L.S. (1982). Effects of monoamine oxidase inhibitors on performance during differential reinforcement of low response rate. *Psychopharmacology, 78*, 214-218.

O'Donnell, J.M., & Seiden, L.S. (1983). Differential-reinforcement-of-low-rate 72-second schedule: Selective effects of antidepressant drugs. *J. Pharmacol. Exp. Ther., 224*, 80-88.

O'Donnell, J.M. (1987). Effects of clenbuterol and prenalterol on performance during differential reinforcement of low response rate in the rat. *J. Pharmacol. Exp. Ther., 241*, 68-75.

Ogren, S.O., Fuxe, K., Agnati, L.F., Gustafsson, J.A., Jonsson, G., & Holm, A.C. (1979). Reevaluation of the indoleamine hypothesis of depression. Evidence for a reduction of functional activity of central 5-HT systems by antidepressant drugs. *J. Neural Transm., 46*, 85-103.

Oswald, I., Brezinova, V., & Dunleavy, D.L.F. (1972). On the slowness of action of tricyclic antidepressant drugs. *Br. J. Psychiat., 120*, 673-677.

Paykel, E.S., Price, J.S., Gillan, R.U., Palmai, G., & Chesser, E.S. (1968). A comparative trial of imipramine and chlorpromazine in depressed patients. *Br. J. Psychiat., 114*, 1281-1287.

Peroutka, S.J., & Snyder, S.H. (1980a). Long-term antidepressant treatment decreases spiroperidol-labeled serotonin receptor binding. *Science, 210*, 88-90.

Peroutka, S.J., & Snyder, S.H. (1980b). Regulation of serotonin-2 (5-HT-2) receptors labeled with [3-H]spiroperidol by chronic treatment with the antidepressant amitriptyline. *J. Pharmacol. Exp. Ther., 215*, 582-587.

Pickel,V.M., Joh, T.H., & Reis, D. (1977). A serotonergic innervation of noradrenergic neurons in locus coeruleus. Demonstration by immunocytochemical localization of the transmitter specific enzymes tyrosine and tryptophan hydroxylase. *Brain Res., 131*, 197-214.

Pinder, R.M., & van Delft, A.M.L. (1983). Pharmacological aspects of mianserin. *Acta Psychiat. Scand. (suppl.), 302*, 59-71.

Porsolt, R.D. (1981) Behavioral despair. In *Antidepressants: Neurochemical, Behavioral, and Clinical Perspectives* (S.J. Enna, J.B. Malick and E. Richelson, eds.), pp.121-139, Raven Press, New York.

Porsolt, R.D., Le Pichon, M., & Jalfre, M. (1977). Depression: a new animal model sensitive to antidepressant treatments. *Nature (Lond.), 266*, 730-732.

Post, R.M., Ballenger, J.C., & Goodwin, F.K. (1980). Cerebral fluid studies of neurotransmitter function in manic and depressive illness. In *The Neurobiology of Cerebrospinal Fluid* (Wood, ed.), pp.685-717, Plenum Press, New York.

Riblet, L.A., Gatewood, C.F., & Mayol, R.F. (1979). Comparative effects of trazodone and tricyclic antidepressants on uptake of selected neurotransmitters by isolated rat brain synaptosomes. *Psychopharmacology, 63*, 99-101.

Riblet, L.A., & Taylor, D.P. (1981). Pharmacology and neurochemistry of trazodone. J. Clin. Psychopharmacol., 1 (Suppl.), 17S-22S.

Sanders-Bush, E., Breeding, M., & Roznoski, M. (1987). 5-HT-2 binding sites after mianserin: Comparison of loss of sites and brain levels of drug. Eur. J. Pharmacol., 133, 199-204.

Sanders-Bush, E., Breeding, M., Knoth, K., & Tsutsumi, M. (1989). Sertraline-induced desensitization of the serotonin 5-HT-2 receptor transmembrane signalling system. Psychopharmacology, 99, 64-69.

Sanger, D.J. (1988). The alpha-2-adrenoceptor antagonists idazoxan and yohimbine increase rates of DRL responding in rats. Psychopharmacology, 95, 413-417.

Schildkraut, J.J. & Kety, S.S. (1967). Biogenic amines and emotion. Science, 156, 21-30.

Scott,J.A., & Crews, F.T. (1985). Increase in serotonin-2 receptor density in rat cerebral cortex slices by stimulation of beta-adrenergic receptors. Biochem. Pharmacol., 34, 1585-1588.

Scott, J.A., & Crews, F.T. (1986). Down-regulation of serotonin-2, but not of beta-adrenergic receptors during chronic treatment with amitriptyline is independent of stimulation of serotonin-2 and beta-adrenergic receptors. Neuropharmacology, 25, 1301-1306.

Schweitzer, J.W., Schwartz, R., & Friedhoff, A.J. (1979). Intact presynaptic terminals required for beta-adrenergic receptor regulation by desipramine. J. Neurochem., 33, 377-379.

Seiden, L.S., Dahms, J.L., & Shaughnessy, R.A. (1985). Behavioral screen for antidepressants: The effects of drugs and electroconvulsive shock on performance under a differential-reinforcement-of-low-rate schedule. Psychopharmacology, 86, 55-60.

Shopsin, B., Friedman, E., & Gershon, S. (1976). Parachlorophenylalanine reversal of tranylcypromine effects in depressed patients. Arch. Gen. Psychiat., 33, 811-819.

Shopsin, B., Gershon, S., Goldstein, M., Friedman, E., & Wilk, S. (1975). Use of synthesis inhibitors in defining a role for biogenic amines during imipramine treatment in depressed patients. Psychopharmacol. Commun., 1, 239-249.

Stanley, M., & Mann, J.J. (1983). Increased serotonin-2 binding sites in frontal cortex of suicide victims. Lancet, 1, 214-216.

Stefanini, E., Fadda, F., Medda, L., & Gessa, G.L. (1976). Selective inhibition of serotonin uptake by trazodone, a new antidepressant agent. Life Sci., 18, 1459-1466.

Stockmeier, C.A., Martino, A.M., & Kellar, K.J. (1985). A strong influence of serotonin axons on ß-adrenergic receptors in rat brain. Science (Wash,DC), 230, 323-325.

Sulser, F. (1983) Mode of action of antidepressant drugs. J. Clin. Psychiatry, 44, 14-20.

Sulser, F., Bickel, M.H., & Brodie, B.B. (1964). The action of desmethylimipramine in counteracting sedation and cholinergic effects of reserpine-like drugs. J. Pharmacol. Exp. Ther., 144, 321-330.

Ueki, S. (1982). Mouse-killing behavior (muricide) in the rat and the effect of antidepressants. In New Vistas in Depression (S. Langer, R. Takahashi, T. Segawa and M. Briley), vol.40, Advances in the Biosciences, pp.187-194, New York, Pergamon Press.

Vetulani, J., & Sulser, F. (1975). Action of various antidepressant treatments reduces reactivity of noradrenergic cyclic AMP-generating system in limbic forebrain. Nature, 257, 495-496.

Waldmeier, P.C. (1981). Stimulation of central serotonin turnover by beta adrenoceptor agonists. Naunyn-Schmiediberg's Arch. Pharmacol., 317, 115-119.

Waldmeier, P.C., Bauman, P.A., Hauser, K., Maitre, L., & Storni, A. (1982). Oxaprotiline, a noradrenaline uptake inhibitor with an active and an inactive enantiomer. Biochem. Pharmacol., 31, 2169-2176.

Wolfe, B.B., Harden, T.K., Sporn, J.R., & Molinoff, P.B. (1978). Presynaptic modulation of beta-adrenergic receptors in rat cerebral cortex after treatment with antidepressants. J. Pharmacol. Exp. Ther., 207, 446-457.

20 Unitary Concepts in Behavioral Biology

Trevor Archer
University of Göteborg

Lars-Göran Nilsson
University of Umeå

Knut Larsson's "window onto nature itself" gave some tremendous insights and opened new horizons in psychobiology. The present treatise of the neuroendocrine axis, incorporated in this volume in his honor, provides only a cursory view of the proliferation of basic research that he stimulated, as described by the topics covered here. The initial topic of this volume deals primarily with particular aspects of maternal behavior and ontogenetic processes controlling development. Mayer, Factor, and Rosenblatt (Chapter 2) demonstrated (a) that aggressive behavior in the maternal rat occurs at the same time as components of fundamentally maternal behavior, pup retrieving, nestbuilding, and anogenital licking behavior, and (b) are modulated by estradiol and progesterone. Significantly, maternal behavior, but not aggressive behavior, was affected by prolactin. Hård and Engel (Chapter 4) present a neuropharmacological review of ultrasonic vocalization in rat pups; undoubtedly, this behavior must ultimately affect that of the rat dams, and as they point out, vocalization reaches a peak around 9-11 days and diminishes from about 16 days of age. Therein lie some components of an important pup-mother interaction and the neuroendocrine management. The chapters by Leon (Chapter 3) and Harding (Chapter 5) describe different neuroendocrine axes of developmental process. These two chapters offer an interesting comparison of early learning: olfaction, on the one hand, and male song birds, on the other. For each respective species, rats as opposed to zebra finches and red-winged blackbirds, the particular modality under examination was shown to be intimately controlled by complex relationships of specific stimuli with selective neurohormonal factors at particular brain sites. Both chapters relate to the enormous plasticity of the developing brain and the critical changes necessary for completion of an adequate developmental process, independent of the exact species and behavior under investigation; here, one may discover remarkable commonalities.

In keeping with Knut Larsson's main interests regarding the neurobiology of sexual behavior, several chapters deal with various aspects of sexual behavior and sexual preference. Meyerson (Chapter 7) presents an exhaustive review on the neuropharmacology of lordosis behavior, whereas Södersten (Chapter 8) reviews the role of testosterone and its metabolism in male rat sexual behavior. The chapter by van de Poll (Chapter 6) offers a different dimension to those of Meyerson and Södersten, in that stimulus events and conditioning factors are examined by introducing procedures developed to study internal stimulus control (sometimes referred to as state-dependent or drug-dependent learning). Van de Poll's chapter bears some relationship upon those of Harding and Leon by focussing the role of gonadal hormones on elements of functional adaptation and plasticity. The approach of van de Poll is extremely worthwhile and innovative. However, only a small portion of his research has been described in the chapter in this volume. To do him justice, one requires some further information on the neuroendocrine-anatomical developments that are currently being produced in his laboratory. It may be useful to compare Meyerson's and Södersten's chapters with that of Ahlenius, Hillegaart, and Larsson (Chapter 9) which reviews monoaminergic influences on the sexual behavior of male rats.

The chapters by Komisaruk (Chapter 11), Everitt (Chapter 10), Beyer and González-Mariscal (Chapter 13), and Micevych (Chapter 12) represent a somewhat different approach to the neurobiology of sexual behavior than those chapters, mentioned above, that deal mainly with the pharmacoendocrine analysis of behavior. These latter chapters represent an extremely diverse contribution that collectively encompass a wide functional variation, over the neuropharmacology-endocrine axis at specific neuroanatomical sites, and even involve structure-activity data on synthetic and naturally-occurring gonadal hormones. The multidimensional approach is well illustrated in Everitt's chapter on the neuroendocrine and psychological mechanisms underlying male sexual behavior. He presents exciting evidence relating stimulus-reward associations in the basolateral amygdala with dopaminergic control of the incentive (an estrous female rat) under the modulation of hormone-dependent hypothalamic processes. Komisaruk's chapter, reviewing the effects of vaginal stimulation on neuroendocrine, behavioral, and analgesic processes, relates a very complete story. His extremely painstaking and consequential approach may provide some exciting insights for the eventual solution of problems involving pain control. The chapter by Beyer and González-Mariscal, in turn, reviews the current state of progress in the area of psychosexual development and brain modulation of sexual differentiation. Two important aspects may be noted by an uninitiated but interested observer: (1) at the receptor level, the concept of "neuromodulatory pregnancy" involving the activation of "trophic" events seems to implicate the involvement of second messenger processes within the chosen functional model, i.e., lordosis behavior. (2) From the point-of-view of drug development, the structure-activity relationships of compounds related to progesterone may well offer new leads in the search for therapeutic compounds having as their mechanism of action the modulation of ion channels at particular brain sites. In this vein, one may note another approach employing the synthesis of so-called "antihormonal" agents to provide serenic drugs that reduce aggressive behavior (cf. Brain, Simon, Hasan, Martinez, & Castano, 1988). We are, indeed, fortunate in this section to have procured a fourth excellent chapter, that of Micevych, which offers new dimensions on processes underlying sexual differentiation regulating reproductive behavior in the female rat. He provides developmental, autoradiographic, and behavioral evidence for the influence and interactions of cholecystokinin (CCK) and gonadal hormones in the modulation and ontogeny of lordosis in male and female rats.

The chapters by Hwang and Broberg (Chapter 16), Carlsson (Chapter 17), and Heimann (Chapter 14) present psychobiological approaches to some aspects of clinical psychology. Heimann presents a brief review of imitative behavior in neonates, as observed from a sociobiological perspective. Contrary to the earlier Guillaume-Piagetian ideas, this review indicates that an imitative capacity develops relatively quickly during the initial weeks following birth (e.g., Heimann, 1989; Heimann, Nelson, & Schaller, 1989). One valuable consideration pertains to the evidence for individual differences in neonatal imitation and it is interesting to regard the adaptive perspective within which this succinct review is presented.

Carlsson has derived a quite comprehensive analysis that essentially utilizes psychophysiological techniques to achieve a biobehavioral understanding of particular disease states. He describes several examples concerning disorders of affect that have been the focus of much attention within psychosomatic medicine. These disorders include complaints with a wide range of symptoms, generally with an important kernel derived from pain-related and phobic behaviors. He provides an interesting analysis of clinical biofeedback effects with an eye towards a possible mechanism. Such a mechanism should extend beyond the simple and now specific "placebo" explanation to one that seeks an understanding at cognitive, emotional, and motivational levels. The chapter by Hwang and

Broberg deals with problems arising from social inhibition. They offer a short review of psychobiological research on inhibited children, together with some of the results from a broad study of social inhibition and its links with other aspects of social behavior. As a consequence of the "Göteborg Child Care Project", the possible role of out-of-home care in modulating the penchant for 'outgoing' or 'inhibited' traits was investigated and some implications, even clinical, were discussed.

The chapters by Brain and Al-Hazmi (Chapter 18) and Marek and Seiden (Chapter 19) offer an approach that is essentially behavioral pharmacology in its orientation. The former representing a wide analysis of a given pharmacologically active compound, ethyl alcohol, in different behavioral models, using mice, and the latter presenting a drug screen, the Differential-Reinforcement-of-Low-Rates (DRL-72s)—72 seconds, through which a very large series of antidepressant and potential antidepressant compounds have been filtered. In spite of a few conflicting sources of results, the Seiden and co-workers (e.g., Seiden, Dahms, & Shaughnessy, 1985) have developed a method for capturing potential anti-depressant-actions. This test has a surprisingly low rate of false positives (compounds active in the screen but not clinically) and false negatives (*not* active in the screen but with some degree of efficacy). One aspect of the DRL-72s antidepressant analysis that should be noted is the alteration of interresponse time frequencies as a function of drug treatment. Another aspect pertains to the possible theoretical analysis of the 'depressed' nature of a hungry or thirsty rat trying to obtain reinforcements in an extremely difficult schedule. Perhaps a comprehensive analysis of several behavioral parameters and biological markers, directly following each experimental session in the skinner box, would develop an explanation for this effect. Thus, cognitive, emotional, and motivational variables may be examined before and after the DRL session by application of a spatial learning, open-field, and preferred-taste intake procedure, respectively. Concomitantly, neurotransmitter (serotonin and noradrenaline) and endocrine (e.g., corticosterone) measures would certainly strengthen the analysis and provide a truly unitary neuroendocrine-behavioral analysis with clinical relevance.

Aspects of neuroendocrine involvement in affective disorders

Within the general ambition of pursuing a general understanding of some aspects of behavioral biology, it is possible to identify human syndromes that may be tested in animal models. For example, Hellhammer and his co-workers (e.g., Hellhammer & Gutberlet, 1988) have examined certain affective states that have, in humans, influenced both male fertility and gonadal function. In the examination of personality attitudes, sexual steroid hormones, gonadotropins, and seminal parameters in the male partners of 218 barren couples, it was found that significant correlations existed between a low sperm count, low serum testosterone levels, and a competitive personality profile (Hellhammer et al., 1985a,b; Hubert et al., 1985). A comparison was made to depressed patients and these did not show evidence of infertility. Hellhammer and Gutberlet (1988) have analyzed other data to indicate that infertile men may show similarities with 'coronary-prone' behavior. On the basis of these indications they designed to compare behavioral, neuroendocrine, and histological parameters of rats that had previously experienced two types of treatment: 1) *Learned-helplessness,* by which they were exposed to shocks over which they had no control, and 2) *Activity-wheel-stress,* in a standard type running-wheel (see Hellhammer et al., 1984a, 1984b; Rea & Hellhammer, 1984, for details). The former group, subjected to uncontrollable shock, were notably passive but showed no evidence of alterations to testosterone levels, luteinizing hormone, or testicle histology. The group subjected to activity-stress, however, showed a lowered blood supply to the gonads and a decrease in testosterone levels in serum and testes. Further, histological evidence indicated a degeneration of spermatids at the testosterone-sensitive stages of spermatogenesis. It is interesting to note that this condition was associated with an elevated noradrenaline turnover in several brain regions, including the neocortex, hippocampus, striatum, thalamus, hypothalamus, midbrain, pons-medulla region, and cerebellum. These data of Hellhammer et al. seem to confirm, in rats, investigations indicating a reduction of plasma testosterone after prolonged stress, e.g., intensive parachute training (Davidson, Smith & Levine, 1978), surgery (Carstensen, Amer, Wide, & Amer, 1973), and military training (Kreutz, Rose, & Jennings, 1972).

The relationship between stress, arousal, aggression, free testosterone, and fertility in males is certainly a very complex one. Testosterone in saliva (which seems to correlate very well with free serum testosterone, e.g., Wang, Plymate, Nieschlag, & Paulsen, 1981) was found to increase under conditions of sexual arousal and aggression (Pirke, Kockott, & Dittmar, 1974; Rubin, Reinisch, & Haskett, 1981) but decreased under conditions of military stress (Kreuz, Rose, & Jennings, 1972). Hellhammer, Herbert, and Schürmeyer (1985b) indicated that salivary testosterone was elevated soon after the onset of erotic and sexual stimulation (film material) but declined after stressful dental surgery films. Surprisingly, aggressive films did not affect the levels of testosterone. There is evidence of increased testosterone in young violent males (DeBold & Miczek, 1985). These diverse indications make it difficult to reconcile personality traits, stress, stimulus objects, aggression, testosterone levels, and infertility, given the positive correlation between testosterone and spermatogenesis (for interesting reviews on certain aspects, see Brain, 1981; Dixon, Huber, & Kaeserman, 1984). For present purposes it may suffice to note that three types of hypotheses have attempted to explain the interaction between stress and infertility: 1) stress-related problems lead to infertility, 2) infertility triggers psychosocial distress, and 3) there exists an interaction causal relationship between infertility and stress/distress (Wright, Allard, Lecours, & Sabourin, 1989).

In this context it is of interest to note the involvement of this type of prolonged stress in animal models of chronic stress and antidepressant treatment. Table 1 presents a simple relationship between some changes in various parameters after the chronic application of stressors and some antidepressant drugs (essentially of the typical tricyclic class of compound).

It should be borne in mind that the neuroendocrine axis plays an important role in interactions of alcohol, stress, and aggression. Several other factors modulate this interaction, including situational and/or environmental variables, social influences, personality, genetics, and cultural predispositions (Brain, 1986; Dixon, 1982; Valzelli, 1981). Alcohol abuse is implicated more than any other drug, with occasions of violence and aggression, and is associated with homicides, suicides, physical and sexual assaults (Cherek, Steinberg, & Manno, 1985; Pernanen, 1976). A large number of investigations over different species have shown that basal testosterone and moderate doses of alcohol have a profound interactive effect on aggressive behavior (e.g., Blanchard & Blanchard, 1987; DeBold & Miczek, 1985; Winslow & Miczek, 1985). A cursory indication seems to be that low-to-moderate doses of alcohol can substantially increase the aggressive behavior of dominant or highly aggressive males, whereas higher doses produce the rate-depressing effect observed over a wide range of behaviors (Winslow, DeBold, & Miczek, 1987). Both noradrenaline and serotonin appear to be involved in the eventual assignation of a dominance/submissive status (Pucilowski & Kostowski, 1981). The neuroendocrine control of alcohol-aggressive behavioral functions in humans becomes greatly dependent on the precise social setting (e.g., high vs. low provocation) in addition to the underlying personality traits (Taylor, 1967; Taylor & Leonard, 1983). Lindman and co-workers (e.g., Lindman 1982, 1985; Lindman, Lindfors, Dahla, & Toivola, 1987) studied various aspects of mood change in aggressive and non-aggressive individuals drinking alcohol. Lindman, Järvinen, and Vidjesko (1987) found that, although basal testosterone levels in saliva were higher in aggressive than in nonaggressive individuals, alcohol did not necessarily alter the behavior of these individuals but rather that the aggressive trait was a fairly reproducible trait with a given predisposition (but see also Boyatzis, 1975; Taylor & Gammon, 1975). It is interesting to note that alcohol may not, indeed, suppress testosterone levels as has formerly been suggested (Ellingboe, 1987; Mello & Mendelson, 1978) but rather enhance them (Lindman et al., 1987). This may, of course, be dependent upon measures of salivary or serum testosterone. Considerations of alcohol and aggression are pertinent to any analysis of the neuroendocrine axis in mood change and, therefore, allow derivation of unitary concept.

The type of stressor employed by Hellhammer et al. (e.g., 1984a), found to reduce testosterone levels in the plasma of human males (see above), does seem to have its own affective component. For example, in situations of continuous stress exposure, as, for in

Table 20.1. Changes in Various Parameters after Chronic Application of Stressors and Antidepressant Drugs (AD)

Effect	Observed after chronic stress	Observed after chronic AD
Attenuation of clonidine hypothermia	immobilization (Platt et al., 1983) & footshock (Danysz, not published)	Von Voigtlander et al., 1978; Gorka & Zacny, 1981; Danysz, submitted
Attenuation of decrease in SS in reaction to acute shock	footshock (Zacharko et al., 1983)	Zacharko & Anisman, 1984
Increase in locomotor response to amphetamine	footshock (Herman et al., 1984)	Spyraki & Fibiger, 1981
Decrease in immobility time in the "behavioral despair" test	immobilization (Platt & Stone, 1982)	Kitada et al., 1981
Increase in serotonergic syndrome	immobilization (Kennett et al., 1985)	Stolz & Marsden, 1982
Attenuation of acute stress-induced suppression of escape and avoidance reactions	footshock (Weiss et al., 1975)	Petty & Sherman, 1979
Attenuation of ulcer formation	restraint & cold (Pare, 1986)	Aguwa & Ramanujam, 1984
Tolerance to morphine analgesia	immobilization (Benedek & Szikszay, 1985)	Kellstein et al., 1984
Decrease in b receptors binding and/or sensitivity in the brain	footshock (Stone & Platt, 1982; Stone et al., 1984)	Vetulani & Sulser, 1977; Selinger-Barnette et al., 1980
Decrease in 3H-clonidine binding (alpha-2) in the cortex	immobilization (Lynch et al., 1983) (1)	Pilc & Vetulani, 1982

SS = self-stimulation
(1) it should be pointed out that an increase in alpha-2 adrenoceptors number has been reported, as well, by Cohen et al., (1986) and Torda et al., (1983)

259

stance, under front line or behind enemy line battle conditions, or in particular conditions of prisoner-of-war camps, depressive symptoms and suicidal tendencies are believed to provide rare instances of maladaptive, self-destructive behavior. The few, more-or-less controlled studies in existence report that behavior patterns placing a premium on survival are most dominant (Haslam, 1984; Solomon & Benbenishty, 1986). Thus, frontline and battle exposure treatment was found to correlate well with lowered rates of posttraumatic stress disorders and elevations of performance, fitness, and mood. The implication from these findings for a broader understanding of the neuroendocrine axis refers essentially to a type of cost-benefit analysis in which an adaptive, antidepressive benefit seems to be weighted against the possible reality of decreased sperm counts and testosterone levels in the male.

Prenatal stress and ontogeny

A different aspect of stress/distress and neuroendocrine reactions with eventual influences upon function involves the effects of prepartum stress on mother-infant relationships and problem-solving in the offspring of stressed mothers. Muir and Pfister (1988) demonstrated differential relationships between corticosterone, oxytocin and prolactin, and novelty stress in nulliparous female rats. In a subsequent study, Pfister and Muir (1989) found that novelty stress during pregnancy caused elevations in the corticosterone levels of lactating dams (postpartum days 6 and 21), whereas oxytocin did not have the same effect. They concluded that through effects on corticosterone, but not prolactin, stressors may affect the lactation process. Certainly, much evidence suggests that stress during pregnancy affects various aspects of offspring-dam relations (Grota & Ader, 1969; Muir, Pfister, & Ivinskis, 1985; Rosenblatt, 1965) and behavioral parameters in the offspring (Archer & Blackman, 1971; Barlow, Knight, & Sullivan, 1979; Pfister, 1979). Pfister and his colleagues (e.g., Muir et al., 1985) have produced comprehensive analyses of the cognitive performance of the offspring of mothers subjected to prepartum stress. In a highly complex study employing the Hebb-Williams problem-solving maze, Muir et al. (ibid.) found that performance, as indexed by the number of errors made, was negatively affected in the stress offspring and confirmed earlier results (e.g., Pfister, 1980; Young, 1963). Surprisingly, however, Muir et al. (1985) found that an enriched postpartum environment failed to critically improve the worsened performance in the maze test as a result of prepartum stress. It seems that heightened arousal and activation of the pituitary-adrenal axis in these offspring did not counteract the effects of prepartum stress. Perhaps the concomitant manipulation of brain noradrenaline (NA) levels and the pituitary-adrenal axis may be necessary; in avoidance learning situations, this particular neuroendocrine axis has much consequence for performance (Archer, Ögren, Fuxe, Agnati, & Eneroth, 1981; Ögren, Archer, Fuxe, & Eneroth, 1981). Further, NA levels appear to modulate the facilitatory effects of an enriched environment which in itself affects the concentration of nerve growth factor in forebrain regions (Mohammed, Jonsson, & Archer, 1986).

Neurotransmitter-hormonal interactions in learning tasks

Much evidence suggests that the neural-gonadal hormone axis modulates several characteristics of learned behaviors. In an exhaustive review, Van Haaren, Van Hest, and Heinsbroek (1990) found that, in aversively-motivated learning situations, the presence of an aversive causes greater behavioral inhibition in intact male rats (i.e., not castrated) than in intact females (i.e., not ovariectomized). They indicate, further, that if gonadectomy is performed when the animals have been allowed to reach an adult stage of development, the active avoidance performance of both males and females is unaffected (e.g., Scouten, Grotelueschen, Beatty, 1975), whereas neonatal castration or androgenization of female rats reversed the behavioral differences observed in shuttlebox situations (e.g., Beatty & Beatty, 1970). In addition, adult castration may alter free operant avoidance to resemble that of female rats

Table 2

Total Error (Mean and Standard Error) Scores For All Groups On Each Of The Six Problems
Composing The Test Battery

Group		Aversively Motivated Problems				Appetitively Motivated Problems		
	N	Visual Discrimination	Maze	Inclined Plane Discrimination	N	Detour A	Detour B	Detour C
Control	86	20.8 ± 0.5	10.0 ± 0.4	7.4 ± 0.3	60	5.8 ± 0.3	6.8 ± 0.8	5.0 ± 0.4
Neocortex								
Frontal	11	20.8 ± 2.6	7.5 ± 0.6	11.0 ± 1.3	11	14.5 ± 1.8*	10.6 ± 2.1*	12.7 ± 1.7*
Parietal	10	36.0 ± 4.1*	30.9 ± 4.8*	18.9 ± 1.7*	10	12.0 ± 1.7*	9.9 ± 2.2*	8.5 ± 1.6*
Occipitotemporal	9	57.6 ± 1.5*	36.0 ± 7.5*	12.6 ± 2.1	10	5.6 ± 1.1	13.5 ± 1.5*	6.3 ± 1.3
Punctuate (12 lesions)	10	28.7 ± 2.0*	23.9 ± 3.4*	10.5 ± 1.1	8	12.0 ± 1.9*	8.1 ± 1.9	7.4 ± 1.1
Punctuate (16 lesions)	9	61.8 ± 4.2*	44.1 ± 9.0*	20.6 ± 3.9*	7	16.7 ± 1.4*	18.6 ± 1.0*	13.1 ± 2.0*
Basal ganglia								
Caudoputamen	13	21.7 ± 2.0	13.2 ± 1.1*	10.0 ± 1.0	13	11.1 ± 1.1*	10.7 ± 2.2	8.9 ± 1.2*
Globus Pallidus	7	43.0 ± 4.6*	16.3 ± 1.8*	13.3 ± 1.2*	12	9.3 ± 1.0*	13.1 ± 2.0*	7.8 ± 1.3*
Entopeduncular nucleus	12	27.6 ± 2.1*	16.2 ± 3.3*	8.7 ± 0.9	12	14.8 ± 1.4*	17.8 ± 1.1*	13.3 ± 1.6*
Substantia nigra	8	31.3 ± 2.9*	15.5 ± 1.7*	19.9 ± 2.9*	9	11.9 ± 2.2*	17.7 ± 1.2*	13.6 ± 2.0*
Limbic forebrain								
Frontocingulate cortex	9	17.1 ± 1.7	20.0 ± 3.3*	18.9 ± 2.9*	17	11.8 ± 1.6*	14.5 ± 1.5*	8.1 ± 1.5
Posterior cingulate	10	15.9 ± 1.9	31.0 ± 2.5*	19.2 ± 2.1*	13	13.9 ± 1.4*	10.5 ± 2.0	10.8 ± 1.6*
Dorsal hippocampus	10	22.7 ± 2.1	33.6 ± 4.4*	11.8 ± 1.5	10	15.6 ± 1.9*	16.6 ± 2.2*	18.2 ± 0.8*
Amygdala	7	18.3 ± 2.3	13.3 ± 1.8*	6.3 ± 0.8	7	5.9 ± 0.7	5.6 ± 1.6	4.7 ± 1.0
Thalamus								
Anterior complex	7	13.6 ± 1.9	33.9 ± 3.6*	18.1 ± 1.3*	7	16.9 ± 1.4*	14.3 ± 2.7*	16.3 ± 2.2*
Lateral complex	7	23.7 ± 2.4	37.1 ± 3.5*	20.0 ± 2.2*	11	7.0 ± 0.8	10.5 ± 2.1	8.4 ± 1.7*
Ventrolateral complex	8	27.5 ± 3.9*	17.6 ± 3.5*	24.5 ± 2.3*	8	11.1 ± 2.5*	12.0 ± 1.7*	10.9 ± 2.3*
Mediodorsal complex	9	15.7 ± 2.1	20.3 ± 3.6*	12.8 ± 1.3*	9	15.0 ± 2.1*	8.0 ± 2.1	9.8 ± 1.8*
Parafascicular complex	9	20.0 ± 1.4	21.0 ± 4.1*	17.2 ± 2.1*	9	17.2 ± 1.2*	16.9 ± 1.2*	7.7 ± 1.8
Ventral complex	6	27.3 ± 2.7*	13.0 ± 0.6*	10.8 ± 1.8	7	5.6 ± 1.2	3.7 ± 0.9	4.4 ± 1.4
Hypothalamus (lateral)	10	29.8 ± 3.4*	39.1 ± 5.6*	12.2 ± 1.1	7	15.6 ± 2.3*	14.1 ± 2.0*	16.9 ± 2.0*
Reticular formation								
Rostral midbrain area	9	17.5 ± 1.8	15.0 ± 1.6*	30.5 ± 2.9*	8	11.4 ± 2.2*	11.8 ± 2.0*	4.9 ± 2.1
Pontine area	6	47.7 ± 4.0*	38.7 ± 6.8*	26.8 ± 5.2*	13	10.3 ± 1.2*	8.6 ± 1.1*	6.0 ± 1.1*
Other Brainstem areas								
Midbrain central gray	9	29.0 ± 1.7*	15.8 ± 3.1*	18.7 ± 2.0*	10	10.3 ± 1.4*	8.8 ± 1.8	6.1 ± 1.2*
Median raphe	6	48.8 ± 4.0*	44.4 ± 9.0*	24.5 ± 4.4*	11	14.6 ± 1.9*	18.8 ± 0.8*	17.0 ± 1.4*
Lateral midbrain area	7	21.1 ± 1.5	7.0 ± 3.5	10.0 ± 0.6	11	5.5 ± 1.0	6.4 ± 1.8	5.6 ± 1.2

*Differs from respective sham-operated control subgroup, Mann-Whitney U, $p < 0.05$.
Reproduced with permission from Thompson et al. (1986).

261

(Davis, Porter, Burton, & Levine, 1976). However, neither neonatal nor adult gonadal hormone manipulation appear to affect passive avoidance/inhibitory conditioning to any reliable extent (Phillips & Deol, 1977; van Oijen, van de Poll, & de Bruin, 1980), whereas resistance-to-extinction in taste-aversion experiments was more prolonged in male than female rats (Chambers, 1976). It is interesting to note that in positively motivated tasks male rats showed greater lever contact rates (in a Skinner box), but that female rats performed more efficiently (i.e., obtained more reinforcements at lower response rates) on schedules requiring differential reinforcement of low rate responding (Beatty, 1973; Zeier, Baettig, & Driscoll, 1978). On the other hand, male rats appear to perform better on male learning tasks (Einon, 1980; Juraska, Henderson, & Muller, 1984), and treatments altering the gonadal hormones appear not to produce the effects observed in avoidance tasks. In sum, the basal, circulating gonadal hormone profiles of male and female rats do certainly affect the performance of learned behaviors though perhaps not necessarily association processes. These constraints will have some critical bearing on the design of experiments to examine neurobiological substrates.

The importance of the neuroendocrine axis in behavioral processes has been established by an enormous number of studies employing learning techniques, generally some form of avoidance learning (e.g., McGaugh, 1983; McGaugh, Introini-Collison, Nagahara, & Cahill, 1989). One highly successful illustration is shown by the interaction of adrenoceptors with opioid peptides in specific brain regions. Pre- or post-training administration of opioid peptides cause defects in learning performance (e.g., Izquierdo, 1979; McGaugh, 1988), effects generally blocked by naloxone (Martinez et al., 1988). The disruptive effects of β-endor-phins are antagonized by low doses of adrenaline (Izquierdo & Dias, 1985) whereas those produced by high doses of adrenaline were antagonized by naloxone (Introini-Collison & McGaugh, 1987). Selective lesioning of forebrain noradrenaline terminals blocked the facilitatory effects of naloxone (Introini-Collison & Baratti, 1986; Kapp & Gallagher, 1983), as did intra-amygdaloid injections of propranolol (Introini-Collison, Nagahara, & McGaugh, 1989). This evidence and other data suggest some negative influence of amygdaloid-caudate opioid systems on basically *procedural* memory tasks whereas other results suggest a negative influence of hippocampal-amygdaloid opioids on *declarative* memory (Collier & Routtenberg, 1984). In each case, this influence may be modulated by noradrenergic fibers. The classification of *procedural* and *declarative* memory pertains to that of Squire and Cohen (1980) in the context of the radial eight-arm maze. Table 2 presents a rudimentary conceptualization of *procedural* and *declarative* memory in the septo-hippocampalamygda-loid and amygdaloid-caudate areas with regard to adrenergic-opioid mechanisms. According to an hypothesis presented earlier (Archer, 1990), one may consider evidence for some interference of declarative memory ('knowing what') following fornix-fimbria lesion, stimulation of opiate receptors in hippocampal regions or certain types of noradrenaline denervations. On the other hand, procedural memory ('knowing how') may be disrupted by treatments influencing noradrenergic/opiate pathways in amygdaloid-striatal regions or caudate lesions. In recent publications, Heimer, Alheid, and co-workers (cf. Alheid & Heimer, 1988; Alheid, Heimer, & Switzer, 1990; Heimer, de Olmos, Alheid, & Zaborszky, 1990, in press) have described three descending corticofugal pathways: (1) the striatopallidal system, with afferents projecting to the putamen and caudate regions; (2) the extended amygdala, with axons originating in the hippocampus, entorbinal cortex; and (3) the magnocellular forebrain complex, with afferents from the same areas that innervate the extended amygdala. In Heimer's terms of neuroanatomical analysis (cf. Heimer et al., 1990, in press), one may speculate that the striatopallidal system may be primarily involved in tasks of *procedural* memory whereas the extended amygdala together with the magnocellular forebrain complex may modulate *declarative* memory.

REFERENCES

Alheid, G.F., & Heimer, L. (1988). New perspectives in basal forebrain organization of special relevance for neuropsychiatric disorders; the striatopallidal, amygdaloid, and corticopetal components of substantia innominata. *Neuroscience, 27,* 1-39.

Alheid, G.F., Heimer, L., & Switzer, R.C. (1990). The basal ganglia. In G. Paxinos (Ed.), *The Human Nervous System* (pp. 483-582). San Diego: Academic Press.

Archer, J.E., & Blackman, D.E. (1971). Prenatal psychological stress and offspring behavior in rats and mice. *Dev. Psychobiol., 4,* 193-248.

Archer, T. (1990). Learning techniques to study memory processes in opiate and/or adrenergic systems. In R.C.A. Frederickson, J.L. McGaugh, and D.L. Felten (Eds.), *Peripheral Signaling of the Brain: Neuronal-immune and Cognitive Function,* in press. Toronto: Hogrefe & Huberg Publ.

Archer, T., Ögren, S.O., Fuxe, K., Agnati, L., & Eneroth, O. (1981). On the interactive role of noradrenaline and corticosterone in two-way active avoidance in the rat. *Neurosci. Lett., 27,* 341-346.

Barlow, S.M., Knight, A.F., & Sullivan, F.M. (1979). Plasma corticosterone response to stress following chronic oral administration of diazepam in the rat. *J. Pharm. Pharmacol., 31,* 23-26.

Beatty, W.W. (1973). Effects of gonadectomy on sex differences in DRL behavior. *Physiol. Behav., 10,* 177-178.

Beatty, W.W., & Beatty, P.A. (1970). Hormonal determinants of sex differences in avoidance behavior and reactivity to electric shock in the rat. *J. Comp. Physiol. Psychol., 73,* 446-455.

Blanchard, R.J., & Blanchard, D.C. (1987). The relationship between ethanol and aggression: studies using ethological models. In B. Olivier, J. Mos, and P.F. Brain (Eds.), *Ethopharmacology of Agonistic Behavior in Animals and Humans* (pp. 145-341.. Dordrecht: Nijhoff.

Boyatzis, R.E. (1975). The predisposition toward alcohol-related interpersonal aggression in men. *J. Stud. Alcohol, 36,* 1196-1207.

Brain, P.F. (1981). Diverse actions of hormones on "aggression" in animals and man, In I. Valzelli and I. Mongese (Eds.), *Aggression and Violence: A Psychobiological and Clinical Approach* (pp. 99-149). Milan: Edizioni.

Brain, P.F. (1986). *Alcohol and Aggression.* London: Croom Helm.

Brain, P.F., Simon, V., Hasan, S., Martinez, M., & Costano, D. (1988). The potential of antiestrogens as centrally-acting antihostility agents: recent animal data. *Int. J. Neurosci., 41,* 169-177.

Carstensen, H., Amer, I., Wide, L., & Amer, B. (1973). Plasma testosterone, LH and FSH during the first 24 hours after surgical operations. *J. Steroid Biochem., 4,* 605-611.

Chambers, K.C. (1976). Hormonal influences on sexual dimorphism in rate of extinction of a conditioned taste aversion in rats. *J. Comp. Physiol., 90,* 851-856.

Cherek, D.R., Steinberg, J.L., & Manno, B.R. (1985). Effects of alcohol on human aggressive behaviour. *J. Stud. Alcohol, 46,* 32-38.

Collier, T.J., & Routtenberg, A. (1984). Selective impairment of declarative memory following stimulation of dentate gyrus granule cells: a naloxone-sensitive effect. *Brain Res., 310,* 384-387.

Davidson, J.M., Smith, E.R., & Levine, S. (1978). Testosterone. In H. Ursin, E. Baade, and S. Levine (Eds.), *Psychobiology of Stress—A Study of Coping Men* (pp. 57-62). New York: Academic Press.

Davis, H., Porter, J., Burton, J., & Levine, S. (1976). Sex and strain differences in leverpress shock escape behavior. *Physiol. Psychol., 4,* 351-356.

DeBold, J.F., & Miczek,K.A. (1985). Testosterone modulates the effects of ethanol on male mouse aggression. *Psychopharmacol. Bull., 85,* 286-290.

Dixon, A.K. (1982). Ethopharmacology: a new way to analyse drug effects on behaviour. *Triangle, 21,* 95-105.

Dixon, A.K., Huber, C., & Kaeserman, F. (1984). Urinary odors as a source of indirect drug effects on the behavior of male mice. In K.A. Miczek, M. Kruk, and B. Olivier (Eds.), *Ethopharmacological Aggression Research* (pp. 81-91). New York: Alan R. Liss Inc.

Einon, D. (1980). Spatial memory and response strategies in rats: age, sex and rearing differences in performance. *Q. J. Exp. Psychol., 32,* 473-489.

Ellingboe, J. (1987). The effects of ethanol on sex hormones in men and women. In K.O. Lindros, R. Ylikahri, and K. Kiianmaa (Eds.), *Proceedings of the Third Congress of the International Society for Biomedical Research on Alcoholism* (pp. 109-116). Oxford: Pergamon Press.

Grota, L.J., & Ader, R. (1969). Continuous recording of maternal behavior in Rattus norvegicus. *Anim. Behav., 17,* 722-729.

Haslam, D.R. (1984). The military performance of soldiers in sustained operations. *Aviat. Space Environ. Med., 55,* 216-221.

Heimann, M. (1989a). Imitation during the first months of life—what we know and what we don't know. *Nordisk Psykologi, 41,* 193-203 (in Swedish).

Heimann, M., Nelson, K.E., & Schaller, J. (1989). Neonatal imitation of tongue protrusion and mouth opening: Methodolgical aspects and evidence of early individual differences. *Scand. J. Psychol., 90,* 90-101.

Heimer, L., de Olmos, J., Alheid, G.F., & Zaborszky, L. (1990). "Perestroika" in the basal forebrain: Opening the border between neurology and psychiatry. *Prog. Brain Res.,* in press.

Hellhammer, D.H., Rea, M.A., Bell, M., & Belkien, L. (1984a). Activity-wheel stress: Effects on brain monoamines and the pituitary-gonadal axis. *Neuropsychobiology, 11,* 251-254.

Hellhammer, D.H., Rea, M.A., Bell, M., Belkien, L., & Ludwig, M. (1984b). Learned helplessness: Effects on brain monoamines and the pituitary-gonadal axis. *Pharmacol. Biochem. Behav., 21,* 481-485.

Hellhammer, D.H., & Gutberlet, I. (1988). Male infertility: Preliminary evidence for two endocrine mediates of stress on gonadal function. In J.A. Ferrendelli, R.C. Collins, and E,M. Johnson (Eds.), *Neurobiology of Aminoacids, Peptides and Trophic Factors* (pp. 227-230). Holland: Kluver.

Hellhammer, D.H., Hubert, W., Freischem, C.W., & Nieschlad, E. (1985a). Male infertility: relationships among gonadotropins, sex steroids, seminal parameters and personality attitudes. *Psychosom. Med., 47,* 58.

Hellhammer, D.H., Hubert, W., & Schürmeyer, T. (1985b). Changes in saliva testosterone after psychological stimulation in men. *Psychoneuroendocrinology, 10,* 77-81.

Hellhammer, D.H., Rea, M.A., Bell, M., Belkien, L., & Ludwig, M. (1984). Learned helplessness: Effects on brain monoamines and the pituitary-gonadal axis. *Pharmacol. Biochem. Behav., 21,* 481-485.

Hubert, W., Hellhammer, D.H., Freischem, C.W. (1985). Psychobiological profiles in infertile men. *J. Psychosom. Res., 21,* 161.

Introini-Collison, I.B., & Baratti, C.M. (1986). Opioid peptidergic systems may modulate the activity of β-adrenergic mechanisms during memory consolidation processes. *Behav. Neural Biol., 45,* 227-241.

Introini-Collison, I.B., & McGaugh, J.L. (1987). Naloxone and beta-endorphin alter the effects of posttraining epinephrine on retention of an inhibitory avoidance response. *Psychopharmacology, 92,* 299-235.

Introini-Collison, I.B., Nagahara, A.H., & McGaugh, J.L. (1987). Memory enhancement with intra-amygdala post-training naloxone is blocked by concurrent administration of propranolol. *Brain Res., 476,* 94-101.

Izquierdo, I. (1979). Effect of naloxone and morphine on various forms of memory in the rat: Possible role of endogenous opiate mechanisms in memory consolidation. *Psychopharmacology, 66,* 199-203.

Izquierdo, I., & Dias, R.D. (1985). Influence on memory of posttraining and pre-test injections of ACTH, vasopressin, epinephrine, or β-endorphin and their interaction with naloxone. *Psychoneuroendocrinology, 10,* 165-172.

Juraska, J.M., Henderson, C., & Muller, J. (1984). Differential rearing experience, gender and radial maze performance. *Dev. Psychobiol., 17*, 209-215.

Kapp, B.S., & Gallagher, M. (1983). Naloxone enhancement of memory processes: Dependence on intact norepinephrine function. *Soc. Neurosci. Abst., 9*, 828.

Kreuz, L.E., Rose, R.M., & Jennings, J.R. (1972). Suppression of plasma testosterone levels and psychological stress: a longitudinal study of young men in an officer's candidate school. *Arch. Gen. Psychiat., 26*, 479-482.

Lindman, R. (1982). Social and solitary drinking: Effects on consumption and mood in male social drinkers. *Physiol. Behav., 28*, 1093-1095.

Lindman, R. (1985). On the direct estimation of mood change. *Percep. Psychophys., 37*, 170-174.

Lindman, R., Järvinen, P., & Vidjeskog, J. (1987). Verbal interactions of aggressively and nonaggressively predisposed males in a drinking situation. *Aggr. Behav., 13*, 187-196.

Lindman, R., Lindfors, B., Dahla, E., & Toivola, H. (1987). Alcohol and ambience: Social and environmental determinants of intake and mood. In K.O. Lindros, R. Ylikahra, and K. Kiianmaa (Eds.), *Proceedings of the Third Congress of the International Society for Biomedical Research on Alcoholism* (pp. 385-388). Oxford: Pergamon Press.

Martinez, J.L., Weinberger, S.B., & Schuteis, G. (1988). Enkephalins and learning and memory: A review of evidence for a site of action outside the blood-brain barrier. *Behav. Neural. Biol., 49*, 192-221.

McGaugh, J.L. (1983). Hormonal influences on memory. *Ann. Rev. Psychol., 34*, 297-323.

McGaugh, J.L. (1988). Modulation of memory storage processes. In P.R. Solomon, G.R. Goethals, C.M. Kelley, and B.R. Stephens (Eds.), *Perspectives on Memory Research* (pp. 33-64). New York: Springer-Verlag.

McGaugh, J.L., Introini-Collison, I.B., Nagahara, A.H., & Cahill, L. (1989). Involvement of the amygdala in hormonal and neurotransmitter interactions in the modulation of memory storage. In T. Archer and L-G. Nilsson (Eds.), *Aversion, Avoidance, and Anxiety: Perspectives on Aversively Motivated Behavior* (pp. 231-249). Hillsdale, NJ: Lawrence Erlbaum Assoc.

Mello, N.K., & Mendelson, J.H. (1978). Alcohol and human behavior. In L.C. Iversen, S.D. Iversen, and S.H. Snyder (Eds.), *Handbook of Psychopharmacology, Vol. 12*, (pp. 235-317). New York: Plenum Press.

Mohammed, A.K., Jonsson, G., & Archer, T. (1986). Selective lesioning of forebrain noradrenaline neurons at birth abolishes the improved maze learning performance induced by rearing in complex environment. *Brain Res., 398*, 6-10.

Muir, J.L., & Pfister, H.P. (1988). Influence of exogenously administered oxytocin on the corticosterone and prolactin response to psychological stress. *Pharmacol. Biochem. Behav., 29*, 699-703.

Muir, J.L., Pfister, H.P., & Ivinskis, A. (1985). Effects of prepartum stress and postpartum enrichment on mother-infant interaction and offspring problem-solving ability in Rattus norvegicus. *J. Comp. Psychol., 99*, 468-478.

Ögren, S.O., Archer, T., Fuxe, K., & Eneroth, P. (1981). Glucocorticoids, catecholamines and avoidance learning. In K. Fuxe, J-Å. Gustafsson, and L. Wetterberg (Eds.), *Steroid Hormone Regulation of the Brain* (pp. 355-377). New York: Pergamon Press.

Pernanen, K. (1976). Alcohol and crimes of violence. In B. Kissin and H. Begleiter (Eds.), *The Biology of Alcoholism, Vol. 4. Social Aspects of Alcoholism* (pp. 351-444). New York: Plenum Press.

Pfister, H.P. (1979). The glucocorticoid response to novelty as a psychological stressor. *Physiol. Behav., 23*, 649-652.

Pfister, H.P. (1980). *Prenatal psychological stress in the rat: Behavioral and physiological responses to novel environments in offspring.* Unpublished Doctoral Dissertation, University of Newcastle, New South Wales, Australia.

Pfister, H.P., & Muir, J.L. (1989). Psychological stress and administered oxytocin during pregnancy: effect corticosterone and prolactin response in lactating rats. *Int. J. Neurosci., 45,* 91-99.

Phillips, A.G., & Deol, G.S. (1977). Neonatal androgen levels and avoidance learning in prepubescent and adult male rats. *Horm. Behav., 8,* 22-29.

Pirke, K.M., Kockott, G., & Dittmar, F. (1974). Psychosexual stimulation and plasma testosterone in man. *Arch. Sex. Behav., 3,* 577-584.

Pucilowski, O., & Kostowski, W. (1981). Effects of stimulation on the raphe nuclei on muricide behavior in rats. *Pharmacol. Biochem. Behav., 14 (Suppl. 1),* 25-28.

Rea, M., & Hellhammer, D.H. (1984). *Psychoth. Psychosom. 42,* 218-223.

Rosenblatt, J. (1965). The basis of synchrony in the behavioral interaction between the mother and her offspring in the laboratory rat. In B.M. Foss (Ed.), *Determinants of Infant Behavior, Vol. 3* (pp. 3-41). New York: Wiley.

Rubin, R.T., Reinisch, J.M., & Haskett, R.F. (1981). Postnatal gonadal steroid affects on human behavior. *Science, 211,* 1318-1324.

Scouten, C.W., Grotelueschen, L.K., & Beatty, W.W. (1975). Androgens and the organization of sex differences in active avoidance behavior in the rat. *J. Comp. Physiol. Psychol., 88,* 264-270.

Seiden, L.S., Dahms, J.L., & Shaughnessy, R.A. (1985). Behavioral screen for antidepressants: The effects of drugs and electroconvulsive shock on performance under a differential-reinforcement-of-low-rate schedule. *Psychopharmacology, 86,* 55-60.

Solomon, Z., & Benbenishty, R. (1986). The role of proximity, immediacy, and expectancy in frontline treatment of combat stress reaction among Israelis in the Lebanon war. *Amer. J. Psychiat., 143,* 613-617.

Taylor, S.P. (1967). Aggressive behavior and physiological arousal as a function of provocation and the tendency to inhibit aggression. *J. Pers., 35,* 297-310.

Taylor, S.P., & Gammon, C.B. (1975). Effects of type and dose of alcohol on human physical aggression. *J. Pers. Soc. Psychol., 32,* 169-175.

Taylor, S.P., & Leonard, K.E. (1983). Alcohol and human physical aggression. In R.G. Green and E.J. Donnerstein (Eds.), *Aggression: Theoretical and Empirical Reviews, Vol. 2* (pp. 77-101). New York: Academic Press.

Squire, L., & Cohen, N.J. (1980). Preserved learning and retention of pattern-analyzing skill in amnesia: dissociation of knowing how and what. *Science, 210,* 207-210.

Valzelli, L. (1981). *Psychopharmacology of Aggression and Violence.* New York: Raven.

van Haaren, F., van Hest, A., & Heinsbroek, R.P.W. (1990). Behavioral differences between male and female rats: Effects of gonadal hormones on learning and memory. *Neurosci. Biobehav. Rev., 14,* 23-33.

van Oijen, H.G., van de Poll, N.E., & de Bruin, J.P.C. (1980). Effects of retention interval and gonadectomy on sex differences in passive avoidance behavior. *Physiol. Behav., 25,* 859-862.

Wang, Plymate, Nieschlag, & Paulsen (1981).

Winslow, J.T., DeBold, J.F., & Miczek, K.A. (1987). Alcohol effects on the aggressive behaviour of Squirrel monkeys and mice are modulated by testosterone. In B. Olivier, J. Mos, and P.F. Brain (Eds.), *Ethopharmacology of Agonistic Behaviour in Animals and Humans* (pp. 223-244). Dordrecht: Nijhoff.

Winslow, J.T., & Miczek, K.A. (1985). Social status as a determinant of alcohol effects on aggressive behaviour in Squirrel monkeys (Saimiri sciureus). *Psychopharmacology, 85,* 953-958.

Wright, J., Allard, M., Lecours, A., & Sabourin, S. (1989). Psychosocial distress and infertility: a review of controlled research. *Int. J. Fertil., 34,* 126-142.

Young, R.D. (1963). Effect of prenatal maternal injection of epinephrine on postnatal offspring behavior. *J. Comp. Physiol. Psychol., 56,* 929-932.

Zeier, H., Baettig, K., & Driscoll, P. (1978). Acquisition of DRL-20 behavior in male and female, Roman High- and Low-avoidance rats. *Physiol. Behav., 20,* 791-793.

Subject index